The First (International) Symposium

on

THE CONSTITUTION OF GLASS

with Introduction and Commentary by

Adrian C. WRIGHT

Society of Glass Technology
Sheffield
2012

Historical Papers in Glass Science & Technology, Volume 1.

Copyright © 2012, Society of Glass Technology.

First Published 2012.

Society of Glass Technology
9 Churchill Way
Chapeltown
Sheffield,
S35 2PY
UK

ISBN 978-0-900682-64-3

Professor William Ernest Stephen Turner

Contents

Foreword

This is the first volume of a proposed series, entitled *Historical Papers in Glass Science & Technology*, most of which will present facsimile reproductions of important early papers on glass science and technology published in the *Journal of the Society of Glass Technology*, and possibly elsewhere. Each volume will include a commentary by the Volume Editor(s), placing the papers in their historical context, together with a short summary of the important conclusions from each paper.

The *Journal of the Society of Glass Technology* was first published in 1917, the year following the founding of the Society of Glass Technology by Professor William Ernest Stephen Turner in 1916, and the last volume (XLIII) appeared in 1959, after which it was replaced by two separate publications, *Glass Technology* and *Physics and Chemistry of Glasses*. More recently, in 2006, the latter journals were combined with the Deutsche Glastechnische Gesellschaft (DGG) journal *Glass Science & Technology* (formerly *Glastechnische Berichte*) to form the new *European Journal of Glass Science and Technology* parts *A: Glass Technology* and *B: Physics and Chemistry of Glasses*.

BRIAN W. MCMILLAN
SGT PRESIDENT
October, 2010

Preface

During research for a future monograph,* the Author came across several papers from a long-forgotten *Symposium on the Constitution of Glass* organised by the Society of Glass Technology in May 1925. This seems to have been the very first meeting, with an international participation, dedicated specifically to the constitution of glass, and is of exceptional interest because it provides a unique summary of the concepts concerning the constitution and structure of glass that were current during the mid 1920s.

All but one of the Symposium papers were published in Volume 9 (1925) of the *Journal of the Society of Glass Technology* and, in 1927, they were collected together, with the General Discussion and two further papers, in a volume entitled *The Constitution of Glass*, edited by W.E.S. Turner,† which is similarly forgotten and has long been out of print. Given the historical importance of the Symposium, it has been decided to republish an updated version of the 1927 volume to which has been added an historical introduction and a short commentary on each paper, to place it in context. The opportunity has also been taken to include two further papers published in 1930, thus extending the history of glass constitution/structure research up to the point just prior to Zachariasen's famous 1932 paper, *The Atomic Arrangement in Glass* [W. H. Zachariasen, *J. Amer. Chem. Soc.* **54** (1932), 3841.], on what subsequently became known as the random network theory.*

An English translation of the paper *The Polymorphism and Annealing of Glass* by A. A. Lebedev [*Trudy Gos. Opt. Inst.* No. 10, 2 (1921).] is included in an Appendix to this monograph in lieu of his missing Symposium paper and is based on an American version by the National Translations Center. A few corrections have been made to the original American translation, the most important being the substitution of the word *alloy* for (the incorrect) *melt* in the

---

* A. C. Wright, *An Historical Introduction to the Constitution and Structure of Glass*, in preparation.
† W. E. S. Turner, *The Constitution of Glass* (Soc. Glass Technol., Sheffield, 1927).

text quoted on page 17 of this monograph. In addition, all of the references have been checked, completed/corrected as necessary, and converted to a modern format.

The Author would like to thank Sergei V. Nemilov, for searching the archives of the S.I. Vavilov State Optical Institute in Sankt Petersburg and for information concerning A. A. Lebedev, and Natalia M. Vedishcheva, for help in locating references, for translations from the Russian and for providing Fig. 2. Grateful thanks are also due to the Inter-Library Loans Section of the University of Reading Library, for their help in locating and obtaining photocopies of many of the early papers discussed in Sections II and IV and/or cited by A. A. Lebedev, and to Liane Anders-Gorczyza of the Deutsche Glastechnische Gesellschaft Library for information concerning the missing Berlin translation of Lebedev's 1921 paper into English, and for a photocopy of Lebedev's 1927 paper published in *Die Glas-Industrie*.

ADRIAN C. WRIGHT.
October, 2010.

Preface to 1927 Volume

The present volume consists of a collection of papers which have already appeared in the pages of the *Journal of the Society of Glass Technology*. Eight of them were contributed to a General Discussion in May, 1925, on the subject of *The Constitution of Glass* and to them have been added, in order to complete the volume, an additional paper by Mr. V. H. Stott, in 1926, and a concluding paper by Dr. W. Rosenhain, in 1927, the latter propounding views which may serve as a new starting-point in the investigation of substances in the vitreous state.

The subject of the constitution of glass is one in which, hitherto, there has been no adequate guiding theory. It is obvious, therefore, that it may be approached from different directions, a fact which provided warrant for publishing, under a single title, a series of papers of very diverse character. And at least there is some advantage in allowing individual authorities to give free rein in presenting their own personal views. Despite the fact that the present volume lacks the systematic presentation which a single author or compiler may have given to it, there are few aspects of the subject which do not find reference in one or other of the papers. It is hoped that the publication of the volume may stimulate additional investigators to enter a field so fascinating and so comparatively virgin.

An index has been added which will assist readers in ascertaining the subjects considered.

W. E. S. TURNER.
May, 1927.

AGENDA[(1)]

First Session

Monday 25th May, starting at 7:30 p.m.
(Royal Society of Arts Lecture Theatre)

1. Minutes.
2. Applications for Membership.
3. To receive and discuss the following papers:-
   I.   Prof. W. E. S. Turner (University of Sheffield):
         The Nature and Constitution of Glass.
   II.   Prof. G. Tamman (University of Göttingen, Germany):
         On Glasses as Supercooled Liquids.
   III.   Dr. F. Eckert (Glaswerke Ruhr, Essen, Germany):
         Some Remarks on the Constitution of Glass.
   IV.   Dr. A. Q. Tool & E. E. Hill (Bureau of Standards, Washington, USA):
         On the Constitution and Density of Glass.
   V.   Prof. H. Le Chatelier (Sorbonne University, Paris):
         On the Viscosity and Allotropy of Glass.

Second Session

Tuesday 26th May, starting at 2:30 p.m.
(Chemistry Lecture Theatre, University College)

To receive and discuss the following papers:-
   VI.   Sir W. H. Bragg, FRS (Royal Institution, London):
         The Structure of Quartz and Silica.
   VII.   V. H. Stott, M.Sc. (National Physical Laboratory, Teddington):
         The Viscosity of Glass.
   VIII. Dr. G. W. Morey & Dr. N. L. Bowen (Geophysical Lab., Washington, USA):
         The Melting Relations of the Soda-Lime-Silica Glasses.
   IX.   Dr. A. A. Lebedeff (Optical Institute, University of Leningrad, Russia):
         Polymorphic Transformations in Glass.
   X.   Dr. G. W. Morey & Dr. R. W. G. Wyckoff (Geophysical Lab., Washington, USA):
         X-Ray Studies of Soda-Lime-Silica Glasses.

Seventh Annual Dinner

(Hotel Cecil, Strand, 7:00 for 7:15 p.m.)

I. INTRODUCTION

"What do we know about the molecular complexity of glass?
What do we know about the presence of compounds in glasses?"
William Ernest Stephen Turner (1925)[2]

On Monday 25th and Tuesday 26th May 1925, as the major part of its 81st Meeting, the Society of Glass Technology (SGT) organised the first (International) *Symposium on the Constitution of Glass*, at which members of the Faraday Society, the Optical Society and the Physical Society were also present, together with others interested in the subject. It seems likely that the Symposium was initiated by the Society's founder, Professor William Ernest Stephen Turner (Frontispiece), as a forum for discussion between leading scientists from Europe, Russia and the United States of America. There was also a great need among glass technologists for a more fundamental understanding of the nature of glass, to allow them to develop new compositions with superior properties, particularly for the optical industry. Unfortunately, probably due to the expense and time-consuming nature of international travel during the 1920s, none of the overseas participants were able to attend in person. However, several of their papers were read by others and were discussed by those present.

The advance Agenda for the 81st Meeting[1] is given opposite and includes a number of famous names from the glass and crystallographic communities. The first session was scheduled for the evening of 25th May and the second during the afternoon of 26th May, followed by the Seventh Annual Dinner, at which the Society was to welcome a number of important guests, as may be seen from the guest list[1] overleaf.

The dinner[3] was attended by 84 members and guests, and the Chair was occupied by the Society's President, Mr. Thomas Courtney Moorshead. The toast of *The Society of Glass Technology* was proposed by Sir Richard T. Glazebrook, a former Director of the National Physical Laboratory, who claimed a hereditary connection with glass through his grandfather's involvement in the manufacture of flint glass. Sir Richard recalled that in 1912, before the founding of the Society of Glass Technology, the General Board of the National Physical Laboratory had drawn attention to the general need for the improvement of glass for optical and other scientific purposes,* but that this requirement was now laid fully on the shoulders of the Society. On a recent trip to America he had learned of the work performed there on optical glass during the World War (I), and questioned whether the Society had sufficiently appreciated the importance of the pure science on which glass technology is based. Had he been able to attend the Symposium lectures, it was possible that he would have been amply satisfied with the Society's intentions. Although the manufacture of glass was some thousands of years old, it seemed that very little was yet known as to its real constitution, but it was clear from the Symposium papers and from the names of those involved that this question was being addressed.

SEVENTH ANNUAL DINNER
(Hotel Cecil, Strand)
Guest List[1]
Frank J. Boam, Esq. (President, Refractories Association).
Sir William H. Bragg, K.B.E., M.A., D.Sc., F.R.S.
(Director, Royal Institution).
Sir Richard T. Glazebrook, K.C.B., M.A., D.Sc., F.R.S.
(Former Director, National Physical Laboratory).
Sir Richard A. Gregory, D.Sc. (Editor of *Nature*).
Bertie Pardoe-Thomas, Esq. (Master of the Glaziers Company).
F. E. Smith, Esq., C.B.E., F.R.S. (President of the Physical Society).
Thomas Smith, Esq., M.A. (President of the Optical Society).
Harry J. Thwaites, Esq. (Master of the Glass Sellers Company).[†]

------

* At the outbreak of World War I, concern was expressed as to the superiority of the German optical glass industry, mainly as a result of the work of Carl Zeiss, Ernst Abbé and Otto Schott in Jena. As a result, W. E. S. Turner became involved in problem solving for the local glass industry, an activity that led first to the establishment of the Department of Glass Technology at the University of Sheffield in 1915, and then to the founding of the Society of Glass Technology in 1916.
† Prevented by indisposition from being present.

In his response to the toast, the President thanked Sir Richard Glazebrook, and expressed his satisfaction as to the success of the Symposium. He remarked that the addition of physical techniques to the traditional medium of chemical analysis would enable the solution of many problems, both scientific and technological, and highlighted the potential role of physicists and chemists in the control of glass manufacturing processes.

The Society's guests were toasted by Mr. E. A. Coad-Pryor, who commented that the Society had never entertained such a galaxy of stars as on that particular occasion, and praised the lucidity of Sir William Henry Bragg's presentation on the $\alpha \rightleftharpoons \beta$ transformation of quartz. Industry was beginning to realise that science was not merely a bundle of facts, but might best be described as organised common-sense and, as an example of former attitudes, he quoted a director of a large glass factory as saying:

"Oh, yes, we keep a chemist, and he does very good work indeed; but in the laboratory, of course; naturally, we never allow him to go out on the works."

E. A. Coad-Pryor (1925)[3]

However, he concluded, senior members of the glass trade were looking forward to a time when their industry would be willing to progress by the organisation of its common-sense!

Sir William Henry Bragg responded on behalf of the Society's guests. He made the point that there was pure science on the one hand and applied science on the other, but that there ought not to be a big gap between the two. As for science and its relationship to glass, glass technologists had thousands of years of accumulated knowledge behind them. However, scientists now had the benefit of new techniques to help them catch up with the technology. Sir William ended with the hope that science and industry would in the future be able to walk together hand in hand.

As may be seen from the final programme[4] on page 5, the order of the Symposium papers was somewhat different from that proposed in the Agenda. The papers by Tammann,[5] Tool & Hill,[6] Morey & Bowen[7] and Wyckoff & Morey[9] were presented by others, whilst that by Le Chatelier[11] was taken as read and Eckert's manuscript[12] arrived too late for a translation to be prepared in advance of the Symposium. With the exception of that by Lebedev (given as Lebedeff in the Agenda), who was unable to send a manuscript, all of the papers were published first in the *Transactions* section of the *Journal of the Society of Glass Technology* and then later in a volume of collected papers edited by Turner,[13] which provides a unique record of the contemporary understanding of the constitution of glass. (Note that the order of the authors for Ref. 9 is reversed with respect to that in the Agenda.)

The reason why Lebedev did not contribute to the Symposium is unclear. A search of the archives of the S.I. Vavilov State Optical Institute[14] has revealed no reference either to the Symposium or to an associated manuscript. However, it is unlikely that Lebedev was officially prevented from participating, since he later spent nine months (November 1930 to August 1931)* in London, working in Sir William Henry Bragg's laboratory, where he performed structural studies using X-ray and electron diffraction.[15,16] During this visit he also became the first to use a magnetic lens for focussing the beam in an electron diffraction experiment.[17]

Lebedev's contribution to the Symposium was to have been entitled *Polymorphic Transformations in Glass*. This suggests that there would have been significant overlap with his classic and frequently-quoted 1921 paper *The Polymorphism and Annealing of Glass*,[18] which reintroduced what later became known as the ***crystallite***† **hypothesis** concerning the structure of glass. It is also likely that his paper would have included further material from his subsequent paper *On the Annealing of Optical Glass*[21] published in 1924, not long before the Symposium. However, whereas an updated version of the latter paper was later (1926) published in French,[22] today the former is only known in Russian, even though the volume list-

---

* These dates are taken from Lebedev's autobiography, which is kept in the archives of the Personnel Department of the V.I. Vavilov State Optical Institute.
† The term ***crystallite*** was not used by Frankenheim[19] (*cf.* Section II), nor by Lebedev in his early (1920s) papers, and its first application to glass structure seems to have been by Rosenhain[20] in 1927.

FINAL SYMPOSIUM PROGRAMME[4]
First Session
(Royal Society of Arts Lecture Theatre)

Chair: Mr. T.C. Moorshead (President of the SGT)

I.   Prof. W.E.S. Turner (University of Sheffield):
The Nature and Constitution of Glass.[2]
(Discussion: T.C. Moorshead, E.A. Coad-Pryor, V. Dimbleby
& W.E.S. Turner)

II. Prof. G. Tamman (University of Göttingen, Germany):
On Glasses as Super-cooled Liquids.[5] (Presented by Mr. F.F.S. Bryson.)

III. Dr. A.Q. Tool & E.E. Hill (Bureau of Standards, Washington, USA):
On the Constitution and Density of Glass.[6]
(Presented by Mr. E.A. Coad-Pryor.)

IV. Dr. G.W. Morey & Dr. N.L. Bowen (Geophysical Lab., Washington, USA):
The Melting Relations of the Soda-Lime-Silica Glasses.[7]
(Presented by Dr. M.W. Travers.)
(General Discussion: T.C. Moorshead, W.E.S. Turner, M.W. Travers,
R.S. Clay, P.E. Shaw, E.A. Coad-Pryor & W.M. Hampton)

Second Session
(Chemistry Lecture Theatre, University College)

Chair: Prof. F.G. Donnan (President of the Faraday Society)

I.   Sir W.H. Bragg, FRS (Royal Institution, London):
The Structure of Quartz and Silica.[8]
(Discussion: F.G. Donnan, W.E.S. Turner, T.M. Lowry, R.S. Clay, H.D.H.
Drane & E.A. Coad-Pryor)

II. Dr. R.W.G. Wyckoff & Dr. G.W. Morey (Geophysical Lab., Washington, USA):
X-Ray Studies of Soda-Lime-Silica Glasses.[9]
(Presented by Miss V. Dimbleby.)
(Discussion: F.G. Donnan & W.H. Bragg)

III. V.H. Stott, M.Sc. (National Physical Laboratory, Teddington):
The Viscosity of Glass.[10]
(Discussion: F.G Donnan, W.E.S. Turner, L.N.G. Filon, A.O. Rankine,
E.A. Coad-Pryor & V.R. Main)

ing on the back cover of the 1924 paper indicates that an English translation was published in Berlin in 1923.* (Extended abstracts of both papers were also published in the *Abstracts and Reviews* Section of the *Journal of the Society of Glass Technology*[23,24]). Hence it was decided to include, as an Appendix to this volume, an English version of Lebedev's 1921 paper based on a hitherto unpublished American translation,[25] in lieu of that missing from the Symposium.

Before considering the Symposium papers in detail, it is important to place them in the context of contemporary ideas concerning the structure of matter, both crystalline and amorphous, and to explain why the approach to the constitution of glass was predominantly molecular. Hence a brief summary of the relevant papers that predate the Symposium is presented in the next section, whilst those published subsequently and up to the end of the decade are outlined in Section IV.

---

* The author has been unable to trace any existing copy of this translation by the (Soviet) Bureau of (Foreign) Science and Technology, neither in Germany nor in Russia.

II. HISTORICAL BACKGROUND

"The ice is music, the glass is noise; the ice is order, the glass is confusion. In the glass, molecular forces constitute an inextricably entangled skein; in the ice they are woven to a symmetric web ..."

John Tyndall (1863)[26]

The first important question to be asked concerning the nature/constitution of glasses was whether they are mixtures, solutions or compounds. Even the alchemists of old were aware that, during glassmaking, the raw materials are transmuted by the fire into something new, which has properties different from those of the basic ingredients; in other words that glasses are the product of a chemical reaction and not simply quenched liquid mixtures. On the other hand, they cannot be simple chemical compounds since, in most systems, they can be formed over a continuous range of composition and not just at specific (crystalline) stoichiometries. Michael Faraday[27,28] was perhaps the first to suggest that the situation is somewhat more complex, in that glasses cannot be considered simply as compounds, but should be regarded *"rather as a solution of different substances one in another, than as a strong chemical compound"*;[27,28] *i.e.* as a mutual solution.

The Russian chemist, Dmitri Ivanovich Mendeleev, was greatly interested in silicates, both crystalline and vitreous, and considered that an understanding of silicate glasses had to be based on that of the corresponding crystalline materials found in nature:[29-31]

"... considerations of the chemical composition and structure of glass cannot be separated from considerations of the chemical composition and structure of silicates that occur in nature."

Dmitri Ivanovich Mendeleev (1864)[29,30]

He argued that the position of silica as the weakest acidic oxide, next in sequence to alumina the weakest basic oxide, led to the formation of *"chemical compounds of indefinite composition"*[29-31] (*i.e.* compounds of undefined stoichiometry), and described such materials as *"alloys formed not from metals but from oxides that include silica and silicates."*[29-31] Mendeleev suggested that the variable composition is the result of the very common isomorphous replacements that occur in silicates, which may lead to a change in crystalline form and/or a transition to the amorphous state, as in the case of glasses, lava, slags, *etc.* The latter do not contain strictly definite compounds, but are nevertheless homogeneous and isotropic, and may crystallise to form a definite chemical compound on cooling. The ability of silicic acid to form a wide variety of salts should be contrasted with the more limited range of salts formed by strong acids; *e.g.* nitrates and sulphates. In respect of the constitution of glass, it is particularly interesting to note that Mendeleev distinguished between two different silicate components:

> *"Consequently, in the composition of every such compound we can distinguish invariable and variable portions. The invariable portion characterises the given substance, whilst the variable portion contains components that can undergo various substitutions without great changes in properties."*
> Dmitri Ivanovich Mendeleev (1864)[29,30]

In modern terminology, the invariable portion can be identified with the silicate framework, and the variable portion associated with network modifying oxides. In this way, Mendeleev regarded glass as a non-crystallising or amorphous alloy of indefinite chemical compounds (silicates) that can only be formed within certain limiting proportions of its constituent oxides.

Early theories of the atomic structure of matter concentrated almost exclusively on the crystalline state, as represented by the well-defined morphology of single crystals.* By the end of the 19th century, crystals were envisaged as comprising regular periodic arrangements of their constituent atoms/molecules. To the chemist, crystalline structures were based on molecules whereas,

---

* A more detailed account of the early history of crystallography is given in Ref. 32.

in 1883, William Barlow[33,34] proposed that crystals were composed of spherical atoms that packed themselves into as small a volume as possible. Barlow's structures were, however, criticised by Sohncke[35] in that the number of neighbours for a given atom did not correspond to its valency, as would be required for a molecular structure:

"By this example we see that from Mr. Barlow's point of view both the notion of chemical valency and of chemical molecule completely lose their present import for the crystallised state."

L. Sohncke (1884)[35]

Sohncke[35] does, however, concede that his objection did not destroy Barlow's theory, *"since chemical valency does not yet belong to perfectly clear and fixed notions, and since the idea of the chemical molecule in a crystal is also not evident and clear."* In his response to Sohncke's comments,[35] Barlow[36] cites the phenomenon of electrolysis. The fact that the ions are liberated at opposite poles whilst *"no apparent alteration takes place in the fluid between the poles"* demonstrates that any particular atom can change its neighbours without destroying the chemical bonding, which favours the view that similar atoms equally near to a given atom are similarly related to it.

Very little had so far been postulated concerning the constitution of glasses, although the above quotation from John Tyndall does suggest a disordered structure, in contrast to the regularity of the crystalline state. In 1835, the German scientist Moritz Ludwig Frankenheim[19] had proposed the first scientific hypothesis of glass structure,[37] *viz.* that glass consists of very small crystals of various sizes such that, on heating, the smallest ones melt earlier than the larger ones thus creating a type of *"lubrication"*, which allows the glass to flow at temperatures much lower than the melting point of the larger crystals and hence provides an explanation of the process of softening. Similarly, Frankenheim[19] considered gels to be composed of aggregates of small crystals with pores in between them, which would account for their low density, easy solubility and elasticity.

Liquid–liquid phase separation had been recognised by Otto Schott, as a consequence of adding P_2O_5 to glass-forming melts:

> *"My melting experiments revealed that SiO_2 and B_2O_3 were compatible, while P_2O_5 exhibits hostile behaviour. While $SiO_2 + B_2O_3$ as a glass-forming double acid accept bases in any proportions, the addition of P_2O_5 almost always causes milky opalescence, apparently because phosphates are as insoluble in the melt as oil in water."*
>
> Otto Schott (Quoted by Eberhard Zschimmer[38])

The subject of phase separation in borate systems, comprising the combination of a metal oxide plus boric acid anhydride (B_2O_3) and concentrations from pure B_2O_3 up to the orthoborate (3:1) composition, was discussed in 1904 by Guertler,[39] who divided these systems into three groups:

(i) All concentrations yield clear homogeneous melts. On quenching, the melts either crystallise or yield a clear glass (Li, Na, K, Rb, Cs, Tl and Ag).

(ii) At high enough temperatures, a clear homogeneous melt is obtained but, on cooling, demixing occurs over a more limited concentration interval to form an emulsion (Cu, Pb, Bi, Sb, As, Ti, Mo, W, and V).

and

(iii) At higher temperatures and concentrations of the metal oxide, a clear melt is obtained but, at lower oxide contents, a two-liquid region occurs in the melt, at least up to 1400°C. A single liquid is also formed at very small oxide contents, which on quenching yields an opalescent glass. For some systems, clear glasses are found at much higher oxide contents, whereas other systems yield metaborate crystals with opalescent glass (Mg, Ca, Sr, Ba, Zn, Cd, Mn, Fe, Co, Ni, La, Ce, Pr, Nd, Sm, and Gd).

The very first crystalline structure to be determined, that of rock-salt (NaCl) published by William Lawrence Bragg[40] in 1913, immediately brought crystallography into conflict with chemists in that rocksalt did not contain any NaCl molecules, but rather the sodium and chlorine atoms were arranged in a *"chessboard pattern"*,[40] with each sodium atom having six chlorine neighbours and *vice versa*. It was objected that molecules are *"the basis of the whole of chemistry"*[41] and W. L. Bragg recalled[41] being begged by a professor of chemistry to find that each sodium atom was just a tiny bit closer to one chlorine atom than it was to the other five! This controversy extended well into the 1920s, with perhaps the last contribution from the famous chemist Henry Edward Armstrong:

'Poor Common Salt!

"Some books are lies frae end to end," says Burns. Scientific (save the mark) speculation would seem to be on the way to this state! ... Prof. W. L. Bragg asserts that "In sodium chloride there appear to be no molecules represented by NaCl. The equality in number of sodium and chlorine atoms is arrived at by a chess-board pattern of these atoms; it is a result of geometry and not of a pairing-off of the atoms."

This statement is more than "repugnant to common sense." It is absurd to the n...th degree, not chemical cricket. Chemistry is neither chess nor geometry, whatever X-ray physics may be. Such unjustified aspersion of the molecular character of our most necessary condiment must not be allowed any longer to pass unchallenged ... It were time that chemists took charge of chemistry once more and protected neophytes against the worship of false gods: at least taught them to ask for something more than chess-board evidence.'

Henry Edward Armstrong (1927)[42]

In contrast to the structure of rocksalt, that of diamond[43,44] was much more readily accepted by chemists, even though it revealed no obvious molecules, because each carbon atom has four neighbours arranged tetrahedrally in space at a distance of 1·522 Å,[44] which is entirely consistent with the tetravalency of carbon. An interesting feature is that Ref. 44 includes two photographs of a ball-and-stick model of diamond, with the sticks linking each carbon atom to its four neighbours; *i.e.* coincident with the C–C (covalent) bonds. However, W. H. & W. L. Bragg comment:

> *"The union of every carbon atom to four neighbours in a perfectly symmetrical way might be expected in view of the persistent tetravalency of carbon. The linking of six carbon atoms into a ring is also an obvious feature of the structure. But it would not be right to lay much stress on these facts at present, since other crystals which do not contain carbon atoms possess, apparently, a similar structure."*
>
> William Henry Bragg & William Lawrence Bragg (1913)[44]

The initial X-ray diffraction study of an amorphous material (liquid benzene) was published by Debye & Scherrer[45] in 1916, only three years after W. L. Bragg's paper on sodium chloride.[40] Debye & Scherrer's paper[45] introduced the well-known Debye equation for the scattering from an isotropic material and used it to (mis)-interpret their diffraction halo from liquid benzene. In general, liquids were assumed to have structures based on molecules and, by the time of the Symposium, three explanations had been advanced for the origin of the diffraction halos from liquids, as summarised in Wyckoff & Morey's Symposium paper:[9]

(i)   Constant distances between atoms in chemical molecules,

(ii)   The average distances apart of the molecules themselves, or,

(iii) More or less transitory groupings of atoms or molecules into crystal-like aggregates.

The first X-ray photograph for a glass known to the Author (vitreous silica) was published in 1917 by Kyropoulos,[46] who obtained X-ray powder diffraction patterns from a series of silica polymorphs and other samples derived from silicic acid. However, he failed to detect any diffraction halo for silica glass apart from broadening of the transmitted beam. A further X-ray diffraction pattern for a "good" glass was given in 1920 by Scherrer[47] and revealed the similarity between the diffraction halo obtained from the glass and those from liquids. Despite this similarity and the concept of glasses as supercooled liquids or mutual solutions, hypotheses concerning the constitution of glasses developed along rather different lines to those for liquids, in that there was increasing support for the idea that multi-component glasses are not homogeneous on a molecular scale; *i.e.* that they are ***nanoheterogeneous***.* Thus, for example, Frink[48] connected the brittleness of glasses with physical and chemical heterogeneity.

As discussed in Section V, prior to the development of X-ray techniques, the concept of an amorphous material was not clearly defined, such that colloidal materials were regarded as amorphous. The relationship between the colloidal, amorphous and crystalline states is discussed in a series of papers by von Weimarn,[49-51] commencing in 1907, which were later cited by Lebedev[18] as favouring a crystalline structure for glass (Footnote on page 312 of the Appendix to this volume). Von Weimarn's research strongly suggested to him that no substances are truly amorphous and that all solids are really crystalline, the amorphous or colloidal condition merely being one where the crystals are so small as to be invisible under a microscope. He also deduced a simple empirical relationship to predict the size of the crystals formed under given conditions, which is inversely proportional to the number of nucleation sites. A number of other authors also considered the relationship between glasses and colloids, such as Kroll,[52,53] who considered high molecular weight phosphate glasses to be colloidal in character. According to Kroll,[52,53] ultraphosphate glasses are not undercooled liquids, but solid materials in the colloidal rather than the crystalline state; *i.e.* colloidal-amorphous sols. Similarly, Filon & Harris[54] supposed that (flint) glass has the nature of a coarse dispersoid or colloid and used the concept of a di-phasic system to explain its photoelastic properties (see later, Section IV). On the other hand, in 1915, Quincke[55] likened the structures of glasses to those of foams, with highly siliceous walls that enclosed more al-

---

* The term ***nanoheterogeneous*** is used to denote composition fluctuations with length scales in the range 1 to 10 nm (10–100 Å), and to distinguish such fluctuations from micro-phase separation.

kaline silicates in the cellular space, rather than the gaseous phase found in true foams. This concept arose from Quincke's hypothesis that the first step in the process of crystallisation is the separation of the liquid into two immiscible liquid phases, one of which is formed in a relatively small quantity and becomes distributed in such a way as to form thin cell walls enclosing drops of the major phase, thus forming a foam-like structure.

The subject of a possible crystalline nature for glass was raised again by Sir Herbert Jackson[56] during the presentation of his Trueman Wood Lecture, entitled *Glass and Some of its Problems*, to the Royal Society of Arts on Wednesday April 30th 1919. One of the problems considered was the question

"Is glass truly amorphous or vitreous or has it any crystalline structure or tendency to crystalline structure?"

Sir Herbert Jackson (1919)[56]

Although no evidence for crystallinity had been found from microscopic examinations of glass surfaces etched with hydrofluoric acid or produced by fracture, Sir Herbert introduced the concept of the submicroscopic crystallinity of annealed glasses based on their phosphorescence under UV light, X-rays and cathode rays. The same glasses when rapidly quenched were practically non-phosphorescent, and the phosphorescence increased with increasing crystallinity, which led to the conclusion that a truly vitreous body would exhibit no phosphorescence.

Later in his lecture, Sir Herbert considered both opalescent and coloured glasses. He observed that many glasses, which crystallise fairly readily, pass through an opalescent phase before finally becoming visibly crystalline. The opalescence is due to the scattering of light by numerous small transparent globules that normally, under the microscope, appear vitreous and transparent with no evidence of crystallinity. He notes that this segregation (*i.e.* phase separation) is produced by adding fluorides, phosphates, *etc.*, to the batch. Coloured glasses, on the other hand, were sometimes divided into two main groups:

(i)   Those in which the colouring matter is diffused in very small particles throughout the glass, and which may be likened to colloidal solutions (sols),

and,

  (ii)   Those in which the colouring substances are in a state resembling that of a solution, and which may be likened more nearly to aqueous solutions of coloured salts.

However, just as with aqueous solutions, there is a continuous gradation from separate particles, visible in the ultramicroscope, down to a true solution. The lecture concluded with a discussion of the factors that influence colour; *e.g.* particle size in the case of gold, and alkali content for glasses containing cobalt. The colloidal nature of gold ruby and related glasses incorporating metallic particles of colloidal dimensions had in fact been appreciated from the work of Zsigmondy,[57-59] and further discussion of the colloidal nature of both ruby glass and opals can be found in a paper published by Liesegang[60] in 1924.

In his 1919 *Note on the Formation of Glass*, Bradford[61] cites the work of von Weimarn[49-51] and argues that there are many indications that the solidification of glass is a crystallisation process. For example, clear, transparent vitreous arsenic trioxide slowly changes at ordinary temperatures into an opaque white crystalline substance resembling porcelain, the growth of visible crystals being due to the tendency of larger crystals to develop at the expense of smaller ones. Bradford[61] also quotes the work of Sir Herbert Jackson to the effect that glasses exist exhibiting all degrees of readiness to crystallise. The development of visible crystals is dependent upon the rate of crystallisation, which for glasses is very small due to their low thermal conductivity and (high) viscosity. This allows a high degree of supercooling and hence many nucleation sites so as to yield a colloidal substance.

On 25th October 1920, the Faraday Society and the Physical Society of London jointly held a General Discussion on *The Physics and Chemistry of Colloids and their Bearing on Industrial Questions*, at which Morris W. Travers presented a paper, *On the Nature of Glass*,[62] which compared glasses to colloidal systems. He argued that, although it was commonly stated that glass, in its rigid form, is still a supercooled liquid, this was based on the assumption that its properties change continuously with temperature. However, Tool & Valasek[63] had recently called attention to a marked discontinuity in the behaviour of glass in the neighbourhood of the softening point in the form of a sudden increase in the rate of absorption of heat. (In fact, this had been reported much earlier by Day & Allen[64] in 1905.) The most obvious alternative suggestion was that glass is a colloid and that, with increasing temperature, it passes through the stages

$$\text{elastic gel} \rightleftharpoons \text{colloidal sol} \rightleftharpoons \text{crystalloidal sol}$$

Travers[62] argues that the hypothesis that glass is a colloid (*i.e.* some form of polyphase liquid system) had frequently been put forward and he quotes the work of Carl Barus,[65,66] who used it to explain the behaviour of glass towards water, and of Quincke[55] (see above). Travers' paper[62] formed part of the Section on *Glass and Pyrosols* and it is interesting to note that, in his opening address to this Section, Sir Herbert Jackson[67] expressed his doubts concerning the relevancy of the subject to a discussion of colloids, even though he knew of suspensions of colouring matter in glass that were colloidal. Only a very brief summary[67] is given of Sir Herbert's address, but it seems to have closely followed his Trueman Wood Lecture.

In 1921, Alexander Alexeevich Lebedev published his famous paper, entitled *The Polymorphism and Annealing of Glass*[18,23] (Appendix 1), in which sharp changes in certain properties (birefringence, refractive index and thermal expansion coefficient) of six multi-component silicate glasses at temperatures in the region 540–600°C were interpreted in terms of a ***polymorphic transformation***. This he assumed to be closely linked to the $\alpha \rightleftharpoons \beta$ transition of crystalline quartz at 575°C and hence the hypothesis was advanced that

"glass is an aggregate of highly dispersed crystals which also include quartz crystals but in all probability not in the pure form but rather in the form of a solid solution with cer-

tain other components; therefore the polymorphic transformation does not occur immediately but in a certain temperature interval in which the glass gradually passes through a series of equilibrium states."
Alexander Alexeevich Lebedev (1921)[18]

In fact, as Sergei V. Nemilov[37] has pointed out, Lebedev's association[18] of the sudden changes in properties with the $\alpha \rightleftharpoons \beta$ transformation of quartz was incorrect, and it is now known that they arise from the process of softening and the glass transition, which at that time were not fully understood. It was just unfortunate that the glass transition region for Lebedev's six glasses happened to roughly coincide with the $\alpha \rightleftharpoons \beta$ transition temperature for crystalline quartz. Nevertheless, it is still of great interest to examine the basis of Lebedev's hypothesis, both because of its importance in defining the future direction of Soviet glass science and because of the insight it affords to contemporary ideas concerning the role of molecules in the structure of the vitreous state of matter.

In Lebedev's original 1921 paper,[18] evidence for the presence of highly dispersed quartz crystals in his glasses is based on the work of Mügge[68] who, in 1907 and before the advent of X-ray crystallography, had suggested that the $\alpha \rightleftharpoons \beta$ transformation of quartz is a consequence of a change of the crystal lattice, rather than a change in the (SiO_2) molecules. Thus, if the observed characteristics of glasses were a result of the $\alpha \rightleftharpoons \beta$ transition, it is necessary to conclude that the glasses contain highly dispersed quartz crystals. These highly dispersed crystals, later termed crystallites (see footnote on page 4), were considered to be formed during the *"freezing"* of the glass rather than being present due to incomplete reaction/melting of the original batch materials, and Lebedev[18] returns to the idea of glass as a silicate alloy, as first proposed by Mendeleev:[29-31]

"i.e. glass is not a supercooled liquid but rather an ordinary alloy consisting only of highly dispersed crystals."
Alexander Alexeevich Lebedev (1921)[18]

That the quartz crystals did not come from the original batch was demonstrated by preparing a silicate glass from ground melted (fused) quartz (silica glass), instead of sand (polycrystalline α-quartz). In this connection, Lebedev observes that fused quartz (*i.e.* vitreous silica) does not exhibit any special characteristics at 575°C. (This is to be expected since T_g for vitreous silica falls within the stability range of tridymite, and not that of quartz.)

Lebedev[18] cites a paper by Wallace,[69] on the crystallisation of glasses from the Na_2SiO_3–$MgSiO_3$ system, in which the author explains the solidification of a glass by the slow rate of crystal growth and not by a small number of nucleating crystals. Lebedev then argues that it is difficult to assume that, between the formation of the visible spherulites observed by Wallace[69] and *"the molecularly dispersed structure that some have attributed to glass"*,[18] there are no intermediate stages of submicroscopic crystals. Hence it is most likely that glasses represent just such an intermediate structure of highly dispersed crystals. At this point, Lebedev notes that other considerations favouring a crystalline structure of glass may be found in the series of papers by von Weimarn[49-51] mentioned above and published from 1907 in *Kolloid-Zeitschrift*.

The second paper by Lebedev,[21,22,24] in 1924, is mainly concerned with the annealing of optical glass but, in the Introduction, he takes the opportunity to update his ideas in the light of publications subsequent to his 1921 paper and to reply to its critics. The most influential paper is that by Sosman,[70] who erroneously discusses the structure of the ambient-pressure silica polymorphs in terms of triangular SiO_2 "atom-triplets" linked by Si–Si bonds to form silica "threads" (*i.e.* chains). Sosman[70] argues that the α⇌β transformations arise from a change in the relative positions of the two oxygen atoms in the SiO_2 atom-triplet and that the different α⇌β transition temperatures for quartz (573°C), tridymite (117 and 163°C) and cristobalite (200 to 275°C) are the result of the interaction between the oxygen atoms in adjacent triplets. In the vitreous state, the silica threads are tangled (*cf.* above quotation from John Tyndall[26]), with variable distances between the oxygen atoms in adjacent threads, and hence the transformation is smeared out over a range of temperature so that no marked effect is discernible at any one temperature. Lebedev, therefore, associates the temperature range over which the property changes occur with this smeared-out transition and concludes that there is in fact no need to invoke the presence of quartz crystals in glasses:

"If it is believed that the reason for the transformation in quartz is a transformation in SiO_2 molecules then, to explain the polymorphic transformation in glasses, there is no necessity to suggest that there are small/fine quartz crystals in glass, because the presence of the SiO_2 molecules themselves is sufficient."

Alexander Alexeevich Lebedev (1924)[21,22]

Thus the fact that X-ray studies of glasses had not revealed the presence of crystals could not be considered as proof that the hypothesis, which suggests the important role of a transformation in SiO_2 molecules for polymorphic transformations in glasses, is incorrect.

Lebedev[21,22,24] goes on to explain changes in the refractive index of optical glass in terms of two glass modifications, a low-temperature α-form and a high-temperature β-form. Within the transition region there is an equilibrium between the two forms, as the α-modification is gradually converted into the β-modification with increasing temperature. The heterogeneous structure of glass is the result of incomplete conversion of the β- into the α-modification on cooling; the greater the rate of quenching, the more the amount of the β-modification remaining in the glass.

The explanation in terms of a transition within the SiO_2 molecules, also answers the criticism of Lafon,[71] who rejects Lebedev's association[18] of the changes in the coefficient of thermal expansion with the α$\rightleftharpoons$β transition of quartz on the basis that the anomaly for his glasses occurred at ~400°C. Lafon[71] argues that the volume effects that occur during the α$\rightleftharpoons$β transition can also occur in vitreous silica, the only difference being that the two phases and the volume changes are less sharply defined in a glass than in a crystal.

A similar criticism was made by Tool & Eichlin[72] to the effect that, for the large number of glasses that had so far been investigated, the endothermic effect observed upon heating occurred in a rather definite narrow temperature region that depended on the glass composition and was located in the range 100–800°C. In addition, glasses containing no silica whatsoever exhibited this *"glass transformation"*. Hence the endothermic and related effects observed in glass could not be related to the α⇌β transformation of quartz, this being supported by the fact that X-ray diffraction experiments had not revealed any evidence of crystallinity.

Tool & Eichlin[72] go on to discuss the conclusion of Sir Herbert Jackson[56] (above) concerning the fluorescence of annealed glass under UV- and X-radiation. They argue that this fluorescence could also be the result of molecular readjustments or the formation of compounds having more complex molecules:

"It is quite possible that all the effects so far observed may be explained by the presence of certain molecules, either simple or complex, or molecular arrangements which may or may not have the character of colloidal aggregates."
A. Q. Tool & C. G. Eichlin (1924)[72]

However, X-ray diffraction measurements performed by R.W.G. Wyckoff had failed to detect any difference between quenched and annealed samples. Tool & Eichlin[72] propose that there is a reversible process by which aggregates or compounds are built up on cooling and disintegrate on heating, and that the endothermic effect on heating is explained by a rapid change in mobility within what may be considered as a boundary range. The extent of the formation of aggregates or compounds depends upon the rate of cooling and so is much less for rapidly-quenched relative to annealed samples.

A further exposition of the colloid hypothesis of glass structure was published by Bary[73,74] in two parts, in January and February 1925, just before the Symposium. Bary assumes that the structure of liquid silica comprises polymer chains of SiO_2 molecules terminated by double-bonded oxygen atoms and that, in the case of alkali and alkaline earth silicates, the chains are terminated by alkali or alkaline earth atoms. At ordinary temperatures, glasses are supposed to consist of a solution of crystalloidal alkali silicates (possibly as orthosilicates) in a polymerised polysilicate of divalent metals (Ca, Mg, Pb, Fe, Zn and Ba), a structure consistent with softening, devitrification, electrical conductivity and the action of water on glass. Alumina and boric oxide could also be incorporated into the polysilicate matrix. The rigidity of the glass is due to the large size of the polysilicate molecules and decreases with increasing temperature as these molecules are depolymerised. The plasticity that occurs above a certain temperature is due to the plasticising effect of the dissolved alkali silicate.

An objection to Lebedev's initial conclusion, concerning the presence of quartz crystals in glasses, had also been raised by Turner,[75] who pointed out that a similar transformation occurs in sodium borate glasses that contain no silica. However, Lebedev[21,22] responds that this requires more attention, since Turner does not exclude the possibility that transformations may occur in other molecules, e.g. B_2O_3. It is thus easy to see why Turner would have been very keen that Lebedev participate in the Symposium, so that these issues could be discussed in detail, especially given the fact that Turner makes the same point in his Symposium paper.[2] Indeed, given Lebedev's ambivalence towards quartz crystallites as expressed in his 1924 paper,[21] the history of glass structure research might have evolved along completely different lines, had such a discussion taken place. Certainly, the great crystallite vs. random network controversy might not have dominated glass structure research to quite the same extent for the following 50 years.

III. SYMPOSIUM PAPERS

"It should be remembered that our ignorance of the constitution of the simplest liquids is almost complete; ... and the far more complicated liquid which is known as glass will probably prove a more difficult problem."

George W. Morey & N.L. Bowen (1925)[7]

The above quotation from the Symposium paper by Morey & Bowen[7] probably represents one of the great understatements of glass structure research! At the time of the Symposium, the structure of liquids was still firmly believed to be molecular in character, and hence the same was assumed to be true for the vitreous state. As to the crystallite hypothesis, unfortunately, at the time of composing their manuscripts, none of the Symposium authors seems to have been aware of Lebedev's comments in the Introduction to his 1924 paper.[21] (These are not included in the extended abstract[24] published in the *Abstracts and Reviews* section of the *Journal of the Society of Glass Technology* in December 1925.) Similarly, the relevance to glass structure, and especially to that of vitreous silica, of W.H. Bragg's paper on α- and β-quartz[8] and of Wyckoff's contemporary structure of β-cristobalite[76,77] (See Section V) appears not to have been realised, even by the authors themselves. Nevertheless, the papers presented at the Symposium do include many concepts that are still current today, and/or have been reintroduced in recent years. The most important of these are highlighted in the following brief summaries of the published manuscripts, some of which have updated titles compared to those given in the Final Programme of page 5.

The Nature and Constitution of Glass

"What I have probably succeeded in doing is to establish how unsatisfactory is our knowledge even of the most fundamental data which are necessary before any theory of the constitution of glass can be proposed even in a tentative manner."

William Ernest Stephen Turner (1925)[2]

Turner's paper[2] serves as an introduction to the Symposium and starts by differentiating between single-component glasses as supercooled liquids and commercial (*i.e.* multi-component) glasses as supercooled solutions. To formulate a working hypothesis of the constitution of glass, and to explain its various properties, particularly in the annealing temperature range, it is necessary to take into account two fundamental factors, *viz.* the molecular complexity of the substance(s) composing the glass and, in the case of binary or multi-component glasses, the chemical compounds present (*cf.* quotation at the head of Section I).

In addition to the well established characteristics of glass, Turner[2] draws particular attention to several properties that had been discovered during the preceding few years and had stimulated renewed interest in the vitreous state:

(i)   the endothermic changes that occur when glass is heated within the annealing temperature range,

(ii) the greatly increased thermal expansion within the same temperature range,

(iii) the changes in density and refractive index that take place when glass is subjected to heat treatment,

(iv) the increase in the electrical resistance on re-annealing a prestrained sample of soda–lime glass

and

(v)   the after effect found in studies of photo-elastic stresses.

Assuming that multi-component glasses are solutions, any discussion of molecular complexity should apply to their individual constituents. Turner supposes that soda–lime–silica glasses comprise a concentrated solution of sodium metasilicate (Na_2SiO_3 or

$Na_2O.SiO_2$), sodium disilicate ($Na_2Si_2O_5$ or $Na_2O.2SiO_2$), calcium metasilicate ($CaSiO_3$ or $CaO.SiO_2$) and possibly other silicates in silica, and argues that these silicates are strongly associated. Hence silicate glasses may be regarded as containing complex molecules. Evidence for the existence of chemical compounds is provided by breaks in property–composition data, e.g. at the $Na_2O.2SiO_2$ composition in the specific volume plot for the sodium silicate system.

Since glasses are formed by fusion at high temperature and maintained as such by more or less rapid quenching, it is to be expected that only the compounds stable at high temperatures will persist to any great extent. Alternatively, there may remain just the dissociation products of compounds, or even the component oxides. Turner[2] postulates that the characteristic properties of silicate and borate glasses are influenced by changes in the molecular complexity of the free (unreacted) silica or boron oxide. He also suggests that the formation of a compound may only be possible once the constituent oxides become less mobile on cooling (*cf.* quotation on page 56).

A theory of the glassy state must apply to both single chemical substances and to *"mixtures of differing composition."* The former (SiO_2, B_2O_3, P_2O_5, *etc.*) correspond to a liquid of very high viscosity and Turner suggests that these materials provide examples of **dynamic allotropy*** such that the glass can be considered as a solution comprising a mixture of molecules of different complexity, as in the case of sulphur. For commercial (multi-component) glasses, it is also necessary to consider the possible formation of chemical compounds and their molecular complexity, in addition to that of the basic glass-forming oxide.

Turner suggests that silicate glasses might be conceived of as a rigid silica *"sponge"* impregnated by silicates, or their dissociation products:

> *"A silicate glass might be conceived of as a sponge of silica containing silicates or their dissociation products as*

---

* The term **allotropy** used by Turner and others is an old-fashioned name for polymorphism. It is little used today and is now normally restricted to elements.

the filling medium; or, possibly, to use the analogy which Sosman[70] has employed to explain silica gels, of a mass of silica threads soaked in the silicates or their products of dissociation."

William Ernest Stephen Turner (1925)[2]

The analogy of a sponge, through which water can percolate to its centre, implies that Turner considered both the silica and the silicates to form a continuous structure, which can be compared to the connected phases formed by spinodal decomposition. Turner also draws attention to a possible similarity between the properties of fused (*i.e.* vitreous) and colloidal (silica gel) silica.

Glasses as Supercooled Liquids

"A glass is a non-crystallised, strongly undercooled fusion."

G. Tamman (1925)[5]

The paper by Tamman[5] is primarily concerned with the processes of glass-formation and crystallisation from the supercooled liquid state, and makes no attempt to provide a structural interpretation. Whereas, on heating, a crystal begins to melt at a definite temperature (*i.e.* the melting point, T_m), which is normally independent of the heating rate, the behaviour on cooling the resulting melt is quite different, in that crystallisation does not immediately take place at T_m but undercooling (supercooling) occurs. The extent to which the melt can be supercooled depends on two factors:

(i)   the number of crystallisation centres (nuclei) formed per unit volume in unit time (the nucleation rate),

and

(ii) the rate at which the nucleated crystals grow into the supercooled liquid (the growth rate).

Both (i) and (ii) are dependent on the extent of the supercooling, and are discussed in detail.

The maximum rate of crystal growth occurs with slight undercooling, whilst the maximum nucleation rate occurs at a somewhat lower temperature, at which point the viscosity very rapidly increases. Hence, to form a glass, the liquid should be quenched as rapadly as possible through this *"dangerous"* temperature region, which Tamman suggests extends to no less that 110° below T_m. He also makes the point that the glassy state is not exceptional but that every substance could be obtained in this state, if only its fusion could be chilled rapidly enough.

Tamman considers the supercooling and glass-forming capacity of binary melts and distinguishes four separate scenarios, which are summarised in Fig. 1:

(1) Both components, A and B crystallise without significant supercooling but, in mixtures containing no great excess of either, crystallisation is strongly retarded (*e.g.* Na_2SiO_3 and $MgSiO_3$).

(2) One component (A) crystallises readily on cooling, but the other does not (*e.g.* K_2SiO_3 and Li_2SiO_3).

(3) Both components crystallise readily and form a compound that also crystallises readily, but mixtures of the compound and either of the components solidify as glasses (*e.g.* $NaPO_3$ and $NaBO_2$).

(4) Both components crystallise with difficulty, but a compound of the two crystallises more readily (*e.g.* SiO_2 and P_2O_5).

Soda-lime-silica glasses, on the other hand, may be regarded as ternary mixtures of Na_2SiO_3, $CaSiO_3$ and SiO_2, of which the first two components crystallise readily.

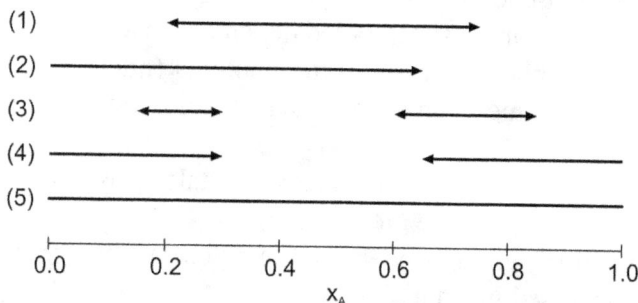

Figure 1. Binary glass formation. Scenarios 1 to 4 are from the work of Tamman, whilst 5 corresponds to the case where glass formation occurs at all values of x_A (*e.g.* B_2O_3 and SiO_2).

Whereas glasses are isotropic, (single) crystals display directional properties; *e.g.* mechanical strength – a single crystal has definite cleavage planes, whilst glasses exhibit conchoidal fracture. The surface etching behaviour for (poly)crystalline and vitreous materials is also different. For the former, maximum etching occurs along the grain boundaries or at specific crystallographic faces. On the other hand, glass yields small pits with curved boundaries.

Some Remarks on the Constitution of Glass

"One finds, as distinct from the condition of a pure solution, all variations up to a more or less ordered aggregation of simple molecules. This fact causes many to regard glass as a super-cooled liquid existing in a colloidal condition, but in a strict sense this may prove correct only in special cases."

Fritz Eckert (1925)[12]

Eckert,[12] who in 1923 had published an extensive review[78] of the physical properties of glasses, starts by differentiating between the state of aggregation of glass and its constitution. In terms of the former, glass is undoubtedly a supercooled liquid but, nevertheless, it exhibits physical and chemical properties that are quantitatively appropriate only to solids. Hence the vitreous state should be regarded as a special condition. Similarly, colloids differ only quantitatively in properties from mixtures or solutions, rather than qualitatively.

Glasses are formed when the molecular structure of the liquid is such as to produce a relatively high viscosity at temperatures where there is still a small possibility of crystallisation. Hence, during quenching through the dangerous temperature range of maximum crystal growth and nucleation rate, so few molecules gather to form crystalline nuclei that the material can still be considered as glass (amorphous). The same conditions of high viscosity retard the chemical reactions occurring in the melt by which, with increasing temperature, compounds are decomposed into simpler compounds and/or dissociated. Thus, in the resulting viscous slowly-reacting liquid, the state of combination/constitution of the constituent molecules is dependent on the total past heat treatment, and the constitution of the resulting glass will depend not only on the initial state but also on the length of time during which it has had to adjust itself as the temperature varies. Indeed, the influence of the past heat treatment may extend as far back as the formation of the batch materials from which the glass is prepared.

The fact that many of the properties of glasses, with the same composition in terms of their constituent oxides, vary with thermal history

> *"... proves that the original view that glass is purely a mixture of oxides is now untenable, ... since the properties of oxide mixtures are purely additive, or practically so, ..."*
>
> Fritz Eckert (1925)[12]

The author considers glass as a concentrated solution of simple and complex silicates, aluminates, borates, *etc.*, mainly in silica, although not an "ideal" homogeneous solution due to the molecular association of the individual constituents (*cf.* quotation at the head of this section) The form of the compounds in the molten liquid when it is more or less rapidly quenched, and how the constitutional properties are altered, depend upon the total, and above all the immediately preceding heat treatment.

On the Constitution and Density of Glass

> *"This preventative action may result either from all the simple molecules or components becoming involved in unwieldy non-crystalline aggregates, or through such a reduction in the mobility of the mass as a whole that any of the simpler components in excess and not included in these aggregates are restrained from the diffusion necessary for crystallisation and segregation."*
>
> A. Q. Tool & E. E. Hill (1925)[6]

In order to discuss the changes in glass arising from heat treatment and annealing, Tool & Hill[6] formulate a tentative hypothesis regarding its constitution, which requires that, under the appropriate cooling conditions, certain practically reversible processes or reactions involving some or all of the constituent molecules advance whilst others, such as segregation or crystallisation, are prevented, thus preserving the vitreous character. This can be seen as a natural extension of the ideas proposed in the earlier paper by Tool & Eichlin[72] outlined in Section II. As indicated by the above quotation, they envisage that the segregation/crystallisation is impeded by the existence in the melt of unwieldy non-crystalline aggregates. Hence crystallisation is suppressed by interference between those processes that lead to (metastable) equilibrium of the glass/melt and those that lead to devitrification.

The character of the various processes, and the rate at which they advance, will change as the glass cools and, at a fixed temperature, the rate should decrease with time until a state of practical (metastable) equilibrium is established. The final state should then be independent of thermal history but, at low temperatures where the changes proceed exceedingly slowly, the condition of the glass is one of approximate (metastable) equilibrium for some higher temperature. Tool & Hill[6] remark that this temperature has been designated the "effective annealing temperature," and that it depends on:

(i)   the condition of the glass before annealing,
(ii) the treatment or annealing temperature,
(iii) the period of treatment,
and
(iv) the mode of cooling.

In general, therefore, glass properties are affected by the total thermal history since the material was last in equilibrium.

Plots of the density at 25°C vs. annealing temperature are given for two glasses, a "medium flint" and Pyrex®. The annealing times increase with decreasing annealing temperature and range from half an hour to 1000 hours. In the case of Pyrex®, the density exhibits a minimum at ~540–570°C, which corresponds to a decrease in density from that for the original glass ($2 \cdot 2381$ g cm^{-3}) of ~$0 \cdot 005$ g cm^{-3}. Since Pyrex® is rich in silica, Tool & Hill[6] suggest that this minimum might be related to the $\alpha \rightleftharpoons \beta$ inversion of quartz near 573°C, but add that this would not necessarily entail the existence of even incipient crystallisation.

Tool & Hill[6] substantiate the conclusion of Eckert[12] that the fact that the properties of a glass are not simply an additive function of those of its independent components implies that the older view that glass may for all practical purposes be considered as a mixture of a number of oxides should be discarded. The condition of a glass is one that is intermediate between the liquid and the solid states. Also it should be regarded as undercooled not only in terms of the process of crystallisation but also with respect to the completion of many processes normal to the vitreous condition; *i.e.* chemical and physical processes including chemical reactions between the constituents, polymerisation, the formation of colloids, interpenetration, diffusion, *etc.*

On the Viscosity and the Allotropy of Glass

"Measurements of the expansion of glass have shown that most glasses exhibit a very marked anomaly in their expansion a little below the point at which they begin to soften. This phenomenon seems to indicate an allotropic transformation of glass analogous to that undergone by molten sulphur when heated above 250°."

Henry Le Chatelier (1925)[11]

The paper on viscosity by Le Chatelier[11] was published significantly in advance of the Symposium, in the March 1925 issue of the *Journal of the Society of Glass Technology (Transactions)*, and discusses the relationship between viscosity and the anomalies in expansion and specific heat close to the softening point. Viscosity is a very important property for glasses and glass-forming melts, *e.g.* in respect of glass blowing, moulding and drawing. It also controls the rate at which bubbles escape from the melt during fining.

Le Chatelier[11] explains the rapid increase in the coefficient of thermal expansion in the neighbourhood of the softening point in terms of an allotropic* transformation, analogous to that under-

---

*   See footnote on page 24.

gone by molten sulphur when heated above 250°C, and concludes that the two allotropic forms should exhibit different viscosity–temperature behaviour. An analysis of viscosity (η) data by English,[79] when plotted as log(logη) *vs.* temperature in degrees centigrade, *t*, yields two straight-line sections with different gradients, which *"confirms the existence of the two allotropic varieties of glass."* However, the problem is that the break between the two straight-line sections lies in the range 900–1100°C, whereas the thermal expansion anomaly occurs at temperatures in the region of 600°C (*cf.* Lebedev[18]); *i.e.* a discrepancy of ~400°C, which remained to be explained from a theoretical point of view.

The Structure of Quartz

"... in fact, the oxygens are arranged about the silicon in a regular tetrahedron ... The result is remarkable for its immediate adjustment to the known tetravalency of the silicon atoms."
Sir William Henry Bragg (1925)[8]

Both W. H. Bragg's Symposium paper,[8] and a more detailed paper published soon after the Symposium by W.H. Bragg & Gibbs[80] in November 1925, discuss the structure of α- and β-quartz. The structure of α-quartz had been studied much earlier in 1914,[81] but with only limited success, and so a new approach was required, which was via the much simpler structure of the β-polymorph.

The earliest crystal structures to be determined (*e.g.* NaCl[40]) were for cubic crystals with the atoms in special positions defined by the crystal symmetry (*i.e.* no arbitrary positional parameters), whereas the structures of non-cubic crystals and those with the atoms in general positions were much more difficult to solve. α-quartz belongs to the trigonal (rhombohedral) crystal system with three SiO_2 "molecules" per unit cell and four unknown atomic positional parameters. The earlier investigation had revealed the silicon atom positions to be related via a 3-fold screw axis parallel to the c_0 axis, but the distance of the silicon atoms from the screw axis and the co-ordinates of the oxygen atom in the asymmetric unit remained to be established.

β-quartz is of higher symmetry than α-quartz and has a hexagonal unit cell with only one arbitrary positional parameter. Hence intensity measurements could be employed to define this last parameter, thus solving the structure of β-quartz. The silicon atoms are surrounded by four oxygen atoms in the form of a regular tetrahedron and each oxygen atom "touches" two silicon atoms, with a Si–Ô–Si bond angle of 155°;[80] *i.e.* the oxygen atom does not lie on the straight line joining the centres of its two silicon atom neighbours.[8] W.H. Bragg seems somewhat surprised at the presence of corner-sharing SiO_4 tetrahedra (*cf.* above quotation), and the use of the word "touches" in Ref. 80 suggests that he viewed the structure as being formed from spherical atoms/ions in contact, rather than being the result of valency (covalent) bonding.

Once the structure of β-quartz had been established, that of α-quartz could be solved, since it was known that the transformation from α- to β-quartz implied small changes in the positions of the atoms in the unit cell, which it was shown resulted from relative rotations of the SiO_4 tetrahedra. The atomic co-ordinates for α-quartz were subsequently published by Gibbs[82] as a separate paper. W.H. Bragg's Symposium paper[8] and that by W.H. Bragg & Gibbs[80] both conclude with a discussion of the twinning of quartz crystals, and an explanation of the electrical properties (piezo-electricity, pyro-electricity and electrical conductivity) in terms of Si^{4+} and O^{2-} ions.

The Viscosity of Glass

"There can be little doubt, therefore, that the sudden increase in the slope of the logarithmic viscosity–temperature curve is due to the same change in the constitution of the glass as that which produces the heat absorption, and that when the glass is in equilibrium a point or region of transition exists in the neighbourhood of 500°."

Vaughan H. Stott (1925)[10]

In the case of the second viscosity paper, the author Stott[10] had the advantage of having read that by Le Chatelier[11] and, on the basis of further viscosity data recorded at the National Physical Laboratory (NPL), argues that some modification of Le Chatelier's conclusions was necessary. As indicated by the above quotation, comparison of a plot of log η vs. t for a "soft laboratory glass" with the heating curve for a small button of the same glass leads to a correlation between the point of inflexion for the former and the region of heat absorption for the latter. Stott[10] concludes that the change occurring in the constitution of glass should occur when the viscosity is in the neighbourhood of $10^{13.7}$ poise. Thus any discontinuities that may occur in viscosity curves at lower viscosity (higher temperature), as found by Le Chatelier,[11] must be ascribed to sources other than that associated with the heat absorption.

Attention is drawn to the difficulties in acquiring accurate viscosity data, and a detailed comparison is presented of the NPL results with previous data. The data of English[79] must be reduced by a factor of three, owing to a confusion between the coefficient of viscosity and the coefficient of viscous traction. This changes the appearance of the log(log η) vs. t curves of Le Chatelier[11] such that, in all cases except for one glass that was liable to devitrification, a smooth curve can be drawn through the data points.

The influence of constitution on the viscosity of the melt at high temperatures ($t \geq 1000°C$) can be seen from the sodium silicate system. The maximum deviation of plots of log η vs. mol% SiO_2 at constant t, from the normal form, occurs at 66 mol% SiO_2, and so Stott[10] suggests that, since the existence of the compound $Na_2O.2SiO_2$ had already been established, it is reasonable to assume that the abnormality in the viscosity curves is due to the presence of this compound in solution in the glass.

The Ternary System Sodium Metasilicate–Calcium Metasilicate–Silica

"There has indeed been much speculation as to the constitution of glass, but no systematic study of the compounds formed by the devitrification of glass, and of the relation of this devitrification or crystallisation to composition and temperature, information which is fundamental to the solution of *the puzzle of the constitution of glass; if, indeed, this puzzle is capable of solution by the methods at our disposal."*

George W. Morey & N.L. Bowen (1925)[7]

At the time of the Symposium, there was little knowledge of the crystalline phases formed by devitrification, even in the most common glass-forming system, soda–lime–silica, and no knowledge of the melting point relationships for these compounds. An investigation had therefore been performed by Morey & Bowen[7] of the equilibrium relationships between the crystalline and liquid phases in the ternary system sodium metasilicate–calcium metasilicate–silica, together with additional measurements on samples from the larger Na_2O–CaO–SiO_2 ternary system containing Na_2O plus CaO in excess of the metasilicate composition. This extended an earlier paper by the same authors[83] on the binary system sodium metasilicate–silica, in which the only intermediate compound identified was sodium disilicate.

Within the region of the Na_2O–CaO–SiO_2 system defined by the triangle $Na_2O.SiO_2$–$CaO.SiO_2$–SiO_2, there are seven compounds giving a total of ten crystalline phases. The polymorphs of SiO_2 (quartz, tridymite and cristobalite) were already known, as were $Na_2O.SiO_2$, $Na_2O.2SiO_2$ and two polymorphs of CaO. SiO_2 (pseudo-wollastonite, α-$CaO.SiO_2$, and wollastonite, β-$CaO.SiO_2$). As a result of their crystallisation studies, Morey & Bowen[7] were able to identify the three further (ternary) compounds, the mixed metasilicates $2Na_2O.CaO.3SiO_2$ ($2Na_2SiO_3$. $CaSiO_3$) and $Na_2O.2CaO.3SiO_2$ ($Na_2SiO_3.2CaSiO_3$), together with $Na_2O.3CaO.6SiO_2$ ($Na_2SiO_3.3CaSiO_3.2SiO_2$). The compound $Na_2O.2CaO.3SiO_2$ occupies the largest field of the ten phases, whereas $Na_2O.3CaO.6SiO_2$ (devitrite) is of particular importance in glass technology, being the primary phase to separate in a large proportion of commercial glasses, even though its stability field is small and far from its own composition.

Morey & Bowen[7] contrast crystalline compounds, having a definite structure and unique properties, with glasses, which they envisage as a continuous series of liquids or solutions whose properties change continuously with composition. An important result from their work is that there are four compounds that exist in or near the composition range that includes all commercial soda–lime–silica glasses, *viz.* $Na_2O.2SiO_2$, $Na_2O.2CaO.3SiO_2$, $Na_2O.3CaO.6SiO_2$ and SiO_2. All of these, with the exception of silica, are largely dissociated in the liquid phase, as is $2Na_2O.CaO.3SiO_2$, which decomposes before its melting point. $Na_2O.2CaO.3SiO_2$ is stable up to its melting point, whereas $Na_2O.3CaO.6SiO_2$ is decomposed long in advance of its melting point. Thus all of the compounds that should be considered as possible molecular species in soda–lime–silica glasses are highly dissociated, even in the pure state, and hence the stoichiometric liquid will probably consist not of molecular groupings of one species, or of two, but of several, and it is likely that the same molecular groupings exist over a large range of composition:

> *"... there can be no doubt that the molecular constitution of glass is a complex one, involving equilibrium between several molecular species. The relative proportions of these species will, of course, be determined by the usual thermodynamic considerations; these proportions will change with temperature in accord with the usual van 't Hoff principle of mobile equilibrium, and any such change will be continuous, the system remaining homogeneous."*
>
> George W. Morey & N.L. Bowen (1925)[7]

The authors comment that it is difficult to conceive of any change in the system, so long as it remains homogeneous, which would give rise to a discontinuous heat effect such as had been observed by other investigators.

X-Ray Diffraction Measurements on Some Soda–Lime–Silica Glasses

(A Preliminary Note)

> *"Neither do they prove, though they strongly indicate it, that entirely clear and apparently homogeneous glasses may contain large amounts of crystals of colloidal dimensions."*
>
> Ralph W.G. Wyckoff & George W. Morey (1925)[9]

In this, the second Symposium paper co-authored by Morey,[9] the
authors present preliminary X-ray diffraction data for a series of
soda–lime–silica glasses. Of the three explanations that had been
advanced for the origin of the broad diffraction halos from liquids
(Section II), Wyckoff & Morey[9] dismiss the first, since a similar
diffraction pattern is obtained for liquid argon. However, the experimental data necessary for a satisfactory quantitative explanation in terms of interference effects between molecular groupings
did not exist.

All of the glasses studied by Wyckoff & Morey showed no sign
of devitrification under the microscope or milkiness in thick section but, nevertheless, exhibited different types of diffraction pattern. In certain cases, one or more broad bands, thought to be characteristic of glasses, were observed but, in others, narrow bands
or lines were obtained that were as sharp as those produced by
crystals of colloidal dimensions. Some glasses even yielded a combination of broad bands and sharper lines. Both the bands and lines
varied considerably in width and frequently, but not always, several glasses had lines in the same positions. Measurements of the
positions of the lines and bands allowed the diffraction patterns
to be classified into five groups; e.g. glasses in the silica corner of
the ternary $Na_2O–CaO–SiO_2$ phase diagram yielded a broad band
similar to that for vitreous silica. Glasses that crystallised with the
greatest difficulty gave rise to broad bands, whereas the sharper
lines were found for those that were more readily crystallisable.
The authors comment that the available data were not yet sufficient
to show definitely whether the broad bands and sharper lines had a
similar origin but also suggest that entirely clear and homogeneous
glasses may contain large amounts of crystals of colloidal dimensions (see above).

General Discussion

*"In any case, the anomaly of glasses is common to all amor-
phous bodies, and has nothing to do with any particular property of silica, still less with the transformation of quartz."*
                              Henry Le Chatelier (1926)[84]

The General Discussion[84] consists of a series of comments contributed subsequent to the Symposium, presumably once those involved had read the published papers. The first comment is by Professor Henry Le Chatelier and concerns the paper by Turner.[2] Le Chatelier did not believe that the investigation of the physical properties of glasses could throw any light on their constitution, or on the nature of the chemical combinations that occur in them since, in solutions, properties are generally additive and therefore independent of the degree of chemical combination. There were, however, two exceptions that might possibly yield useful information: the optical absorption spectrum and magnetic properties. As to the role of definite compounds in glasses, Le Chatelier only knew of one example, $3ZnO.2B_2O_3$, which gives a beautiful glass on melting, but only exists close to this specific stoichiometry. Additional B_2O_3 forms a separate layer and extra ZnO results in devitrification on cooling the melt.

The transformation point for vitreous B_2O_3, measured in Le Chatelier's laboratory, was 240°C, at which point the coefficient of thermal expansion increased from 14×10^{-6} to 210×10^{-6} deg.$^{-1}$ This phenomenon was not connected with the allotropic* changes of state of any particular compound, as had been thought thereto, and was common to all amorphous bodies.[85]

Professor Turner replied that, in regard to the investigation of the constitution of glass via the measurement of physical properties, he did not think it desirable at that stage to exclude the possibility of success, despite the apparent additivity of many properties. However, Professor Le Chatelier's most interesting comments concerned the striking phenomena observed in the softening range of glasses and other amorphous substances. Professor Turner had conceived of a similar idea three or four years previously; viz. that the heat absorption and marked change in thermal expansion were common to all amorphous substances.

A further comment was submitted by Dr. P. E. Shaw, who reported the observation of "streakiness" on the surfaces of thin glass rods. This he explained in terms of the melt consisting of several

---

* See footnote on page 24.

38 THE CONSTITUTION OF GLASS

materials in a state of (imperfect) solution. On quenching to the vitreous state, some of these would be partially adsorbed into the glass surface, thus leading to a "patchiness". Professor Turner found these remarks intensely interesting, but suggested that the "patchiness" might be associated with the process of marvering, and the consequent formation of a toughened surface layer. Dr. Shaw's observations were also relevant to the non-uniform attack of glass surfaces by corroding agents such as steam and water under pressure.

Professor Le Chatelier also responded to the reservation by Stott,[10] concerning the non-homogeneity of his double-logarithmic formula, by arguing that this formula was merely empirical, but had the advantage that the parameters for the two straight-line sections could be related to the chemical composition of the glasses concerned. He also remarked that he abandoned the notion that the point of intersection of the two straight lines corresponds to an allotropic* transformation. The transformation observed for glasses was not peculiar to the vitreous state, but belonged to all amorphous bodies.[84] At low temperatures, vitreous materials have a coefficient of thermal expansion of the same order of magnitude as that of crystalline solids and, at high temperatures, their thermal expansion is similar to those of liquids. Mr. Stott thanked Professor Le Chatelier for his interesting remarks, but had nothing further to add.

The final contribution to the General Discussion was by Dr. Fritz Eckert, who wanted to add some further evidence to support the hypothesis advanced in his Symposium paper concerning the dependence of the properties of glass upon its past heat treatment. It was known that the viscosity of molten glass, particularly at the working temperature, is very dependent on the type of batch constituents; e.g. a glass made with soda ash is "softer" than the same glass (composition) made from saltcake. The properties of the final glass depend on the decomposition temperatures and products of the batch constituents, the melting temperature and the duration of melting, those most influenced by thermal history being the viscos-

---

*   See footnote on page 24.

ity of the melt and the strength (brittleness) of the final glass. The rigid and fluid states are connected by a transition state characterised by a change in the rate of thermal expansion.

Dr. Eckert argues that it is also possible that reactions, which are relatively incomplete at high temperatures, lead to irregular heterogeneities of a size greater than that corresponding to the molecularly-disperse state but small enough that the glass is optically homogeneous. The large strains, undoubtedly existing between these small heterogeneities, he assumed are the cause of the increased brittleness of glass. His hypothesis was that phases exist in equilibrium in glass at any fixed temperature having varying properties, but that this equilibrium is only more or less completely reached before a change in temperature; hence the dependence of glass properties on thermal history. Professor Turner welcomed Dr. Eckert's additional notes and added that work at Sheffield substantiated and extended the latter's views.

IV. TO THE END OF THE DECADE

"As will be gathered from the foregoing, our available trustworthy data such as would make it possible to understand the nature of glasses are few and any structure of theory erected on them is liable to be unstable. Yet it is desirable to make a beginning, even if all that it is possible to do is to speculate, or to proceed by analogy."

William Ernest Stephen Turner (1925)[2]

Prior to making some general comments in Section V, concerning the Symposium papers, it will be helpful to briefly survey a few important papers on the constitution/structure of glass that were published during the years immediately following the Symposium and continue its themes. The obvious cut-off point is 1930, with the publication of Randall, Rooksby & Cooper's X-ray diffraction studies[86-88] supporting the crystallite theory. These three papers represent the last major contribution to glass structure/constitution research prior to Frederick William Holder Zachariasen's famous 1932 paper, entitled *The Atomic Arrangement in Glass*,[89] which is conventionally credited as introducing the **random network theory** of glass structure, and was to have such a profound effect on the future direction of Western glass science.

The French version of Lebedev's 1924 paper, published in January 1926, incorporates extra references to papers that appeared later in 1924 and in 1925. These include several by Le Chatelier[10,90-92] on the viscosity–temperature relationship for different kinds of glass, which Le Chatelier interpreted in terms of the existence of allotropy* in glasses. (Note that Ref. 10 is Le Chatelier's Symposium contribution published in March 1925.) Lebedev's hypothesis concerning the existence of α- and β-modifications of glass is further developed in a paper discussing *Structure Changes in Glass* published in *Die Glas-Industrie* in 1927.[93] Marked changes in glass occurred over a range of ~50°C, which Lebedev terms the "critical range", and heat is absorbed on passage through this re-

---

* See footnote on page 24.

gion (*i.e.* the $\alpha \rightarrow \beta$ transformation is endothermic). The amount of α-modification present at ambient temperature is a maximum for a well-annealed glass, whereas much more of the β-modification is present in a rapidly quenched glass.

Although not published until 1925, the X-ray diffraction study by Seljakow, Strutinski & Krasnikow[94] is mentioned in both versions of Lebedev's 1924 paper,[21,22] but neither version includes a formal reference. Seljokow *et al*[94] found no indication of the presence of crystals in glasses where there were no apparent traces of devitrification. Lebedev[21,22] argues, however, that this does not disprove the hypothesis that explains the polymorphic transformation in glass as one originating within silica molecules (*cf.* quotation on page 19).

One of the extra papers included by Turner[13] in the 1927 volume containing the Symposium papers is that by Stott,[95] entitled *The Viscous Properties of Glass*, published in the *Journal of the Society of Glass Technology (Transactions)* in 1926. In this sequel to his Symposium paper,[10] Stott[95] highlights two distinct problems requiring solution:

(i)   The relationship between viscosity and thermal history for glasses of different composition,

and

(ii) The influence of a *"very wide range"* of shear rate on the apparent viscosity of glass at high temperatures.

As an example of (i), he cites a particular soda–lime–silica glass that had been rendered homogeneous by stirring at 1470°C and yielded a reproducible viscosity–temperature curve at temperatures between 1200 and 1530°C. On maintaining the glass for too long within a critical temperature range below 1200°C, the viscosity became abnormally high, which Stott[95] argues must be associated with some form of heterogeneity. However, there was no evidence of incipient devitrification, nor any difference between the *"homogeneous"* and *"heterogeneous"* glasses under the optical microscope. He suggests that the local variations in viscosity must be very large and notes that submicroscopic heterogeneity in glass at ambient temperatures had been postulated both by Eckert[12] and Filon & Harris.[54]

The second additional paper in Turner's 1927 volume[13] is a long-overlooked but nevertheless extremely important paper by Walter Rosenhain[20] on The *Structure and Constitution of Glass*, which anticipates many of the key ideas found in Zachariasen's well-known paper.[89] As in the latter, Rosenhain[20] begins by noting the then dearth of knowledge concerning the structure and constitution of glass, and subsequently proceeds to discuss the structure of glass in terms of the very much better understood structure of crystalline materials. He points out that a careful study of the papers presented at the Symposium leads to the fundamental generalisation that glass must be regarded as an amorphous solid or under-cooled liquid, but that this just widens the problem to that of the real nature and structure of under-cooled liquids (*cf.* Zachariasen's comments[96] in his 1935 rebuttal of Hägg's criticism[97] of the random network theory.) or quasi-solid amorphous substances. On the other hand, glass behaves *"much as any other super-cooled liquid of complex composition"*, and recent work by Le Chatelier and by Samsoen[85] had indicated that certain phenomena believed to be typical of glass occur generally in amorphous materials.

Considerable progress had been made in recent years in determining the structures of crystalline materials, whereas Rosenhain[20] comments that:

> *"On the experimental side as well as on the theoretical, glass has proved singularly elusive."*
>
> Walter Rosenhain (1927)[20]

One of the conclusions to be drawn from X-ray studies of crystals was that the atom, rather than the molecule, must be regarded as the basic unit of structure. In crystals such as NaCl, molecules clearly do not exist and, in more complex systems, it was doubtful whether there is any significant difference between the bonds within a molecular grouping and those linking such groupings. Materials such as glasses could be regarded essentially as assemblages of atoms in which certain "molecular groupings" could occur with greater or lesser frequency. The more complex the molecule, especially in a mixture or solution of different substances, the greater

is the difficulty of crystallisation and hence the ease of formation
of the vitreous or amorphous state. A simple silicate is much more
likely to crystallise than is a mixture in which a variety of molecular groupings is present.

To explain the difference between crystals and glasses, Rosenhain[20] considers, as a first approximation, an assemblage of atoms
linked together by *"bonds"* (Rosenhain's inverted commas) varying in strength depending on the chemical species involved and the
inter-atomic distance. For any given pair of atoms there is a minimum distance apart and a maximum separation beyond which the
forces that constitute an inter-atomic bond cease to operate. Variations from the equilibrium bond length always lie within narrow
limits and are of similar magnitude whether caused by heating,
leading to fusion, or under stress, resulting in fracture or plastic
yielding. More importantly, Rosenhain[20] realises the necessity to
consider bond-angle bending in addition to bond stretching forces
(*i.e.* directional rather than spherically-symmetric bonding):

*"... it is difficult to avoid the idea that inter-atomic forces are
vectorial and that inter-atomic linkages tend to be formed in
certain definite directions, that is, that the lines joining the
centre of an atom to the centres of other atoms to which it is
linked tend to form certain definite angles with one another
and that linkages in other directions, if they can occur at
all, imply a certain distortion of the atom or its surrounding force-field which implies storage of energy. If this view
is accepted, it follows that even in an irregular assembly of
atoms there will be a tendency for the atoms to form linkages
only in certain relative directions or to turn round in such a
way as to bring the angles between linkages as close to the
normal angle as possible. The formation of linkages in an irregular assembly is thus limited by two factors – inter-atomic distance and angular relations between adjacent atoms."*

Walter Rosenhain (1927)[20]

The criterion for a crystalline structure is that it yields sharp
X-ray reflections, which implies a regular atomic (strictly inter-

planar) spacing over a sufficiently large region of space compared to the X-ray wavelength. For an amorphous material, on the other hand, the atomic arrangement must be irregular, at least over distances for which a regular arrangement would give rise to a sharp reflection. The question was how regular should such an arrangement be? Rosenhain's answer[20] is that the atoms will first arrange themselves such that they exceed their minimum separation, but then they will try to "satisfy" as many of their potential bonds as possible, consistent with the restrictions on bond length and angle outlined in the previous paragraph. His final picture is thus of the crystal with a regular strain-free structure and the glass having an irregular arrangement incorporating both bond-length and bond-angle strain, and the energy associated with any "unsatisfied" bonds:

> *"Here the idea at the root of our considerations is that the total number of atomic linkages is materially smaller than in the stable crystalline arrangement, and that whilst in the undistorted crystal lattice all the effective linkages are, between given kinds of atoms, alike, those in the amorphous assembly will differ widely, both as regards inter-atomic distance and in regard to the angular relations between the atoms."*
>
> Walter Rosenhain (1927)[20]

Having established his concept of the structure of amorphous solids, Rosenhain[20] then applies his ideas to an explanation of various properties, including softening, specific heat, elastic properties, fracture, thermal expansion, pressure effects, atomic diffusion, surface properties, electrolytic replacement of alkali cations, *etc.* For example, on heating a crystalline substance, the amplitude of the atomic thermal oscillations increases until, at a single temperature (the melting point), the structure breaks down with the absorption of the latent heat of fusion (energy required to break bonds) to form the liquid. The situation for a glass or amorphous solid is rather different. Initially, as for a crystalline solid, raising the temperature results in an increase in thermal vibration amplitudes. However, above a certain temperature (the glass transition temperature, T_g?), the most highly strained bonds begin to break and hence the heat necessary to raise the temperature comprises two components, one to increase the amplitude of thermal vibration and the other to supply the small amounts of latent heat necessary to break the most strained bonds at any given temperature. Thus the glass softens over a range of temperatures, as more and more bonds are broken, rather than exhibiting a discrete melting point. The release of strain energy will lead to the evolution of heat, whilst heat will be absorbed during bond rupture, resulting in a slight overall absorption of heat. Rosenhain[20] associates the onset temperature for the breaking of the most strained bonds with the sudden increase in the rate of absorption of heat in the neighbourhood of the softening temperature observed by Tool & Valasek[63] (Section II). Similarly, at the onset temperature, there should be a rapid and marked expansion, followed by a generally higher rate of thermal expansion, as found experimentally, and, above this temperature, the specific heat should be higher than for the corresponding crystalline material, due to the latent-heat contribution.

Towards the end of his paper, Rosenhain addresses the presence of more or less complex molecules in glass, which he assumes must occupy the greater part of the material. He also considers the role of aggregates:

> *"It is also possible, so far as experimental evidence goes, that aggregates of atoms or even of molecules may be present in the form of very minute crystallites*, too small to yield X-ray reflexions."*
>
> Walter Rosenhain (1927)[20]

The bonding within such aggregates, he argues, is too close and regular to give rise to the phenomena characteristic of the amorphous state and so they act as neutral or "dead" zones. In a complex substance, there will similarly be a range of different bonds, varying in both closeness and strength, as shown by the fact that it is pos-

---

*   The first use of the word *crystallite* in a paper on glass structure?

sible to replace one alkali metal by another via electrolysis at temperatures in excess of 400°C, thus rupturing alkali metal–oxygen bonds. On the subject of well-defined compounds in glasses, Rosenhain[20] comments that their presence should be indicated by marked maxima or minima, or at least breaks, in property curves that are in general only rarely encountered, mainly in simpler types of glass.

The discussion of chemical heterogeneity in binary and multi-component glasses during the Symposium seems to have been limited to singe-phase glasses, even though the phenomenon of phase separation had been recognised much earlier (Section II). The subject of phase separation was, however, addressed in 1927 in an important two-part paper by Grieg,[98,99] who reported a sub-liquidus immiscibility region (liquid–liquid phase separation) in the phase diagrams for a number of binary $MO–SiO_2$ (M = Mg, Ca, Sr, Mn, Zn, Fe, Ni or Co) and ternary silicate systems. In each case, the two liquids were in equilibrium with cristobalite and one of them comprised nearly pure silica. Cristobalite was found to crystallise very rapidly from the less siliceous liquid on cooling. On the other hand, there was no evidence of liquid immiscibility in the $BaO–SiO_2$, $Al_2O_3–SiO_2$, $Na_2SiO_3–SiO_2$, $K_2SiO_3–SiO_2$ and $B_2O_3–SiO_2$ systems. As regards liquid immiscibility, silica was remarkably similar to boric oxide, as could be seen by comparison with the much earlier results of Guertler[39] (Section II) for the corresponding borate systems.

Two further X-ray diffraction papers were published in 1929, by Clark and co-workers.[100,101] These authors consider glass as a supercooled solution of various silicates or metal oxides in excess silica, and start by considering pure SiO_2. Unfortunately, whilst acknowledging that the structure of quartz is based on SiO_4 tetrahedra, they analyse their diffraction patterns, for both vitreous silica and a multi-component glass based on the soda felspar composition ($Na_2O.Al_2O_3.6SiO_2$), in terms of the erroneous atom-triplet model of Sosman[70] (see page 18), and invoke a structure for vitreous silica consisting of tangled threads of SiO_2 atom-triplets placed end to end. The position of the two peaks in the diffraction pattern for vitreous silica, and for the felspar glass melted at different temperatures, are interpreted using Bragg's law to yield the repeat molecular spacing along the thread ($7 \cdot 15$ Å for vitreous silica) and the spacing between the oxygen atoms in adjacent threads (*i.e.* the molecular width; $2 \cdot 52$ Å for SiO_2).

As indicated above, X-ray diffraction support for the crystallite theory of glass structure was initially provided in 1930 by the three papers of Randall, Rooksby & Cooper[86-88] and, in view of their importance, a facsimile copy of that published in the *Transactions* section of the *Journal of the Society of Glass Technology*[88] is included with the Symposium papers in this volume, together with subsequent discussion.[102] Randall *et al*[86-88] studied a wide range of glasses (vitreous silica, wollastonite, sodium diborate, potassium borate, selenium, potash and soda felspars, boric oxide, a series of more usual soda–lime–silica and borosilicate glasses, and glucose and sucrose glasses), using both Cu K_α and Mo K_α X-radiation, and a microphotometer was employed to extract quantitative intensity data. All three 1930 publications, however, concentrated on just three of these glasses: vitreous silica, wollastonite and sodium diborate. The new concept introduced by Randall *et al*[86-88] into their data analysis was the fact that, for extremely small crystals (crystallites), the normal Bragg peaks are broadened by an amount that increases with decreasing crystallite size and is given by the Scherrer particle-size equation.[103] Hence their approach was to compare the X-ray diffraction pattern for the glasses with a suitably broadened version of that for the corresponding polycrystalline phase.

The concept of crystal-like ordering had already been applied to the liquid state, where one possible explanation of the origin of the observed X-ray diffraction halos was more or less transitory groupings of atoms or molecules into crystal-like aggregates (Section II). This concept was further developed by Stewart[104-106] who proposed the presence in liquids of temporary ***cybotactic groupings*** of molecules, with an arrangement very probably similar to that which exists in the solid state:

> *"If the X-rays give evidence of periodic molecular grouping, it must not be supposed that these groups are large or that the molecules in any one well defined group remain per-*

manently members of that group. At any one instant these small orderly molecular groups might exist at numerous points in the liquid, the regions between them being not so orderly. The percentage of the volume occupied by fairly orderly groups might be large in some liquids and not so large in others."

G. W. Stewart (1930)[106]

Stewart's ideas[104-106] are quoted by Randall *et al*[87,88] to indicate that, even in liquids, the atoms/molecules are always trying to arrange themselves in the regular fashion associated with the crystalline state. Hence it is to be expected that the same would be true of glasses; *i.e.* the crystal-like groupings in liquids would be frozen in on quenching to the vitreous state. Similarly, further crystallinity could arise in the vitreous state due to incipient devitrification. It is also interesting to note that Stewart's cybotactic groupings were later (1936) invoked in Valenkov & Porai-Koshits' famous paper[107] that led to what became known as the ***modern crystallite theory***.[108,109]

Randall *et al*'s measurements[86-88] for vitreous silica yielded a strong band equivalent to a *d*-spacing of 4·33 Å, plus a very broad, feint band at 1·5 Å, and hence they conclude that the two bands recorded by Clark and co-workers[100,101] were spurious. Their vitreous silica diffraction pattern could be effectively modelled assuming 15 Å α-cristobalite crystallites whereas, for amorphous (precipitated) silica that had been heated for a few hours at 600°C to remove water, the crystallite size was somewhat larger:

"... the resemblance of the observed band and the calculated one shows beyond reasonable doubt that the glass consists of cristobalite crystallites."

J. T. Randall, H. P. Rooksby & B. S. Cooper (1930)[88]

Quartz crystallites were eliminated since α-quartz would yield a band centred at a higher scattering angle than that observed, whereas the crystallites were unlikely to be of tridymite because this is a fairly rare mineral and cristobalite is invariably found when vitreous silica is devitrified at 1200°C. The authors calculate that the average crystallite contains about 20 molecules (SiO_2 composition units) and conclude for the first time that the atomic arrangement in vitreous silica is based on corner-sharing SiO_4 tetrahedra, as in β-cristobalite. They also conclude that

> 'The difference between the crystalline state on the one hand and the "vitreous" and "amorphous" on the other is one of degree, not of kind.'
> J. T. Randall, H. P. Rooksby & B. S. Cooper (1930)[88]

The conclusions for the wollastonite ($CaO.SiO_2$) and sodium diborate ($Na_2O.2B_2O_3$) glasses were very similar. For the former, the best agreement with the glass diffraction pattern was provided by the high-temperature pseudo-wollastonite polymorph, with a crystallite size of 16 Å. In the case of sodium diborate, the glass was compared to the broadened diffraction pattern from a devitrified sample, prepared by allowing the glass to cool with the furnace. On the other hand, the results for soda and potash felspar glasses ($M_2O.Al_2O_3.6SiO_2$; M = Na or K) are rather different. Both glasses gave rise to a single band in nearly the same position as that for vitreous silica, whereas the polycrystalline phases would predict two bands. This suggested that, on heating, the felspar structure had broken down and that it had not reformed on cooling, which is supported by the fact that the devitrification product for potash felspar glass is cristobalite.

As indicated by Randall et al,[87] the centre of the band for vitreous silica is not quite in the same position as the 111 Bragg reflection for polycrystalline α-cristobalite, the former having a d-spacing that is 6·6% larger, which could be explained by theoretical calculations predicting an expansion of the crystalline lattice with decreasing crystallite size below ~500 atoms. Similar discrepancies were observed for other glasses. However, an isotropic expansion of the α-cristobalite lattice by 6·6% would yield a density for the glass that is 17·5% less than that of α-cristobalite and is much too low. {The density of vitreous silica (2·202 g cm^{-3}) is only 5% lower than that of α-cristobalite, as calculated from the X-ray lattice parameters[110] $(2 \cdot 318 \text{ g cm}^{-3})$.}

Whilst the X-ray results suggest an average crystallite size of between 10 and 100 Å, they also reveal that the distribution of crystallite size is not simple, but probably comprises at least two groups, one having a much larger average size than the other. A glass could therefore be conceived as a matrix of the larger crystallites connected together by still smaller units (the smaller crystallites). Since the average crystallite size is close to the limit for reinforced reflection according to Bragg's law, the smaller crystallites (groups of atoms) might be expected to scatter X-rays as independent atoms/molecules, and hence the average crystallite size may not be correct*. The authors also conclude that considerable numbers of groups of atoms smaller than the average exist in glasses and amorphous solids and that the proportion of extremely small particles increases as the average size decreases.

The distribution of crystallite sizes, and the presence of extremely small crystallites, is invoked by Randall et al[87,88] to account for several glass properties (density, "melting", ionic conductivity, thermal expansion and viscosity). It was to be expected that the potential energy of an agglomerate of crystallites would be higher than that for a macroscopic single crystal. Thus, for example, like Frankenheim,[19] Randall et al[87,88] argue that the process of softening can be explained by the fact that the atoms in the smaller units/crystallites can migrate at a much lower temperature than those in a bulk single crystal and that the larger crystallites will "melt" at a somewhat higher temperature than the smaller ones.

The hypothesis "that glasses frequently contain vast numbers of ultramicroscopic crystals" also appears in a commentary by Preston,[111] On the Supposed Diphasic Nature of Glass, in which he challenges the interpretation by Filon & Harris[54] (Section II) of their uniaxial pressure experiments on flint glass. The latter authors explain the resulting double refraction on a photo-elastic basis, but this requires the glass to consist of at least two phases having dif-

---

* It is also necessary to distinguish between the number and volume averages, since the contribution to the X-ray diffraction pattern from a given crystallite size would be proportional to its volume fraction.

ferent stress-optical coefficients. Preston,[111] on the other hand, suggests that the birefringence could arise if anisotropic ultramicroscopic crystals were aligned with their axes in one direction, in which case the observed phenomena would not be photo-elastic in character but the glass would show double refraction of the crystalline variety. This interpretation was supported by a drawn glass tube that showed uniform double refraction along its length and, under the ultramicroscope, was found to be full of minute elongated crystalline particles.

By 1930, the situation *vis à vis* the constitution and structure of glass was that there were several theories, all having one or both of two repeating themes, *viz.* the prediction of chemical nanoheterogeneity for glasses having more than one component, and the presence of some element of molecular clustering and/or crystal-like ordering. In particular, it should be noted that one of the important consequences of the crystallite theory, for binary and multi-component glasses not of crystalline stoichiometry, is that they must be chemically nanoheterogeneous. This concept is diametrically opposed to the much more homogeneous structure predicted by the random network theory in its original form, and as advocated in the papers by Bertram Eugene Warren and co-workers published between 1933[112] and 1941.[113] In the West, the random network theory rapidly replaced the crystallite theory, although the latter, in modified form, together with the concept of a heterogeneous structure, was to dominate Soviet glass science for the next 50 years,[32,114] which perhaps explains why many of the early advances in the area of phase separation in glasses occurred in the former Soviet Union and Eastern Europe, rather than in the West.

V. COMMENTARY

"It must be frankly admitted that we know practically nothing about the atomic arrangement in glasses."

Frederick William Holder Zachariasen (1932)[89]

It is interesting to note that the papers presented during the Symposium concentrated on the (molecular) constitution of glass, but none of them discussed the atomic arrangement either in the molecules themselves or in glasses. Sosman's suggestion[70] concerning the presence of threads of SiO_2 atom-triplets in both crystalline and vitreous silica was shown to be incorrect in 1925, both by W.H. Bragg's Symposium paper on α- and β-quartz[8] and by Wyckoff's X-ray diffraction determination of the structure of β-cristobalite,[76,77] which clearly demonstrated the absence of discernible SiO_2 molecules.*

"In such a structure, as in the diamond or sodium chloride, no intimate association of a few atoms corresponding to a chemical molecule, or ion group, can be observed. If 'molecules' exist in such a crystal they can scarcely be smaller than the entire piece."

Ralph W. G. Wyckoff (1925)[76]

The concept of the crystal comprising a single molecule (*cf.* Julg's monograph *Crystals as Giant Molecules*[117]), which also applies to the structure of diamond, as determined by W. H. & W. L. Bragg in 1913,[43,44] is tantamount to an acceptance of a network structure, since the nature of the valency (covalent) bonding within mole-

---

* The only difference between Wyckoff's and the currently accepted structure[115,116] is that the oxygen atoms are located on the straight line joining their two silicon atom neighbours, thus yielding too short a Si–O bond length (1·541 Å) and a Si–Ô–Si bond angle of 180°. It is now known that the oxygen atoms are displaced such that the Si–O bond length is 1·611 Å at 300°C and the Si–Ô–Si bond angle is 146·7°.[115] However, there is orientational disorder and the oxygen atoms can rotate around the line joining the silicon atoms,[115,116] with the result that the "average" oxygen atom position is in fact that given by Wyckoff.[76,77]

cules, *i.e.* the sharing of electron pairs between atoms, had already been established by Lewis[118] in 1916. However, like W. H. Bragg, Wyckoff[76] seems unconvinced as to the role of valency and the presence of electron sharing (covalent bonds):

> "*It will be noticed that each silicon atom is surrounded by four equally distant oxygen atoms and each oxygen atom by two silicon atoms, but whether this correspondence to the valencies of the two atoms involved has any physical significance or is, from the standpoint of the valencies, more or less a matter of chance, can be at present a subject for speculation only. If electron sharing between atoms is possible and if such electron sharing takes place in high-cristobalite, then one might expect some such structure as the one found here; but as far as existing information shows the atoms of silicon and oxygen may equally well be electrostatically charged, as are undoubtedly the atoms in many inorganic compounds, like sodium chloride or magnesium oxide.*"
>
> Ralph W. G. Wyckoff (1925)[76]

Hence, whereas chemists had objected to the absence of molecules, these crystallographers found it difficult to accept the presence of electron sharing {*i.e.* the type of (covalent) bonding found within molecules} and that not all inorganic crystals were simply packings of spherical ions!

On 3rd June 1930, W. L. Bragg[119] read a paper on *The Structure of Silicates* at a joint meeting of the Society of Glass Technology and the Deutsche Glastechnische Gessellschaft, which was held in London. The corresponding manuscript, published in the *Journal of the Society of Glass Technology (Transactions)* and included as one of the facsimile papers in this volume, summarises a much longer paper of the same title in *Zeitschrift für Kristallographie*[120] and adds a paragraph on vitreous silicates. It also followed an earlier (1927) *Nature* paper,[121] again entitled *The Structure of Silicates*, in which crystalline silicate structures are considered from the point of view of packings of oxygen atoms with metal and silicon atoms in the interstices. Thus the SiO_4 tetrahedron is regarded as four closely packed oxygen atoms with a silicon atom at the centre, the O–O distance of ~2·7 Å being characteristic of a close packing of oxygen atoms. On the other hand, crystalline structures where the oxygen atoms are arranged as an hexagonal or cubic close packed array are exceptional and normally the structures are more open, but nevertheless may be regarded *"as an embroidery of the metal atoms upon an oxygen framework."*[121] W. L. Bragg[121] also comments that the *"absence of the molecule in solids is in general only found in inorganic compounds."*

The paper presented by W. L. Bragg at the London meeting[119] concentrates on the various ways in which corner-sharing SiO_4 tetrahedra are linked together in the crystalline state; *viz.* orthosilicate anions, polysilicate and cyclic anions, chains, sheets and three-dimensional networks. The metallic cations are incorporated into the structure such that there is a local balancing of the electric charge between the cations and the negatively-charged (*i.e.* non-bridging) oxygen atoms. In the vitreous state, W. L. Bragg indicates that he expected there to exist a tendency towards regular arrangements and cites the X-ray diffraction data of Randall (Rooksby & Cooper), for compounds with both crystalline and vitreous phases, that indicate localised imperfect groupings similar to those in the ideal crystal.

W. L. Bragg's *Zeitschrift für Kristallographie* review[120] includes a footnote with comments by Victor Moritz Goldschmidt, who had received a copy prior to publication, concerning the nature of the bonding in silicates, which he compares to fluoroberyllates and germanates. Goldschmidt argues that the cation:anion radius ratio is a more important factor than the character of the interatomic bonds and concludes:

> *"I have the impression that the differences between ionic bonds and valency bonds in crystals in many cases at least are not so important for the mechanism of atomic interaction as one might assume from analogous cases of gaseous molecules and that the rude picture of ionic spheres may not be very far from reality even in cases where the interaction in the crystals is not an ionic interaction sensu stricto."*
>
> Victor Moritz Goldschmidt (1930)[120]

The problem concerning the nature of the bonding in silicates seems to have been due to the assumption that the bonding in solids was purely the result of either electron sharing (intra-molecular, valency or covalent bonds) or the electrostatic attraction between charged ions of opposite sign (ionic bonds) whereas, in most cases, it is intermediate in nature, with a varying degree of ionic/covalent character. Thus the association of the number of neighbours around a given atom with the chemical valency is appropriate for predominantly (single) covalent bonding, whilst the chessboard pattern of W. L. Bragg (NaCl) is characteristic of a predominantly ionic solid. In the case of both crystalline and vitreous silicates, it appears not to have been appreciated that the network modifying cation – non-bridging oxygen atom bonding is more ionic than that within the SiO_4 tetrahedra, although W. L. Bragg does concede:

"The silicates are an assemblage of silicon, oxygen and metal atoms which as crystalline structures occupy a position intermediate between the metallic oxides and the inorganic salts. To single out silicon–oxygen complexes in them, and compare these to acid radicles, is somewhat arbitrary and is perhaps only justified by convenience of description ... It seems preferable, however, to single out the silicon-oxygen link as essentially different to the electrostatic bond between metal and oxygen. We must then think of the silicon–oxygen complexes as units, packed together with metal atoms between."

William Lawrence Bragg (1930)[120]

This separation of the silicate and metal components is reminiscent of Mendeleev's comments[29,30] concerning the invariable and variable portions of silicate structures (*cf.* Section II) and, like Mendeleev, W. L. Bragg considers the role of isomorphous replacements in (crystalline) silicates and the problem of establishing the correct chemical formulæ.

Neither Wyckoff[9] nor W. H. Bragg,[8] who seems to have been unaware of Wyckoff's papers,[76,77] extended their crystallographic studies of β-cristobalite[76,77] and of α- and β-quartz,[8,80] respectively, to suggest that the structure of vitreous silica might also be based on SiO_4 tetrahedra. Indeed, it is sad to reflect that, had Wyckoff had the courage to follow up on his comments concerning a single (giant) molecule and the possibility of electron sharing (covalent bonds) in β-cristobalite, and extended them to vitreous silica, he might well have predicted a disordered corner-sharing SiO_4 tetrahedral network structure several years before the random network theory!

The idea of a random structure for vitreous silica is very briefly mentioned by Gibbs[122] in his 1926 paper on *The Polymorphism of Silicon Dioxide and the Structure of Tridymite*, but he too was still thinking in terms of ions:

"Lastly, fused silica may be regarded as a random distribution of charged ions in thermal agitation, which is far too viscous to crystallise on cooling, even in a very long period; in order that devitrification may occur, the mass has to be held for a long time at a high temperature not far below its melting point."

R. E. Gibbs (1926)[122]

On the other hand, Turner[2] comes tantalisingly close to the idea of a network structure:

"The formation of a compound might only be possible when the constituent oxides lost their mobility and the component atoms became linked up into the continuous arrangement which X-ray spectrographs indicate as occurring in crystal structures."

William Ernest Stephen Turner (1925)[2]

Clearly, the time had come for the concept of a disordered vitreous network, but nobody seems to have realised this until the work of Rosenhain[20] and Zachariasen.[89]

It is important to understand that, in 1925, the concept of an amorphous solid was not clearly defined, since it was only with the improvement of X-ray diffraction techniques that it became possible to distinguish between materials that were truly amorphous and those that were polycrystalline, but with crystals/crystallites too small to be observed in the optical microscope. Prior to X-ray diffraction, the method of identifying amorphous materials was via their physical and chemical properties; *e.g.* from their shape (often rounded or roughly spherical in the case of minerals), isotropic optical properties, lack of cleavage planes, *etc.* Hence, in his review of amorphous minerals published in 1917, Rogers[123] includes colloidal materials as a subset of amorphous minerals, and comments that most amorphous minerals are hardened hydrogels that incorporate water (*e.g.* opal). He quotes an early mineralogist, Breithaupt, who divided amorphous substances into two groups, glasses and gels. Ordinary glasses, Rogers[123] argues, are amorphous but not colloidal, although there are exceptions, such as opalescent glasses. Similarly, it was possible that mineral hydrogels should not, strictly speaking, be classified as colloids, but rather as colloidal in origin, since technically the term colloid implied that a material is in the form of a sol (colloidal solution). Rogers[123] also criticises the tendency for some contemporary mineralogists to describe as colloidal not only hydrogel materials but also microcrystalline substances of colloidal origin, thus confusing the terms amorphous and colloidal. It is, therefore, easy to see why, at the time of the Symposium, there was still confusion as to the relationship between colloidal materials and glasses, and as to whether glasses contained extremely small crystals (crystallites) such that there was a continuous gradation from polycrystalline materials through colloidal systems to glasses (*cf.* quotation from Randall *et al*[88] on page 49).

In addition to the evolution of ideas concerning the constitution and structure of glasses, it is interesting to observe the development of an understanding as to the nature of the glass transition. Several of the Symposium papers,[2,10,11,84] and those by Lebedev,[18,21-24,93] incorporate the idea of a polymorphic (allotropic) transformation to explain the endothermic effect and increase in thermal expansion within the annealing/softening range, which can of course now be correlated with the glass transition. Various authors put forward explanations as to why glasses soften over a range of temperatures, rather than exhibiting a discrete melting point. Thus, according to Lebedev,[18]

> " ... *the polymorphic transformation in glass is essentially not an instantaneous but a slow viscosity change of the medium, and it has a character corresponding to the transformations of impure crystals and solid solutions, i.e. the transformation is completed in a certain temperature interval, so that the glass passes through a continuous series of equilibrium states and each temperature has its corresponding specific equilibrium state.*"
>
> Alexander Alexeevich Lebedev (1921)[18]

Tamman's suggestion[5] that, with sufficiently rapid cooling, any melt could be quenched into the vitreous state has been supported in more recent years via the development of rapid quenching techniques, which have led, for example, to the discovery of metallic and heavy metal fluoride glasses.

Commenting on Le Chatelier's Symposium paper[11], and on those published in *Comptes Rendus*[90,91] in 1924, Preston[111] points out that Le Chatelier's double-logarithmic equation, which had already been criticised by Stott[10] in his Symposium paper (See also the Symposium General Discussion session[84] on page 229.), is physically meaningless and depends mainly on the fact that the French (c.g.s) unit of viscosity (poise) is very small. In English or French (Engineering) units, $\log(\log \eta)$ is imaginary, since η is less than unity and $\log \eta$ is negative! Similarly, the fact that the formula yields a straight line is a mere mathematical accident. Le Chatelier replies[124] that his equation has two practical advantages, that it does give a straight line and that it revealed the relationship between the viscosity of glasses and their chemical composition.

It is interesting to note that the general consensus among the Symposium papers was that binary and multi-component glasses are nanoheterogeneous. For example, Turner[2] likens soda–lime–silica glasses to a sponge, consisting of interpenetrating regions of silicates and silica. As indicated in Section IV, this is at variance with the early interpretation of the Zachariasen–Warren random network theory, which predicted a much more homogeneous structure, with the network modifying cations distributed randomly in holes in the silicate network close to non-bridging oxygen atoms; *i.e.* any heterogeneity was merely statistical. Although always accepted by Soviet glass scientists, it was only much more recently that the idea of nanoheterogeneity in single-phase glasses has become established in the West; *e.g.* as incorporated into the *modified random network model* proposed by Greaves.[125]

During the Symposium, various ideas were expressed concerning the nature of the molecular groupings/complexes in glasses, ranging from non-crystalline agglomerates that inhibit crystallisation on quenching[6] to crystals of colloidal dimensions.[9,12] The concepts of molecular groupings[6,12] and the existence of compounds in glasses[2,7,10,12] can be linked to modern thermodynamic modelling studies using the model of associated solutions,[126,127] as indeed can the idea of chemical equilibria between different compounds in the melt, such that their relative concentrations vary with both temperature and composition. The suggestion that, even at a crystalline stoichiometry, a glass with more than one component includes a range of groupings with different stoichiometries[7] is also supported by thermodynamic modelling.

The constitution of soda–lime–silica glasses within the commercial composition range is discussed by several Symposium authors,[2,5,7] who imagine them as comprising a solution of silicates in excess silica. However, the various authors differ as to which silicate compounds are involved. Turner[2] includes $Na_2O.SiO_2$, $Na_2O.2SiO_2$ and $CaO.SiO_2$, whereas Tamman[5] has only $Na_2O.SiO_2$ and $CaO.SiO_2$. Morey and Bowen,[7] on the other hand, point out that there are four compounds that have compositions near to those of commercial glasses, $Na_2O.2SiO_2$, $Na_2O.2CaO.3SiO_2$, $Na_2O.3CaO.6SiO_2$ and SiO_2, and hence it is interesting to compare these suggestions with the chemical structure predicted by the model of associated solutions.[128] This is shown for the cut with 70.5 mol% SiO_2 in Fig. 2. The first point to note is that the chemical structure is dominated by unreacted SiO_2 and hence is entirely consistent with the concept of a solution of silicates in excess silica. However, calcium metasilicate is absent and sodium metasilicate only appears in very small amounts at very high values of

$x_{Na2O}/(x_{Na2O}+x_{CaO})$. The major silicate mole fractions are from $Na_2O.2SiO_2$, $2Na_2O.CaO.3SiO_2$ (rather than $Na_2O.2CaO.3SiO_2$; *cf.* Morey & Bowen[7]) and $Na_2O.3CaO.6SiO_2$, with a minor contribution from $3Na_2O.8SiO_2$ ($3Na_2SiO_3.5SiO_2$), which was unknown at the time of the Symposium and not identified by Morey & Bowen[83] in their work on the system Na_2SiO_3–SiO_2.

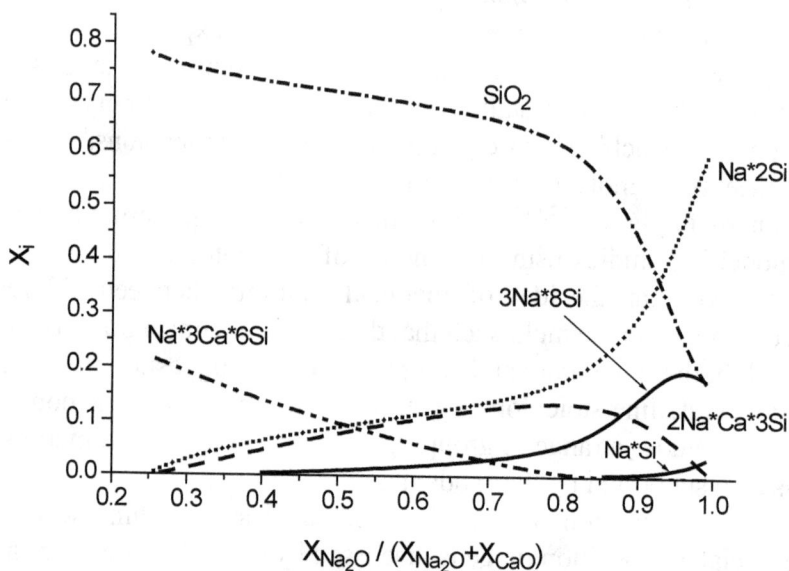

Figure 2. The chemical structure of soda–lime–silica glasses containing 70·5 mol% SiO_2.[128] The sodium silicate and sodium calcium silicate species are denoted iNa.jSi and iNa.jCa.kSi (*i.e.* iNa$_2$O.jSiO$_2$ and iNa$_2$O.jCaO.kSiO$_2$, respectively).

Tool & Hill's hypothesis[6] that, on quenching to form a glass, crystallisation is inhibited by the presence of unwieldy non-crystalline aggregates was reflected by Gunnar Hägg[97] in 1935 who, in criticising the random network theory, postulated the existence in the (supercooled) melt of groupings larger than the basic structural unit, which are so large and/or irregular and so strongly bonded that the growth of crystalline nuclei is inhibited by the resulting frustration, thus leading to glass formation. Although emphatically dismissed by Zachariasen,[96] Hägg's ideas were later adopted and extended by Krogh-Moe,[129-132] who was particularly interested in borate glasses and stressed the identity between the larger groupings (**superstructural units**) in these glasses and those in the corresponding crystalline materials.

An interesting semantic question that is reopened by the discussion of the role of molecular aggregates, chemical groupings and/ or crystallites in glasses, and of the relationship between glasses and colloids, concerns the difference between a mixture and a solution. This was discussed as early as 1899 by Nikolai Nikolaevich Lyubavin[133] in his seven volume treatise *Technical Chemistry*:

> " *It is highly probable that both solutions and glasses are merely intimate mixtures, which differ from typical mixtures such as granite only by the extreme subdivision of their constituent parts, which makes them invisible and difficult to separate.* "
>
> Nikolai Nikolaevich Lyubavin (1899)[133]

The conventional concept of a solution is an atomic/molecular scale dispersion of the solute throughout the solvent or, in the case of a mutual solution, an atomic/molecular scale mixture of the constituents. The individual solute atoms/ions/molecules may of course be solvated and/or be non-uniformly distributed. If, on the other hand, the solute atoms/molecules cluster together, to form molecular aggregates, crystallites or chemical groupings that exclude the solvent, this becomes a sol (colloidal solution*), which can account for some of the ideas expressed during the Symposium concerning the similarity between glasses and colloids. {The individual colloidal particles themselves may of course be (nano)crystalline.} The important point to note is that, in a sol, the individual colloidal particles (the dispersed phase) are still surrounded by a continuous solvent phase, and that there is a well-defined interface between the colloidal particles and the solvent.

The gradual removal of the solvent from a sol may lead to the

---

* For an introduction to colloid science and a definition of the basic terminology, see Ref. 134.

coagulation of the colloidal particles and/or the precipitation of the dispersed phase (*e.g.* in the case of colloidal gold). If, however, there are strong attractive interactions between the elements of the dispersed phase, the whole system can develop a network structure and becomes a gel. On further removal of the solvent, the initially very floppy gel becomes increasingly rigid and, as the process continues, the material may either remain amorphous or crystallise. An example of the former is the preparation of sol-gel amorphous silica by the dehydration of silicic acid solution. After almost total removal of the solvent (water) the final product is almost indistinguishable from normal melt-quenched silica glass.

The foam-like structure suggested by Quincke[55] (Section II) would occur when the solvent phase remains continuous, even at very low solvent content, and solute-rich clusters or colloidal particles occupy the spaces within the cell walls defined by the solvent. Once the continuity of the solvent phase is broken, and the solvent also exists as clusters/groupings, the material ceases to be a simple/colloidal solution, but whether it is regarded as a colloidal mixture or mutual solution depends upon whether there is a relatively sharp interface between the different components or a gradual transition in composition from one to the other. Thus, for any given system, there must be a solvent mole fraction at which the glass ceases to be a simple solution/sol; *i.e.* with one grouping in the chemical structure (the solvent) in clear excess. A special case concerns those glasses that comprise two interpenetrating continuous (connected) phases, *e.g.* Turner's sponge[2] or phase-separated glasses formed by spinodal decomposition, since such a system is not a true mixture. It can however be regarded as a network colloid,[134] if the boundaries are relatively sharp or, alternatively, as a mutual solution where there is a more gradual spatial transition between the phases.

The paper by Rosenhain,[20] included by Turner in his 1927 collection of papers,[13] clearly represented a significant advance and a new approach both to the constitution/structure of glass and to the understanding of glass property–structure relationships. Although his paper is not cited by Zachariasen,[89] Rosenhain[20] predates the latter's ideas concerning the role of strain in an irregular (disordered) structure in determining the phenomenon of softening. He also proposes that there are similar linkages in crystals and glasses, that the atoms vibrate about their equilibrium positions and that the structure is not entirely random due to the minimum distances between atomic centres. On the other hand, Rosenhain[20] almost certainly places too great an emphasis on unsatisfied bonds, relative to bond length and angle strain. In a complex silicate glass it is unlikely that there will be many unsatisfied Si–O bonds, although the environment surrounding the network-modifying cations is likely to be less regular than in the corresponding crystalline phase, perhaps with fewer "bonded" oxygen atoms, due to the less directional nature of the bonding.

Despite identifying the role of bond angle forces, Rosenhain[20] does not use the word network. However, his concept of a disordered structure constrained by bond length and angle forces is surely equivalent to a so-called random network in all but name. Nevertheless, the introduction of the term *glass/vitreous network* had to await Zachariasen's paper,[89] in which the latter then proceeded to discuss the conditions for the network to be relatively strain-free, leading to his well-known criteria ("rules") for glass formation. It is interesting to note that, whilst Zachariasen[89] is considered to be the originator of the random network theory, nowhere in his paper does he actually refer to a random network! The term *random network hypothesis* was in fact first introduced by Warren[112] in his first (1933) paper on the vitreous state, entitled *X-Ray Diffraction of Vitreous Silica*.

As expounded by Lebedev[18] and by Randall *et al*,[86-88] the crystallite theory was not completely defined, in that these authors did not describe in detail the interface between the individual crystallites: *i.e.* whether this took the form of a grain boundary or whether chemical bonding is continuous across the interface with a more or less disordered interconnecting region. Lebedev's analogy of an alloy[18] and Randall *et al*'s concept of the larger crystallites cemented together by the smaller ones[87,88] (very similar in concept to an alloy!) suggest the former, whereas the modern crystallite theory[108,109] envisages disordered interconnecting regions with a much smaller crystallite volume fraction.[108]

The impact of Randall, Rooksby & Cooper's papers[86-88] is evident from the discussion of that published in the *Journal of the Society of Glass Technology (Transactions)*,[88] which appeared in the same journal[102] during the following year. Much of this discussion centred upon the effect that the presence of cristobalite crystallites would have on the thermal expansion of vitreous silica. In particular, the thermal expansion coefficient of cristobalite is very much larger than that of vitreous silica and there is no indication of a sudden expansion of vitreous silica at ~200°C (the $\alpha \rightleftharpoons \beta$ transition temperature for cristobalite) On being questioned concerning the proportion of cristobalite present in vitreous silica, Randall[102] replied that there must be ~80%, a value quoted by Evgenii Alexandrovich Porai-Koshits[108] in his seminal paper presented at the 1953 conference on *The Structure of Glass*. He also indicated that very tiny crystallites would have rather different properties to those of bulk crystals. On the other hand, Turner[102] thought that it was necessary to assume that, if minute crystallites existed, their presence was masked by the amorphous medium in which they were situated (*i.e.* the interconnecting regions between crystallites).

In some ways, history has been very unkind to Lebedev. His only error, caused by an unlucky coincidence of temperatures, was to initially associate the softening of his glasses with the $\alpha \rightleftharpoons \beta$ transition of crystalline quartz. Everything else followed logically from this assumption. Even so, Lebedev's ambivalence towards the existence of quartz crystallites in his glasses is clear from the Introduction to his paper of 1924.[21,22] Unfortunately, however, with the advent of the random network theory, this was ignored by Soviet workers and the great crystallite *vs.* random network controversy became politically motivated,[114] with the result that several leading Soviet glass scientists refused to abandon the idea of crystallites long after Lebedev himself[135] had acknowledged that the two theories were merely different ways of viewing the same structure:

> *"It is easy to see that such a gradual transition may be imagined between these theories that their juxtaposition becomes meaningless. Indeed, in the case of a continuous network it can always be imagined that the ordering of the*

atoms may go so far around individual atomic sites that such regions can be quite justly termed crystallites; on the other hand, if we start with the crystallite theory, we must imagine that the individual crystallites are so small that it is impossible to speak of the presence of quite definite, sharply delineated fine crystals, as obviously between the grains there must be transitional zones with irregular particle distribution, and the lattices of the grains themselves may be considerably distorted because of the very small size of the fine crystals."

Alexander Alexeevich Lebedev (1940)[135]

The crystallite theory was destined to continue for another 30 years, but then died dramatically following the conference entitled *Discussion on the Modern Status of the Crystallite Hypothesis of Glass Structure* that took place in December 1971 to celebrate the 50th anniversary of the publication of Lebedev's original paper![114]

VI. CONCLUDING REMARKS

"The problem of the constitution of glass is one whose so-lution is probably still far distant, and it will long offer a fascinating field of research."

George W. Morey & N.L. Bowen (1925)[7]

The *Symposium on the Constitution of Glass* was undoubtedly an important milestone, coming as it did very early on in the history of glass structure research. As indicated in Section V, many of the ideas expressed are still current today and form the basis of modern theories/concepts of glass structure. Perhaps the most important of these are the concept of the chemical nanoheterogeneity of glasses having more than one component and the idea of chemical equi-libria within glass-forming melts. These were clearly dominant at the time of the Symposium, but Western glass scientists were later deflected towards a more homogeneous structure by the original interpretation of the Zachariasen–Warren random network theory, and especially Warren's two-dimensional schematic diagrams de-noting the structure of sodium silicate[136] and sodium borate[137] glasses, which reinforced Zachariasen's idea[89] that the cages con-taining the network-modifying cations are statistically distributed throughout the network.

A further important meeting with respect to the constitution/struc-ture of glass, the *Conference on the Vitreous State*, from which the quotation from A. A. Lebedev on pages 64-65 is taken, was held in Leningrad during the period 9th to 11th November 1939, and was organised by the Leningrad Branch of the All-Union D.I. Mend-eleev Chemical Society. This was the first in a series that later gave rise to the well-known Consultants Bureau proceedings volumes. In more recent years, these conferences have been superseded by the highly-successful international conferences on *The Structure of Non-Crystalline Materials*, the first two of which were chaired, un-der the auspices of The Society of Glass Technology, at Churchill College, Cambridge, by Philip H. Gaskell in 1976[138] and 1982.[139]

References*

1. W.E.S. Turner, *Advance Notice for the May 1925 Meeting of the Society of Glass Technology and the Seventh Annual Dinner* (Soc. Glass Technol., Sheffield, 1925).

2. W.E.S. Turner, *J. Soc. Glass Technol. Trans.* **9** (1925), 147.

3. Anon, *J. Soc. Glass Technol. Proc.* **9** (1925), 27.

4. Anon, *J. Soc. Glass Technol. Proc.* **9** (1925), 23.

5. G. Tammann, *J. Soc. Glass Technol. Trans.* **9** (1925), 166.

6. A.Q. Tool & E.E. Hill, *J. Soc. Glass Technol. Trans.* **9** (1925), 185.

7. G.W. Morey & N.L. Bowen, *J. Soc. Glass Technol. Trans.* **9** (1925), 226.

8. W.H. Bragg, *J. Soc. Glass Technol. Trans.* **9** (1925), 272.

9. R.W.G. Wyckoff & G.W. Morey, *J. Soc. Glass Technol. Trans.* **9** (1925), 265.

10. V.H. Stott, *J. Soc. Glass Technol. Trans.* **9** (1925), 207.

11. H. Le Chatelier, *J. Soc. Glass Technol. Trans.* **9** (1925), 12.

12. F. Eckert, *J. Soc. Glass Technol. Trans.* **9** (1925), 267.

13. W.E.S. Turner (Ed.), *The Constitution of Glass* (Soc. Glass Technol., Sheffield, 1927).

14. S.V. Nemilov, *private communication* (2008).

15. M.P. Vanyukov, *Alexander Alexeevich Lebedev*, in: *50 let Gosudarstvennogo opticheskogo instituta im. S.I. Vavilova (1918-1968)* [*50 Years of the S.I. Vavilov State Optical Institute (1918-1968)*] (Izd. Mashinostroyeniye, Leningrad, 1968), p. 642.

16. Anon. *Zh. Prikl. Spekt.* **10** (1969), 883.

17. A.A. Lebedeff, *Nature* **128** (1931), 491.

18. A.A. Lebedev, *Trudy Gos. Opt. Inst.* No. 10, **2** (1921); *J. Russ. Phys. Chem. Soc., Physical Sect.* **50** (1921), 57. (*N.B.* These two publications are absolutely identical!)

19. M.L. Frankenheim, *Die Lehre von der Cohäsion* (Schulz, Breslau, 1835), p. 389.

20. W. Rosenhain, *J. Soc. Glass Technol. Trans.* **11** (1927), 77.

21. A.A. Lebedev, *Trudy Gos. Opt. Inst.* No. 24, **3** (1924).

22. A.A. Lebedeff, *Rev. Opt.* **5** (1926), 1.

23. A.A. Lebedeff, *J. Soc. Glass Technol. Abs. Rev.* **6** (1922), 110.

24. A.A. Lebedeff, *J. Soc. Glass Technol. Abs. Rev.* **9** (1925), 270.

25. Anon., *The Polymorphism and Annealing of Glass*, (National Translations Center, Univ. Chicago).

26. J. Tyndall, *Heat A Mode of Motion*, Fourth Edn (Longmans Green, London, 1870).

---

* Reference numbers in bold type indicate papers of which facsimile copies are bound into this volume.

27.   M. Faraday, *Philos. Trans. Roy. Soc. Lond.* **120** (1830), 1.

28.   M. Faraday, *Experimental Researches in Chemistry and Physics* (Taylor & Francis, London, 1859), p. 231.

29.   D.I. Mendeleev, *Stekliannoye Proizvodstvo* [*Glass Production*] (Obshchestvennaya Pol'za, Sankt Peterburg, 1864) [Repr. *Sochineniya* **17** (Izd. Akad. nauk SSSR, Leningrad-Moskva, 1952) , pp. 46-401].

30.   L.G. Melnichenko, in: *Stroyeniye stekla* (Trudy Soveshchaniya po stroyeniyu stekla, Leningrad, 23-27 noyabrya 1953), Ed. A.A. Lebedev, N.A. Toropov, V.P. Barzakovsky & A.A. Appen (Izd. Akad. nauk SSSR, Moskva-Leningrad, 1955), p. 126. [Tr. *The Structure of Glass* (Consultants Bureau, New York, 1958), p. 98.]

31.   V.P. Barzakovsky, in: *Stroyeniye stekla* (Trudy Soveshchaniya po stroyeniyu stekla, Leningrad, 23-27 noyabrya 1953), Ed. A.A. Lebedev, N.A. Toropov, V.P. Barzakovsky & A.A. Appen (Izd. Akad. nauk SSSR, Moskva-Leningrad, 1955), p. 136. [Tr. *The Structure of Glass* (Consultants Bureau, New York, 1958), p. 105.]

32.   A.C. Wright, *An Historical Introduction to the Constitution and Structure of Glass*, in preparation.

33.   W. Barlow, *Nature* **29** (1883), 186.

34.   W. Barlow, *Nature* **29** (1883), 205.

35.   L. Sohncke, *Nature* **29** (1884), 383.

36.   W. Barlow, *Nature* **29** (1884), 404.

37.   S.V. Nemilov, *Glass Phys. Chem.* **21** (1995), 148.

38.   E. Zschimmer, *Die Glasindustrie in Jena* (Diederichs, Jena, 1912).

39.   W. Guertler, *Z. Anorg. Chem.* **40** (1904), 225.

40.   W.L. Bragg, *Proc. Roy. Soc. Lond. Ser. A* **89** (1913), 248.

41.   G.K. Hunter, *Light is a Messenger* (OUP, Oxford, 2004).

42.   H.E. Armstrong, *Nature* **120** (1927), 478.

43.   W.H. Bragg & W.L. Bragg, *Nature* **91** (1913), 557.

44.   W.H. Bragg & W.L. Bragg, *Proc. Roy. Soc. Lond. Ser. A* **89** (1913), 277.

45.   P. Debye & P. Scherrer, *Nachr. Kgl. Gesell. Wiss., Göttingen, Math.-Phys. Kl.* (1916), 16.

46.   S. Kyropoulos, *Z. Anorg. Allgem. Chem.* **99** (1917), 197. (*N.B.* Correction on p. 249.)

47.   P. Scherrer, in: *Kolloidchemie Ein Lehrbuch*, R. Zsigmondy Third Edn (Otto Spamer, Leipzig, 1920), p. 387.

48.   R.L. Frink, *Sprechsaal* **45** (1912), 690.

49.   P.P. von Weimarn, *Kolloid-Z.* **2** (1907), 76.

50.   P.P. von Weimarn, *Kolloid-Z.* **3** (1908), 282.

51.   P.P. von Weimarn, *Kolloid-Z.* **4** (1909), 123.

52.   A.V. Kroll, *Z. Anorg. Chem.* **76** (1912), 387.

53.   A.V. Kroll, *Z. Anorg. Chem.* **77** (1912), 1.

54.   L.N.G. Filon & F.C. Harris, *Proc. Roy. Soc. Lond. Ser. A* **103** (1923), 561.

55. G. Quincke, *Ann. Phys.* **46** (1915), 1025.
56. H. Jackson, *J. Roy. Soc. Arts* **68** (1920), 134.
57. H. Siedentopf & R. Zsigmondy, *Ann. Phys.* **10** (1903), 1.
58. R. Zsigmondy, *Zur Erkenntnis der Kolloide* (Gustav Fischer, Jena, 1905).
59. R. Zsigmondy, *Kolloidchemie Ein Lehrbuch*, Third Edn (Otto Spamer, Leipzig, 1920).
60. R.E. Liesegang, *Sprechsaal* **57** (1924), 233.
61. S.C. Bradford, *J. Soc. Glass Technol. Trans.* **3** (1919), 282.
62. M.W. Travers, in: *The Physics and Chemistry of Colloids and Their Bearing on Industrial Questions* (HMSO, London, 1921), p. 62.
63. A.Q. Tool & J. Valasek, *Scientific Papers, U.S. Bureau of Standards* **15** (358) (1920), 537.
64. A.L. Day & E.T. Allen, *Carnegie Inst. Wash. Publ.* **31** (1905), 15.
65. C. Barus, *Amer. J. Sci.* **41** (1891), 110.
66. C. Barus, *Amer. J. Sci.* **6** (1898), 270.
67. H. Jackson, in: *The Physics and Chemistry of Colloids and Their Bearing on Industrial Questions* (HMSO, London, 1921), p. 61.
68. O. Mügge, *Neues Jahrb. Mineral. Monatsh.* (1907), 181.
69. R.C. Wallace, *Z. Anorg. Chem.* **63** (1909), 1.
70. R.B. Sosman, *J. Franklin Inst.* **194** (1922), 741.
71. P. Lafon, *Compt. Rend. Hebd. Acad. Sci.* **175** (1922), 955.
72. A.Q. Tool & C.G. Eichlin, *J. Opt. Soc. Amer. & Rev. Sci. Instrum.* **8** (1924), 419.
73. P. Bary, *Rev. Gén. Colloides* **3** (1925), 1.
74. P. Bary, *Rev. Gén. Colloides* **3** (1925), 43.
75. W.E.S. Turner, *J. Soc. Glass Technol. Trans.* **7** (1923), 132.
76. R.W.G. Wyckoff, *Amer. J. Sci.* **9** (1925), 448.
77. R.W.G. Wyckoff, *Z. Kristallogr. Kristallgeom. Kristallphys. Kristallchem.* **62** (1925), 189.
78. F. Eckert, *Jahrb. Radioakt. Elektronik* **20** (1923), 93.
79. S. English, *J. Soc. Glass Technol. Trans.* **8** (1924), 205.
80. W.H. Bragg & R.E. Gibbs, *Proc. Roy. Soc. Lond. Ser. A* **109** (1925), 405.
81. W.H. Bragg, *Proc. Roy. Soc. Lond. Ser. A* **89** (1914), 575.
82. R.E. Gibbs, *Proc. Roy. Soc. Lond. Ser. A* **110** (1926), 443.
83. G.W. Morey & N.L. Bowen, *J. Phys. Chem.* **28** (1924), 1167.
84. H. Le Chatelier, P.E. Shaw, W.E.S. Turner, V.H. Stott, & F. Eckert, *J. Soc. Glass Technol. Trans.* **10** (1926), 95.
85. M. Samsoen, *Compt. Rend. Hebd. Acad. Sci.* **182** (1926), 517.
86. J.T. Randall, H.P. Rooksby & B.S. Cooper, *Nature* **125** (1930), 458.
87. J.T. Randall, H.P. Rooksby & B.S. Cooper, *Z. Kristallogr.* **75** (1930), 196.
88. J.T. Randall, H.P. Rooksby & B.S. Cooper, *J. Soc. Glass Technol. Trans.* **14** (1930), 219.
89. W.H. Zachariasen, *J. Amer. Chem. Soc.* **54** (1932), 3841.

90. H. Le Chatelier, *Compt. Rend. Hebd. Acad. Sci.* **179** (1924), 517.
91. H. Le Chatelier, *Compt. Rend. Hebd. Acad. Sci.* **179** (1924), 718.
92. H. Le Chatelier, *Ann Phys.* **3** (1925), 5.
93. A.A. Lebedeff, *Glas-Industrie* **35** (1927), 6. [*J. Soc. Glass Technol. Abs. Rev.* **11** (1927), 143.]
94. N. Seljakow, L. Strutinski & A. Krasnikow, *Z. Phys.* **33** (1925), 53.
95. V.H. Stott, *J. Soc. Glass Technol. Trans.* **10** (1926), 424.
96. W.H. Zachariasen, *J. Chem. Phys.* **3** (1935), 162.
97. G. Hägg, *J. Chem. Phys.* **3** (1935), 42.
98. J.W. Greig, *Amer. J. Sci.* **13** (1927), 1.
99. J.W. Greig, *Amer. J. Sci.* **13** (1927), 133.
100. C.W. Parmelee, G.L. Clark & A.E. Badger, *J. Soc. Glass Technol., Trans.* **13** (1929), 285.
101. G.L. Clark & C.R. Amberg, *J. Soc. Glass Technol., Trans.* **13** (1929), 290.
102. H.H. Macey, J.T. Randall, S.R. Hind, W.J. Rees, A.T. Green, V. Stott, F. Winks, W.E.S. Turner & B.P. Dudding, *J. Soc. Glass Technol. Trans.* **15** (1931), 54.
103. P. Scherrer, *Nachr. Kgl. Gesell. Wiss. Göttingen Math.-Phys. Kl.* (1918), 98.
104. G.W. Stewart & R.M. Morrow, *Phys. Rev.* **30** (1927), 232.
105. G.W. Stewart, *Phys. Rev.* **31** (1928), 174.
106. G.W. Stewart, *Rev. Mod. Phys.* **2** (1930), 116.
107. N. Valenkov & E. Poray-Koshitz, *Z. Kristallogr.* **95** (1936), 195.
108. E.A. Porai-Koshits, in: *Stroyeniye stekla* (Trudy Soveshchaniya po stroyeniyu stekla, Leningrad, 23-27 noyabrya 1953), Ed. A.A. Lebedev, N.A. Toropov, V.P. Barzakovsky & A.A. Appen (Izd. Akad. nauk SSSR, Moskva-Leningrad, 1955), p. 30. [Tr. *The Structure of Glass* (Consultants Bureau, New York, 1958), p. 25.]
109. S. Urnes, in: *Modern Aspects of the Vitreous State*, Ed. J.D. Mackenzie (Butterworths, London, 1960), p. 10.
110. W.A. Dollase, *Z. Kristallogr.* **121** (1965), 369.
111. F.W. Preston, *J. Soc. Glass Technol. Trans.* **13** (1929), 19.
112. B.E. Warren, *Z. Kristallogr.* **86** (1933), 349.
113. B.E. Warren, *J. Amer. Ceram. Soc.* **24** (1941), 256.
114. N.M. Vedishcheva & A.C. Wright, in: *Borate Glasses, Crystals & Melts: Structure & Applications*, Ed. Y.B. Dimitriev & A.C. Wright (Soc. Glass Technol., Sheffield, 2001), xvii.
115. A.F. Wright & A.J. Leadbetter, *Philos. Mag.* **31** (1975), 1391.
116. M.G. Tucker, M.T. Dove & D.A. Keen, *J. Phys. Condens. Matter* **12** (2000), L425.
117. A. Julg, *Crystals as Giant Molecules (Lecture Notes in Chemistry* **9)** (Springer-Verlag, Berlin, 1978).
118. G.N. Lewis, *J. Amer. Chem. Soc.* **38** (1916), 762.

119. W.L. Bragg, *J. Soc. Glass Technol. Trans.* **14** (1930), 295.
120. W.L. Bragg, *Z. Kristallogr.* **74** (1930), 237.
121. W.L. Bragg, *Nature* **120** (1927), 410.
122. R.E. Gibbs, *Proc. Roy. Soc. Lond. Ser. A* **113** (1926), 351.
123. A.F. Rogers, *J. Geol.* **25** (1917), 515.
124 H. Le Chatelier, *J. Soc. Glass Technol. Trans.* **13** (1929), 24.
125. G.N. Greaves, *J. Non-Cryst. Solids* **71** (1985), 203.
126. B.A. Shakhmatkin & N.M. Vedishcheva, *Fiz. Khim.Stekla* **24** (1998), 333.
127. B.A. Shakhmatkin, N.M. Vedishcheva & A.C. Wright, *Proc. XIX Intern. Congr. Glass, Invited papers* (Soc. Glass Technol., Sheffield, 2001), p. 52.
128 N.M. Vedishcheva, B.A. Shakhmatkin & A.C. Wright, *Phys. Chem. Glasses* **46** (2005), 99.
129. J. Krogh-Moe, *Phys. Chem. Glasses* **1** (1960), 26.
130. J. Krogh-Moe, *Acta Crystallogr.* **15** (1962), 190.
131. J. Krogh-Moe, *Phys. Chem. Glasses* **3** (1962), 1.
132. J. Krogh-Moe, *Phys. Chem. Glasses* **3** (1962), 101.
133. N.N. Lyubavin, *Tekhnicheskaya Khimiya* (*Technical Chemistry*), Vol. 2 (Univ. Publ. House, Moscow, 1899) p.588.
134. D.H. Everett, *Basic Principles of Colloid Science* (Roy. Soc. Chem., London, 1988).
135. A.A. Lebedev, Izv. Akad. Nauk. SSSR [Bull. Acad. Sci. URSS] **4** (1940), 584.
136. B.E. Warren & J. Biscoe, *J. Amer. Ceram. Soc.* **21** (1938), 259.
137. J. Biscoe & B.E. Warren, *J. Amer. Ceram. Soc.* **21** (1938), 287.
138. P.H. Gaskell, Ed., *The Structure of Non-Crystalline Materials* (Taylor & Francis, London, 1977).
139. P.H. Gaskell, J.M. Parker & E.A. Davis, Ed., *The Structure of Non-Crystalline Materials 1982* (Taylor & Francis, London, 1983).

FACSIMILE PAPERS*

---

* The papers are presented in their order of publication in the *Transactions* section of the *Journal of the Society of Glass Technology*, the original page numbers being given in brackets.

II.—On the Viscosity and the Allotropy of Glass.

By HENRY LE CHATELIER.

(*Translated from the French by Walter Butterworth, jun., M.A.*)

THE viscosity is certainly the most important of the properties of glass. On it depend all the methods of manipulation : blowing, moulding, drawing out. It is, on the other hand, the source of numerous difficulties in manufacture; high viscosity resists the escape of gaseous bubbles from the body of the founding mass; it compels the fining to be carried out at a high temperature and to be much prolonged, which makes this operation very costly; further, it gives rise to chilling or tempering as it gets cold, a serious drawback which has to be corrected by a delicate annealing operation. The great interest which the study of this phenomenon presents will be readily understood, and it is surprising how inadequate are the data hitherto available in regard to it. The difficulty of the problem has long discouraged investigators; we are much indebted to Messrs. Washburn and Shelton and to Mr. English for the very important contributions which they have just made almost simultaneously to the study of this question.*

I propose to discuss here the results of these investigators. I shall first endeavour to co-ordinate the measurements made so as to facilitate the use of them by the glass manufacturer. I shall group the figures obtained by means of a simple formula which seems to give the precise law of the phenomenon, and I shall then utilise this formula to study a very interesting theoretical problem, that of the allotropy of glass.

Measurements of the expansion of glass have shown that most glasses exhibit a very marked anomaly in their expansion a little below the point at which they begin to soften. This phenomenon seems to indicate an allotropic transformation of glass analogous to that undergone by molten sulphur when heated above 250°. If this is so, the existence of the two allotropic varieties of glass should be marked by a difference in the law of variation of their viscosity with the temperature. In the case of sulphur, it was the enormous variation in its viscosity which led to the discovery of the existence of an allotropic transformation.

* E. Washburn and G. Shelton, " The Viscosities of Soda–Lime Glasses " (*University of Illinois Bulletin*, 1924, Paper No. 140); S. English, " The Effect of Composition on the Viscosity of Glass " (*J. Soc. Glass Tech.*, 1924, **8**, 205).

The determination of the viscosity of glass presents peculiar experimental difficulties on account of the enormous differences in value of this property in the range of temperature in which glass is treated. This range is one of a thousand degrees, say from 1500°, the fining temperature, to 500°, the final annealing temperature. In this interval of temperature the viscosity varies in the ratio of 1 to 1 milliard and necessitates the employment of different methods of measurement according to the temperature region explored.

For temperatures above 750°, Messrs. Washburn and English both used modifications of the well-approved method of Margules, in which the liquid under investigation was contained in the annular space between two concentric cylinders, one of which was fixed whilst the other was free to rotate. The torsion couple required to rotate the free-moving inner cylinder at a determined rate was measured. This cylinder, made of porcelain in Washburn's method and of platinum–iridium in that of English, was necessarily closed at the ends and supported by a rod of porcelain; thus, it was not a geometrical cylinder, and the customary formulæ for a cylinder could not be applied to it. The graduation of the apparatus was made empirically by means of sugar solutions of various viscosities previously examined by the methods which normally serve for determining the coefficient of viscosity.

Below 800° down to 500°, Mr. English used, for determining the viscosity, the method of measuring the rate of elongation of a rod under definite loads. He verified that at the limiting temperature of 800° the two methods gave very approximately the same results.

The exactitude of the determination of the law of variation of the viscosity as a function of the temperature depends on the exactitude of two distinct measurements, that of the temperature and that of the viscosity. The margin of error in measurements of temperature can be estimated at 10° in the neighbourhood of 1500°, and at 2° in the neighbourhood of 500°, regard being had only to the causes of error due to the measuring apparatus. There is, further, the very great difficulty of procuring absolute uniformity of temperature in the heart of the mass of glass, especially when the viscosity is high. It is quite possible to get total errors of 10° at the lowest temperatures and of 20° at the highest.

The precision of the measurement of the viscosity is only of very secondary importance, since by reason of the rapid variation of this property with the temperature, the errors in the value of this latter have a quite preponderating influence on the exactitude with which the law can be determined. At 500°, a difference of 10° causes a variation of 100 per cent. in the viscosity, and at

14   JOURNAL OF THE SOCIETY OF GLASS TECHNOLOGY.

1500° a difference of 20° causes a variation of 15 per cent. in this coefficient. It is therefore sufficient to measure these coefficients within 25 per cent. at the low temperatures and within 5 per cent. at the high, which is relatively easy. The exactitude with which the law of this phenomenon can be established depends, therefore, solely on the exactitude in the measurement of the temperatures.

The Law of the Phenomenon.

The numerous experiments made on the viscosity of liquids in the neighbourhood of the temperature of the atmosphere have led to the representation of the law of variation of the viscosity as a function of the temperature by an exponential, that is, if a curve is drawn having temperatures as abscissæ and the logarithms of the viscosity as ordinates, the result is a straight line. The authors of the present researches on glass used the same method of representation, but their diagrams show a very pronounced curvature. Thus, a simple exponential does not adequately represent the law of variation of viscosity over a temperature range of 1000°. There should be nothing surprising in this. When a curve is drawn for a small range of values of the variable, the arc of the actual curve can be confounded with its tangent, and the function is represented as a linear function. As a greater range of values of the variable comes to be considered, different functions should be taken to represent the phenomenon; first tangents, then curves osculatory to the actual function, which may for long, perhaps for ever, remain unknown.

As the simple logarithmic formula does not satisfactorily represent the experiments of Messrs. Washburn and English, I set out to find a formula corresponding more closely with the facts. With this in view, I had the idea of applying the method of finite differences. Thus, drawing up a table of the logarithms of the viscosity for temperatures spaced at intervals of 100°, I tried whether these differences would not obey a simple law. I perceived that they increased in geometrical progression. This led me to try a double logarithmic formula

$$\text{Log. log. } \eta = M . t + P$$

that is to say, by plotting the temperatures as abscissæ and the double logarithms as ordinates, we should get a straight line. To verify the appropriateness of this formula, I plotted the corresponding curves for some of the experiments of Mr. Washburn. The agreement is very satisfactory, as Fig. 1 demonstrates, which refers to the three glasses Nos. 1, 2 and 8. It must be noted that the numerical values of Mr. Washburn's memoir have been more evenly spaced on the curves by graphic interpolation, sometimes even by extrapolation; the actual results of the measurements are not given. For the experiments we have chosen, the agreement is as perfect as possible.

For some glasses, on the contrary, it is impossible to represent Mr. Washburn's results by the double logarithmic formula, as is

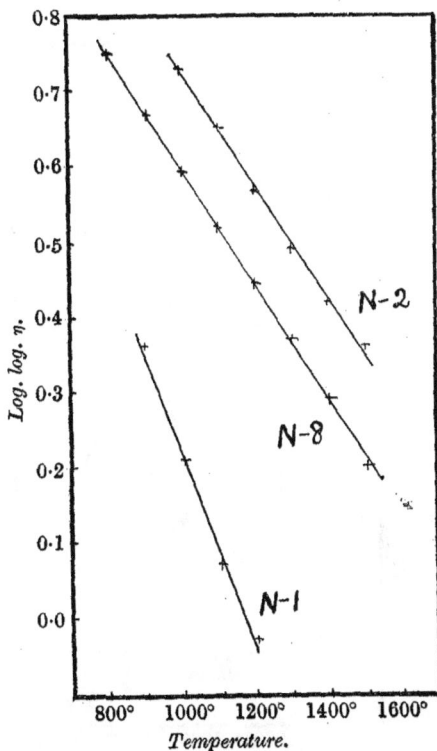

FIG. 1.

seen in Fig. 2 for glasses 9 and 10. The departures from the average direction reach more than 200° at the lower temperatures. The explanation of this divergence is very simple. These abnormal glasses devitrify at the lower temperatures, as indeed the authors of the experiments remark. Thus, crystals of silicate of lime swim in the molten glass and consequently increase considerably its proper viscosity. If fine crystalline powders, as, for example, clay, are placed in suspension in water, an extremely viscous mass is obtained. This is a phenomenon quite distinct from the viscosity of glass properly so-called.

To give an idea of the extreme limits of variation of the viscosity of glass I shall give for the glass No. 8 the value of the viscosity at temperature intervals of 100° and the ratio of the viscosities for the same interval of temperature. The viscosities are expressed in poises, that is, they give in dynes the force required to displace through one cm. per second a plane of area 1 sq. cm. which is 1 cm. distant from a fixed plane.

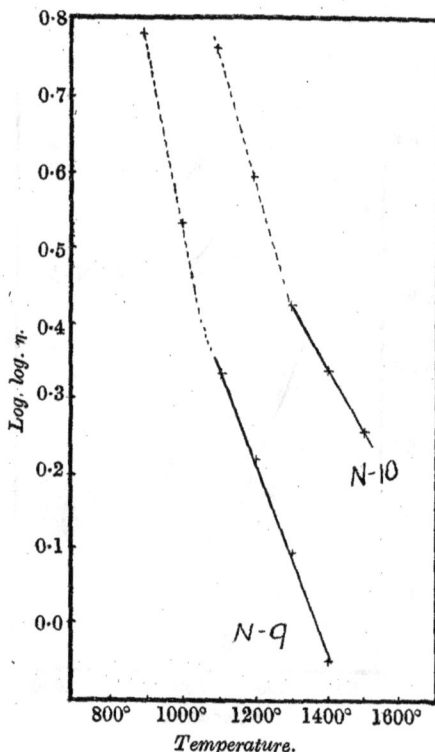

FIG. 2.

Effect of Temperature on the Viscosities and Viscosity Ratios for Washburn's No. 8 Glass.

Temperatures...	1500°	1400°	1300°	1200°	1100°	1000°	900°	800°
Viscosity	44	91	21×10^1	59×10^1	20×10^2	87×10^2	49×10^3	38×10^4
Ratios		2·09	2·33	2·76	3·39	4·37	5·63	7·77

I give in the adjoining table the parameters of the logarithmic formula for the different glasses investigated by Messrs. Washburn and Shelton. To this end, I have written the logarithmic formula in a particular form intended to put in evidence magnitudes which directly concern the glass manufacturer :

$$\text{Log. log. } \eta = M\frac{t - 1000}{1000} + N.$$

M represents the speed of variation of the viscosity as a function of the temperature. The smaller M is, the more extensive is the range of fusibility of the glass.

N represents the viscosity at the temperature of 1000°, that is, in the middle of the working zone of the glass. The greater N is, the harder is the glass.

I have put down in the two last columns of the table the temperature at which the viscosity of the glass equals 10,000 poises and the value in poises at the temperature 1000°.

I have separated the devitrifying glasses from those which are stable under the experimental conditions. The parameters for the devitrifying glasses have been calculated only from the measurements made at the higher temperatures, where the devitrification should be less important. Nevertheless, these coefficients are very uncertain, for one is never sure that some very fine crystals are not present in the molten glass, and the presence of these crystals is sufficient completely to falsify the measurements.

Number of Glass.	Molecular Composition.			M	N	Temperature at Viscosity equal to 10,000 Poises.	Viscosity in Poises at 1,000°.
	SiO$_2$.	Na$_2$O.	CaO.				
1	1·03	1·00		1·40	0·21	720°	42
4	1·55	1·00		0·95	0·49	880	1,100
17	1·52	0·73	0·27	0·88	0·51	800	1,749
16	1·83	0·53	0·47	0·92	0·60	990	9,550
3	2·40	1·00		0·73	0·56	960	4,620
7	2·23	0·67	0·33	0·80	0·57	970	5,630
8	2·33	0·64	0·36	0·76	0·59	1,000	8,620
13	2·76	0·60	0·40	0·78	0·66	1,075	37,200
12	2·63	0·53	0·47	0·77	0·63	1,035	21,000
14	2·64	0·42	0·58	0·60	0·69	1,160	79,400
2	4·85	1·00		0·74	0·71	1,150	153,000

Glasses undergoing Devitrification.

9	1·22	0·81	0·19	1·10	0·55	950	3,520
6	1·50	0·48	0·52	1·18	0·58	980	6,310
5	1·66	0·35	0·65	1·25	0·67	1,050	53,700
15	2·06	0·45	0·55	0·90	0·65	1,040	29,500
10	2·25	0·31	0·69	0·84	0·67	1,084	47,800

Alumina-containing Glass.

12A	Glass 12+5% Al$_2$O$_3$			0·80	0·67	1,090	50,000

Contrary to what might have been hoped for, no clear relation emerges from this table between the chemical composition of the

18   JOURNAL OF THE SOCIETY OF GLASS TECHNOLOGY.

glass and its range of fusibility. Yet the viscosity is certainly a function of the chemical composition, and very probably it ought to be a continuous function. This point, of capital importance in the industrial manufacture of glass, calls for further research.

The Allotropy of Glass.

The extrapolation of the results of these experiments, for temperatures from 500° to 600°, which are the annealing temperatures of glass, leads to consequences incompatible with what is known of the conditions of this annealing. Numerous experiments have shown that in this temperature region the viscosity of glass doubles for a diminution of temperature of 8°, or, which comes to the same thing, it varies in the ratio of 1 to 4,000 for a temperature interval of 100°, whereas the extrapolation of the results reproduced above gives, between 500° and 600°, a ratio always less than 100, that is, 40 times less than the experimental result. We must therefore conclude, either that the formula does not apply to temperatures less than 800°, or that glass undergoes a change of state at these low temperatures. The second hypothesis is correct.

Anomalies in the properties of glass in the neighbourhood of 600° have long been observed; anomalies in specific heat * and anomalies in expansion.† In the course of research conducted in my laboratory, Mr. Lafon ‡ noticed rapid increases in the coefficient of expansion of all glasses in the neighbourhood of the softening point. The increase is even enormous in the case of certain enamels. This led me to think that glass must undergo an allotropic change of state near this temperature, analogous to that of sulphur at about the temperature 250°. The law of variation of the coefficient of viscosity might be very different for these two varieties of glass, as it is for the two varieties of sulphur. I announced this conclusion in a communication made before the Académie des Sciences in regard to the experiments of Mr. Washburn.§ Shortly after the publication of this note, Mr. S. English‖ gave in this Journal

---

* Eichlin, " Absorption of Heat in Glass " (J. Amer. Optical Soc., 1920, 4, 340; 1924, 8, 419).

† Pietenpoll, " Expansion of Glass " (Chem. Met. Eng., 1920, 23, 876); Eckert (Jahrbuch, 1923, 20, 93); Lebedev, " Polymorphisme et Recuit du verre " (Trans. Optical Inst. Petrograd, 1921, Vol. 2).

‡ Lafon, " Anomalies dans la dilatation du verre " (Compt. rend., 1922, 175, 955).

§ H. Le Chatelier, " Sur la viscosité du verre " (Compt. rend., 1924, 179, 517).

‖ S. English, " The Effect of Composition on the Viscosity of Glass " (J. Soc. Glass Tech., 1924, 8, 205).

a very important memoir on the viscosity of glass, containing numerous measurements extending from 500° to 1400°. Applying to these new experiments the geometrical representation of the double logarithmic formula, it is at once evident that the curve is made up of two straight lines of different inclination, the transition from one variety to the other occurring at temperatures spaced out between 900° and 1100°, according to the nature of the glasses. This, therefore, confirms the existence of the two allotropic varieties of glass.

The following numerical table gives, for different glasses studied by Mr. English, the number of the experiment, the molecular composition of the glass, the parameters M and N of the logarithmic formula, and, finally, the temperature of transition from one variety to the other. The variety of glass which is stable at low temperatures is denoted by the symbol α and the variety stable at high temperatures by the symbol β.

Number of the Glass.	Molecular Composition.				Variety α.		Variety β.		Temperature of Transition.
	SiO_2.	Na_2O.	MgO.	Al_2O_3.	M	N	M	N	
I. (1)	3	1	—	—	1·25	0·45	0·91	0·54	700°
II. (329)	3	0·7	0·3	—	1·08	0·57	0·73	0·62	900°
III. (443)	3·6	1	—	0·2	1·11	0·55	0·61	0·58	950°

To show the degree of exactitude with which the double logarithmic formula co-ordinates the results of the experiments, the experimental points relating to these glasses are given on the graph Fig. 3. All these points fall very clearly into the straight lines whose parameters are given in the numerical table.

It will, perhaps, be objected that, as the two straight lines form a fairly obtuse angle, an equally satisfactory representation of the experiments could be obtained by a curve of large radius of curvature. To come closer to grips with this question, I give the comparison between the temperatures calculated and those measured directly. It would appear to be difficult to get a more satisfactory agreement by a continuous curve. The calculation is made for glass No. III, using the parameters given above.

GLASS III.

Variety α.

Experiment	500°	552°	600°	650°	695°	805°
Calculation	496°	551°	601°	652°	689°	809°
Difference	4	1	1	2	6	4

Variety β.

Experiment	1020°	1100°	1180°	1275°	1388°
Calculation	1010°	1100°	1180°	1274°	1370°
Difference	10	0	0	1	18

For the variety which is stable at low temperatures, the agreement is more satisfactory than could be hoped for, by reason of the uncertainty in the measurements of the temperatures. For the variety which is stable at high temperatures, the disagreement at the temperature of about 1400° is rather greater, but should

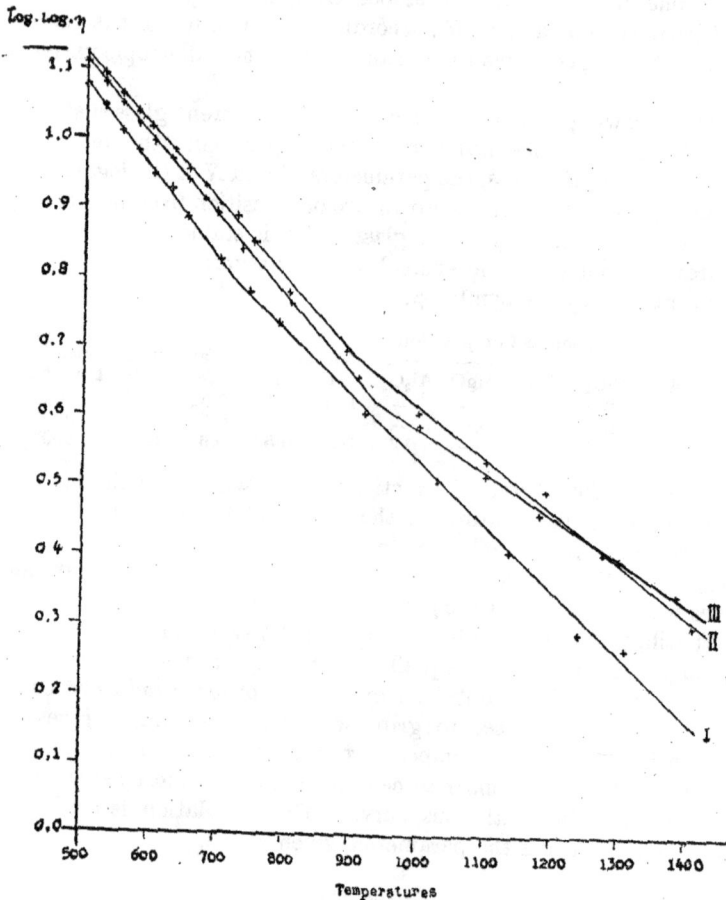

FIG. 3.

not cause surprise, because of the great difficulty of all measurements made at such high temperatures.

Glass, therefore, exhibits a phenomenon identical with that which we know already in the case of sulphur. There is, however, a difference. Sulphur in changing its state undergoes at the same time a considerable variation in the absolute value of its viscosity, whereas for glass only the law of variation of the viscosity as a function of the temperature is modified.

The table below gives the calculation of the coefficients M and N for each of the two varieties of glass investigated by Mr. English.

Number of Glass.	Molecular Composition.						Variety α.		Variety β.		Temp. of Transition.
	SiO₂.	Al₂O₃.	B₂O₃.	Na₂O.	CaO.	MgO.	M	N	M	N	
1	3·0			1·00			1·12	0·50	0·90	0·54	700°
3	3·0			0·90	0·10		1·27	0·46	0·81	0·55	800
4	3·0			0·85	0·15		1·24	0·50	0·79	0·55	875
5	3·0			0·80	0·20		1·24	0·50	0·78	0·55	875
6	3·0			0·75	0·25		1·25	0·50	0·75	0·57	900
7	3·0			0·70	0·30		1·19	0·55	0·73	0·58	900
8	3·0			0·65	0·35		1·15	0·58	0·70	0·59	950
9	3·0			0·60	0·40		1·23	0·56	0·70	0·60	950
10	3·0			0·55	0·45		1·09	0·62	0·68	0·62	1000
11	3·0			0·50	0·50		1·13	0·63	0·65	0·64	1000
26	3·0			0·85		0·15	1·10	0·54	0·78	0·58	850
329	3·0			0·70		0·30	1·06	0·58	0·73	0·61	900
32	3·0			0·55		0·45	0·99	0·58	0·74	0·65	950
390	3·0			0·45		0·55			0·72	0·70	
443	3·6	0·20		1·00			1·11	0·66	0·61	0·58	900
446	4·7	0·50		1·00			1·00	0·63	0·53	0·65	950
600	4·0			1·00			1·35	0·46	0·82	0·60	700
601	3·8		0·20	1·00			1·10	0·55	0·75	0·52	1050
602	3·9		0·40	1·00			1·20	0·52	0·41	0·35	1200
603	3·7		0·50	1·00			1·44	0·46	0·45	0·33	1150
604	3·3		0·60	1·00			1·64	0·40	0·50	0·35	1100
605	3·3		0·90	1·00			2·07	0·29	0·48	0·29	1000
606	2·6		1·30	1·00			2·40	0·18	0·45	0·25	1000
607	1·5		1·50	1·00			3·50	0·20	0·25	0·14	900

These experimental results still leave some uncertainty in regard to the allotropic transformation of glass. They do not agree satisfactorily with the measurements made of the anomaly of expansion and of that of the specific heat. For these two phenomena the point of discontinuity varies from one glass to another between 500° and 700°, say on the average 600°, whereas the meeting point of the two viscosity curves lies between 900° and 1100° for the different glasses, say on the average 1000°. There is thus an interval of 400° between two anomalies which should coincide.

There is, besides, a second difficulty, which will to some extent account for the first. There is no reason a priori why the two allotropic varieties of glass should have in the neighbourhood of the transformation temperature the same absolute value of viscosity, which seems, however, to follow from the form of the curves. There should be, as for sulphur, a very rapid change of viscosity during the period of transformation which should extend over a certain interval of temperature. If the transformation takes place near 600°, as the measurements of expansion indicate, starting from this temperature there should be a very sudden fall in the viscosity, and for an interval of 25° to 50° a branch of the curve approaching the vertical, forming a point of inflection and joining up the two straight lines. This is a matter which remains unexplained for the moment.

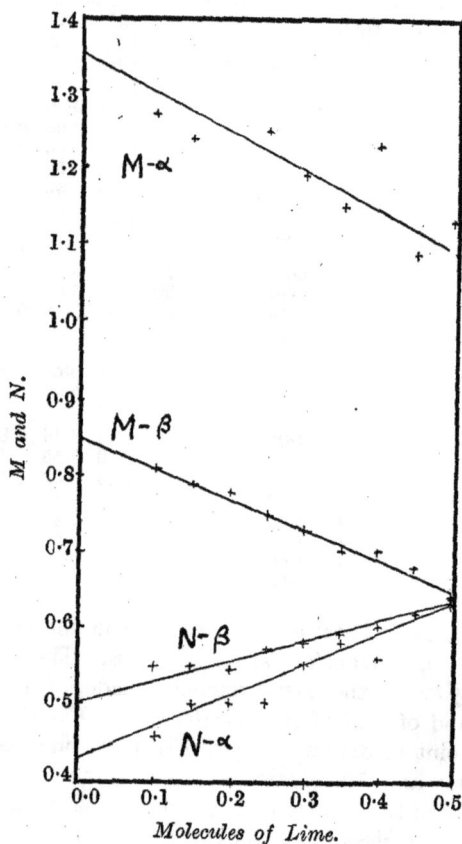

Molecules of Lime.

FIG. 4.

The value of the viscosity and its variation necessarily depend on the composition of the glass. Consequently it should be possible to connect the parameters M and N of the formula with the chemical composition of the glass by definite laws, that is, by algebraic formulæ. I have tried to make this calculation by starting from the numerical results of the above table.

With this aim, I traced for Mr. English's glasses with a lime base the diagrams of Fig. 4, in which the abscissæ represent the number of molecules x of lime contained in that quantity of the glass $3SiO_2, x . CaO, (1 - x)Na_2O$ which includes three molecules of silica, and for ordinates the calculated values of the parameters M and N. We thus find that the values of these parameters can be represented by the following simple formulæ :

Glasses with Lime Base.

$$M_a = 1.35 - 0.5\,x \qquad\qquad M_\beta = 0.85 - 0.4\,x$$
$$N_a = 0.43 + 0.4\,x \qquad\qquad N_\beta = 0.50 + 0.25\,x$$

Similarly for glasses with a magnesia base we should get the following values :

Glasses with Magnesia Base.

$$M_a = 1.14 - 0.3\,x \qquad\qquad M_\beta = 0.80 - 0.2\,x$$
$$N_a = 0.56 + 0.15\,x \qquad\qquad N_\beta = 0.53 + 0.3\,x$$

These results are very interesting for glass manufacturers, for they enable us to foresee in what direction the viscosity of a given glass will vary when part of the soda in the glass is replaced by equivalent proportions of lime.

In the course of his experiments, Mr. English determined the value of the viscosity corresponding with two very important phases in the treatment of glass, that of the temperature of gathering and that of the annealing of the solid glass after working. We may add to these the value of the viscosity corresponding with the fining period. I give, in the following table, the value of the double logarithm corresponding to these three degrees of viscosity. To find the corresponding temperature, it is only necessary to draw on the double logarithmic diagram horizontal lines passing through the corresponding ordinates. Their intersection with the curves gives the temperatures required. The table of viscosities is as follows :

	Fining.	Gathering.	Annealing.
η	100	300	$6 \cdot 10^{12}$
Log. η	2	2.5	12.8
Log. Log. η .	0.320	0.398	1.108

By drawing corresponding lines on Fig. 3, we get the following temperatures for glasses I, II, III :

Glass.	Fining.	Gathering.	Annealing.
I.	1230	1150	475
II.	1410	1300	510
III.	1450	1300	500

Thus the experiments of Messrs. Washburn and Shelton and of Mr. English have caused much progress to be made in our knowledge. It would be desirable to carry them further to clear up two points which are still doubtful.

(i) From the theoretical point of view, the discrepancy must be explained which seems to exist as to the temperature region where the allotropic transformation of glass takes place. There is an interval of 400° between the temperature corresponding with the change of expansion or of specific heat and that corresponding with the discontinuity in the variation of viscosity.

(ii) From the practical point of view, it would be very interesting to take more measurements of glasses of variable silica content, to establish the law of variation of the coefficients of the formula as a function of these proportions of silica, at least between 2 and 3 molecules of silica to one molecule of protoxide.

SORBONNE UNIVERSITY,
  PARIS.

XIII.—*The Nature and Constitution of Glass.*

By Prof. W. E. S. Turner.

(Read at the London Meeting, May 25th, 1925.)

The term glass employed in scientific language has a wider meaning than when used in the industry of glass-making. It denotes a condition rather than a substance and applies to substances of simple chemical composition as well as to those of complex character. A glass has certain well-marked characteristics. It has apparently some of the properties of a solid, but scientifically, we know that it is either a supercooled liquid in the case of a single substance, or a supercooled solution in the case of the ordinary glasses of commerce. In such condition the viscosity is very high; but in keeping with the fact that we have to do with the liquid state is the power of distributing stresses or self-annealing even at ordinary temperature, which instrument makers have informed me has been observed by them to occur in initially strained glasses kept for a period of years.

Besides the characteristics which have been recognised for years as belonging to glasses, others have been discovered as the result of tests on glasses of varied composition within the past few years which have served to quicken interest in this type of substance or in the vitreous state. Of these newer observations I would particularly cite :

(1) The endothermic changes which occur when glass is heated within the annealing temperature range.*

(2) The greatly increased thermal expansion of glass within the same temperature range.†

---

* A. Q. Tool and J. Valasek, *Bureau of Standards Sci. Papers, U.S.A.*, No. 358, 1920; A. Q. Tool and C. G. Eichlin, *J. Opt. Soc. Amer.*, 1920, **4**, 340.
† C. G. Peters and C. H. Cragoe, *Bureau of Standards Sci. Papers, U.S.A.* No. 393, 1920; W. B. Pietenpol, *Chem. and Met. Eng.*, 1920, **23**, 876; A. Lebedeff, *Trans. Opt. Inst. Petrograd*, 1921, **2**, No. 10; P. Lafon, *Compt. rend.*, 1922, **175**, 955.

88

(3) The changes in density and in refractive index which occur when glass is subjected to heat treatment.*

(4) The increase in the electrical resistance which M. J. Mulligan found to occur when a sample of soda–lime glass, originally strained, was re-annealed.†

(5) The after effects found to occur in glass by L. N. G. Filon, H. T. Jessop and F. C. Harris in their studies of photo-elastic stresses.‡

All these phenomena have been noted as the outcome of definite physical measurements. There are others, however, which have resulted from the recent systematic study in the author's Department of the processes of glass melting and glass working, phenomena which have been discussed frequently by experienced glassmakers but have never previously emerged into the purview of scientific determination. It has been definitely confirmed that remelted glass has a different plasticity from the glass produced from raw materials, although the chemical composition is not appreciably different. The presence of a considerable percentage of moisture in the mixture of raw materials from which a glass is made also markedly affects the plasticity—as measured by glass-making operations, such as tube-drawing—although the glass itself contains no moisture capable of detection by the ordinarily used methods of analysis. Remelting of these abnormal glasses with small amounts of alkaline salts restores the initial plasticity without producing marked alteration in the chemical composition.

These new facts, namely, those which have been the result of physical measurements and those resulting from systematic study of the glass-melting processes, all suggest further avenues for research in our progress towards a solution of the nature of glass.

Whatever methods are pursued in this search after the nature of glass, it seems to me that two fundamental factors must necessarily play a part in any attempts to formulate a working hypothesis of the constitution of glass, namely, the molecular complexity of the substance or substances composing the glass, and, in the case of glasses which are composed of two or more substances, of the chemical compounds existing in glass. I propose in this paper to attempt to answer two questions, namely, What do we know about the molecular complexity of glass?, and, What do we know about the presence of compounds in glasses?

* A. Q. Tool and C. G. Eichlin, *J. Amer. Cer. Soc.*, 1925, **8**, 1; F. Twyman and F. Simeon, *J. Soc. Glass Tech.*, 1923, **7**, 199; A. Lebedeff, *loc. cit.*

† *Trans. Roy. Soc. Canada*, 1924, **18** (3), 120.

‡ *Proc. Roy. Soc.*, 1922, [A], **101**, 165; 1923, [A], **103**, 561; 1924, [A], **106**, 718.

The Molecular Complexity of Glass.

On the view that ordinary glasses are solutions, any discussion of molecular complexity should apply to their individual constituents. For the moment, however, and especially because definite measurements of the molecular condition of the substances constituting the glass based on sound theory are lacking, we may be permitted to apply tests perhaps unusual and unorthodox in the hope of extracting even meagre information. Let us suppose that even a complex glass may be treated as an individual substance. Assuming it as such, it would be possible to apply the method of Ramsay and Shields based on molecular surface energy to throw some light on its molecular state.

Travers * states that he applied the test to the surface tension measurements of A. A. Griffiths † and found the complex glass studied to be monomolecular. The calculations, however, were not reproduced.

The recently published surface tension and density data of E. W. Washburn, G. R. Shelton, and E. E. Libman ‡ make a more thorough test possible and I have utilised them as fully as possible. Applying the Ramsay and Shields equation, the value of the molecular weight, M, in the relationship

$$k = \frac{\gamma_1 (Mv_1)^{2/3} - \gamma_2 (Mv_2)^{2/3}}{t_2 - t_1}$$

is that corresponding to the approximate formula selected arbitrarily, it is true, for each glass. Washburn and his co-workers stated the glass compositions only in percentages and not in molecular proportions. For the purpose of trial, I have converted the percentage compositions into molecular proportions assuming $6SiO_2$ present in each case, and the molecular weight, M, used in the formula, corresponds to these molecular compositions. A selection from Washburn's data has had to be made, since, in the case of the glasses Nos. 4, 9 and 16, the density recorded for the temperature 1454° is greater than at 1206°, a rather unlikely fact; whilst the variation of surface tension with temperature in the case of glass No. 6 is so much greater than with any other glass as to suggest the likelihood of error in measurement.

Using the remaining data available, the values of k have been calculated and are given in the following table :—

* J. Soc. Glass Tech., 1921, **5**, 131.
† Phil. Trans., 1920, **220**A, 587.
‡ Univ. of Illinois Bull., No. 140, 1924, **21**, 1.

Glass No.	Molecular composition.			Approx. molecular weight.	Specific volume.		k.
	SiO_2.	Na_2O.	CaO.		At 1206°.	At 1454°.	
3	6	2·50	—	515	0·4132	0·4150	2·11
5	6	1·25	2·39	571	0·3832	0·3861	0·67
7	6	1·87	0·94	529	0·4329	0·4425	0·29
8	6	1·66	0·92	514	0·4425	0·4545	1·60
10	6	0·83	1·84	514	0·4329	0·4484	0·57
13	6	1·30	0·87	489	0·4329	0·4386	0·87
15	6	1·33	1·62	534	0·4032	0·4082	1·76

In the working formula it is not altogether justifiable, in case the molecules are associated, to assume that the value of M is the same over the fairly wide temperature interval from 1206° to 1454°. From remarks already made, and especially in view of the difficulties involved in measurements at high temperature, the degree of accuracy of the surface tension and the density values is probably not high. Remembering these facts, it still remains that in practically all cases the value of k is distinctly, in some cases far, below the average for the so-called normal liquid, namely, 2·12, and it is at least a reasonable view that the value of M present in the first term, namely, $\gamma_1(Mv_1)^{2/3}$, is not sufficiently great, or M in the second term not sufficiently smaller, to raise the value of k to that for the normal liquid. In other words, there is at least some warrant for believing that the average molecular weight of the molecules in the fluid glass at 1206—1454° is high.

As glasses other than pure substances are solutions, the tests applicable to solutions should be more correctly used. The application by Vogt of the Van't Hoff's rule, namely,

$$C = \frac{0·0198\ T^2}{L},$$

to calculate the molecular lowering (C) and its comparison with observed depressions of freezing point, led him to conclude that in silicate melts the silicates such as diopside and olivine had simple molecular weights corresponding to $CaMgSi_2O_6$ and Mg_2SiO_4, respectively. As the values of T (the melting point) and of L (the latent heat of fusion of 1 gram) were not accurately known, the results could scarcely be relied on and similar tests by other workers have not led to Vogt's conclusions.

In the absence of direct test we are driven back on indirect evidence. Of the oxides which are concerned in the production of glasses, namely, SiO_2, B_2O_3, P_2O_5 and As_2O_3, we know that As_2O_3 has a complex molecule both at high temperatures * and in solution in nitrobenzene,† the molecular complexity in the latter solvent increasing with concentration. P_2O_5 in the state of vapour

---

* H. Biltz, Z. physikal. Chem., 1896, 19, 385.   † Biltz, loc. cit.

has a complex molecule.* In the case of boric oxide, B_2O_3, no determination appears to have been attempted and the same remark certainly applies to silica. The very high melting point of silica (1700°) as compared with that of carbon dioxide is very strong evidence of extreme complexity in the solid state (as is also the evidence of X-ray spectrographs), and the high temperature needed

Relation between ratio $\dfrac{M.\ Wt.\ (obs.)}{Simple\ Formula\ Weight}$ *and Dielectric Constant for* $N(C_3H_7)_4I$ *dissolved in various solvents.*

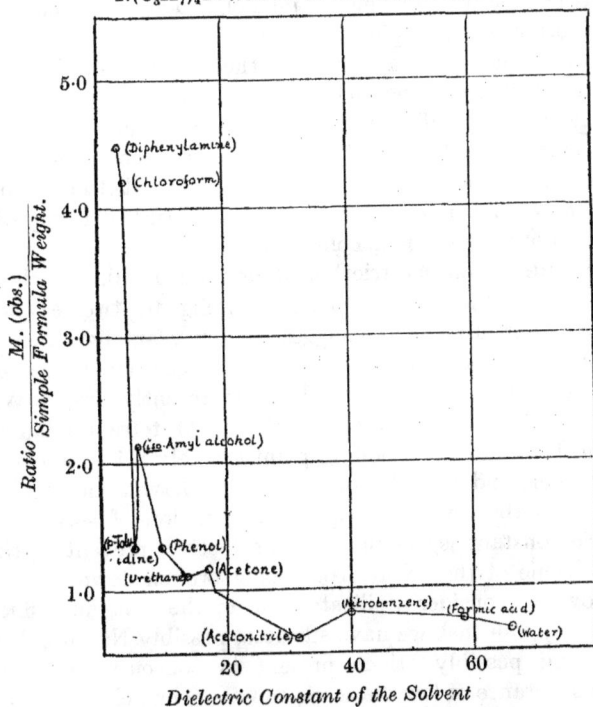

Dielectric Constant of the Solvent

FIG. 1

for the volatilisation of fused silica suggests that the amorphous, glassy form is composed of complex molecules.

We can by analogy, however, surmise that the silicates in solution in silica as glasses have very complex molecules. In a series of investigations both of organic and inorganic substances, particularly of salts, I have shown that the molecular complexity of the solute is intimately connected with the dielectric constant of the solvent; that in solvents of high dielectric constant, such as water or form-

* W. A. Tilden and R. E. Barnett, *Trans. Chem. Soc.*, 1896, **69**, 154.

amide, a salt is apparently dissociated, whilst when dissolved in solvents of low dielectric constant such as benzene or chloroform, the same substance shows evidence of strong association.* Take the case, for example, of tetrapropylammonium iodide. The attached curve (Fig. 1) shows the relationship between degree of association at a dilution 25 milligram-molecules per 100 c.c. of solvent (i.e. $V = 4$) and the dielectric constant of the solvent at the temperature of the molecular weight determination, the solvents employed in the test being water, formic acid, nitrobenzene, acetonitrile, acetone, urethane, phenol, *iso*-amyl alcohol, *p*-toluidine, chloroform and diphenylamine.

Various influences may disturb the relationship between the association factor and the dielectric constant, but there is no doubt at all that the general principle holds good, namely, that association occurs in solvents of low dielectric constants and, for electrolytes, dissociation in solvents of high dielectric constant, the extensiveness of association or dissociation varying with the magnitude of the dielectric constant.

Some salts, such as triethylsulphonium iodide in chloroform solution, have molecular weights running to twelve times the formula weight, and I have suggested, ten years ago, a connection between molecular association and the formation of colloids. A similar suggestion has been made more recently, by P. Walden.†
He showed that a 14 per cent. solution of tetra-amylammonium iodide in benzene had a boiling point scarcely raised above that of pure benzene, and the salt must, therefore, have a molecular weight many times the formula weight. The action of solvents of low dielectric constant is, he suggests, protective, preventing the complex molecule of the solute from being broken down.

If, now, we consider a silicate glass of the soda–lime–silica type, we may suppose that we have silicates, possibly Na_2SiO_3, $Na_2Si_2O_5$, $CaSiO_3$ and possibly others present in concentrated solution in silica, a substance of low dielectric constant, namely, 4·35, according to M. Pirani.‡ Sodium metasilicate is certainly ionised or dissociated in aqueous solution,§ and it is reasonable to assume, from the foregoing evidence, that it is strongly associated when dissolved in silica. By analogy, the same holds good for the other silicates in solution.

---

* *Trans. Chem. Soc.*, 1911, **99**, 880; 1912, **101**, 1923; 1914, **105**, 1751. See also *Molecular Association*, by W. E. S. Turner, published by Longmans, London, 1915.

† *Kolloid-Z.*, 1920, **27**, 97.

‡ Dissertation, Berlin, 1903.

§ F. Kohlrausch, *Ann. Physik*, 1892, **47**, 757.

Hence, we may regard silicate glasses as containing complex molecules.

The Existence of Chemical Compounds in Glass.

The next question relates to the existence of compounds in glasses of more than one component oxide.

Our knowledge of the existence of such compounds is limited and has been gained almost entirely by investigations other than on the properties of the glasses themselves.

When silica is heated with sodium carbonate or with calcium carbonate, reaction occurs at a comparatively low temperature. Proof of chemical reaction is forthcoming in the liberation of all the CO_2 from silica-rich mixtures with sodium carbonate at a temperature of 800°, at which temperature the dissociation of the carbonate is but very small.[*] Quite a number of silicates of metallic elements which may be components of glass have been prepared and crystallised and their properties examined. Much of our knowledge of the existence of silicates has been gained from a study of equilibrium diagrams of two, three, or more component systems. For example, Na_2SiO_3 and $Na_2Si_2O_5$, K_2SiO_3 and $K_2Si_2O_5$, Li_2SiO_3 have thus been indicated by Jaeger and by G. W. Morey and N. L. Bowen; $CaSiO_3$ and Ca_2SiO_4 by G. A. Rankin and F. E. Wright; Pb_2SiO_4 and $PbSiO_3$ by Hilprecht.

Attempts to deduce the presence of compounds actually present in glasses have been very few, and the results seldom convincing. One of the earliest attempts was that of Leydolt [†] who etched glass with HF and concluded that the crystal-like figures developed by this process were due to pre-existing crystals in the glass. The subject was the basis of periodical controversy, being raised by A. Daubrée,[‡] Weatherill,[§] Reinitzer,[||] H. Jackson, and E. M. Rich,[¶] R. L. Frink,[**] and E. W. Tillotson,[††] the last-named investigator proving that the form of the etch figures was independent of the chemical composition of the glass, and was determined by the composition of the solution. There were no crystals pre-existing in the glass as such.

A study of the durability towards chemical reagents of soda-

* See J. W. Cobb, *J. Soc. Chem. Ind.*, 1910, **29**, 69, 250, 399, 608, 799.
† *Wien. Akad. Ber.*, 1852, **8**, 261; *Compt. rend.*, 1852, **34**, 565; *Pogg. Ann.*, 1852, **86**, 104; 1869, **136**, 494.
‡ *Compt. rend.*, 1856, **45**, 792.   § *Amer. J. Sci.*, 1866, **41**, 16.
|| *Dingler's Polytech.*, 1866, **262**, 322.
¶ *J. Soc. Chem. Ind.*, 1901, **20**, 555.
** *Trans. Amer. Cer. Soc.*, 1909, **11**, 299; 1912, **14**, 662; 1913, **15**, 715; *Trans. 8th Inter. Cong. App. Sci.*, 1912, **5**, 57.
†† *J. Ind. Eng. Chem.*, 1917, **9**, 937.

lime–silica glasses inclined some investigators to believe in the existence of a compound $6SiO_2$, CaO, Na_2O. Possibly the practice of
Benrath, Tscheuschner, and later glass technologists of setting
out the limits of composition of durable commercial glasses by molecular formulæ encouraged the idea, but it appeared to be supported
by the tests of Mylius and Foerster on the action of water on vessels
of soda–lime–silica glass, minimum corrosion being found to occur
with the glass of approximately the $6SiO_2$, CaO, Na_2O composition.

This particular composition has frequently been referred to
in literature as that of the " normal glass." Zulkowski,* considered
that the results of Mylius and Foerster demonstrated that there
must be a true double silicate of the composition indicated above,
and proceeded to ascribe to it the graphic formula :—

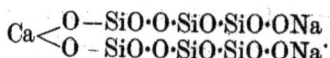

$$Ca\!<\!\begin{array}{l} O-SiO\cdot O\cdot SiO\cdot SiO\cdot ONa \\ O-SiO\cdot O\cdot SiO\cdot SiO\cdot ONa \end{array}$$

Reactions between SiO_2, Na_2CO_3 and $CaCO_3$ in proportions other
than the 6 : 1 : 1 molecular ratio were supposed to yield mixtures
of the " normal glass " together with silicates of sodium and of
calcium. The resistance to the action of water, Zulkowski considered,
was measured by the extent to which the " normal glass " had been
produced during the founding process. In a somewhat similar
manner, constitutional formulæ † were devised for glasses containing
alumina, boric oxide and phosphoric oxide. In the case of glasses
containing silica and boric oxide together with soda and lime,
alternative formula were prepared, one, namely,

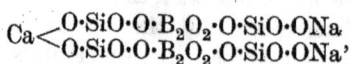

$$Ca\!<\!\begin{array}{l} O\cdot SiO\cdot O\cdot B_2O_2\cdot O\cdot SiO\cdot ONa \\ O\cdot SiO\cdot O\cdot B_2O_2\cdot O\cdot SiO\cdot ONa \end{array}$$

based on the view that a borosilicate was formed, the other on the
supposition that boric oxide yielded a " normal glass " like silica, of
formula $6B_2O_3$, CaO, Na_2O.

Although a strict investigation of the durability of soda–lime
glasses containing more CaO than represented by the 6 : 1 : 1 ratio
has yet to be carried out, it may be pointed out that in the series of
glasses $100SiO_2$, $20Na_2O$, $xCaO$ investigated by Peddle ‡ the value
of x ranged from 0 to 40 and improvement in the durability was
continuous with change of composition. Further, in the case of
the magnesia glasses based on the 6 : 1 : 1 ratio,§ we have found
that the glass $6SiO_2$, $1\cdot1MgO$, $0\cdot9Na_2O$ is more durable than the
$6SiO_2$, $1\cdot0MgO$, $1\cdot0Na_2O$ glass, a result not in accordance with
Zulkowski's contention.

    * Chem. Ind., 1899, 22, 280.   † Chem. Ind., 1900, 23, 108.
    ‡ J. Soc. Glass Tech., 1920, 4, 36.
    § Dimbleby, Muirhead, and Turner, ibid., 1922, 6, 101.

The known existence of a number of silicates of elements which enter into the composition of glasses must at least suggest that these same silicates are present in solution in the glass. Doubt exists, however, on this point. In solution, these compounds may be dissociated either partially or completely. R. B. Sosman* pointed out that amongst the binary compounds of oxides of high melting point, those were generally most stable in which the oxides were combined in their simplest ratios, such as 1 : 1 or 2 : 1. The existence of such simple ratios he regarded as indicating that the oxide molecules retained their individuality in silicates; that the latter were molecular and not atomic compounds, and exhibited their characteristics in this class of compound just as did water of crystallisation which showed the same absorption bonds in the spectrum as when in the free state. The formation of a compound might only be possible when the constituent oxides lost their mobility and the component atoms became linked up into the continuous arrangement which X-ray spectrographs indicate as occurring in crystal structures. On plotting the specific volume for mixtures of $CaSiO_3$ and $MgSiO_3$ against the percentage composition, Sosman found that the curve for the glassy mixtures was continuous, whereas that for the crystalline mixture contained a marked break corresponding to the diopside composition, namely, $CaSiO_3 \cdot MgSiO_3$.

Tammann † likewise concluded that there is a distinction to be drawn between the carbon compounds and the silicates, the latter being largely dissociated when molten. The molecular heats of a number of silicates could be calculated from Neumann's Law.

The fact that for a number of glasses, the relationship between percentage oxide content and certain physical properties, such as thermal expansion, density, tensile strength, compression strength, etc., is approximately linear is distinctly suggestive of survival of the individuality of the oxides, just as the density of dilute aqueous salt solutions can be calculated from Valson's Law of Moduli and that of molecular refraction in aqueous solution by Oudemann's Law, the validity of which depend on the nearly complete dissociation of electrolytes in dilute solution, so that the characteristic properties observed are due to the ions. The general rule, which, although only partly true, indicates that the relation between composition of glasses and their resistance to attack by water is parallel to the solubility of the constituent oxides in water also suggests a state of dissociation as already existing in glasses.

The study of compound formation in solution still leaves us without a very satisfactory theory of the action and interaction of the

* *Trans. Faraday Soc.*, 1917, **12**, 258.
† *Z. anorg. Chem.*, 1922, **125**, 301.

solvent and solute one on the other and of the extent to which
degree of combination occurs. Various properties of mixtures
of many simple chemical substances, such as of salts in water, in
alcohol, etc., have been investigated and it has been usual to inter-
pret many of the breaks which occur in the property-composition
diagram as indicative of compound formation. Such interpretation
is clearly reasonable when the break occurs at a composition corresponding to that of a known molecular compound which can be
crystallised from the solution. A big advance has been made,
within the last generation, of our knowledge of the constitution
of metallic alloys as the outcome of similar studies applied to mixtures of metals, and a variety of properties, including specific
volume, specific electrical conductivity, hardness, electromotive
force and heat of solution, have been applied with interesting
and valuable results. The study of glasses by similar means still
awaits serious development although the beginnings of such study
have occurred. Thus, in regard to the simple system Na_2O–SiO_2,
F. Kohlrausch * showed that successive additions of silica to an
aqueous solution of sodium hydroxide reduced the specific conductivity until a composition approximately Na_2O, $2SiO_2$ was
reached, after which the conductivity remained approximately
constant. In 1909, R. Schaller † showed that the upper devitrification
temperature in a series of glasses of the system Na_2SiO_3–SiO_2 fell
rapidly to about 720°, and the curve then rose to a maximum point
at the molecular composition Na_2O, $2SiO_2$. Mylius ‡ pointed out
that sodium chloride and white of egg first react in sodium silicate
solution when the SiO_2 content of the silicate exceeds that corresponding to Na_2O, $2SiO_2$. In 1918, E. W. Tillotson § found that
a sharp break occurred in the specific refraction-composition curve
for the same system at a point corresponding to the disilicate, and
V. H. Stott‖ indicated a similar break which he found when he
re-plotted the results of Peddle's refractive index measurements.
All these facts agree to show that sodium disilicate, both in aqueous
solution and in glasses, is sufficiently stable as to influence the
value of the physical properties observed; and as all these observations were made several years before G. W. Morey and N. L. Bowen ¶
isolated the disilicate Na_2O, $2SiO_2$ from meltings of the two oxides
and described its properties, it would suggest that a study of the

* Ann. Physik., 1892, 47, 757.   † Z. angew. Chem., 1909, 22, 2369.
‡ Sprechsaal, 1908, 140; see also R. Schaller, Article "Glas," Abegg's
Handbuch der anorg. Chem., 1909, III, 2, p. 373.
§ J. Ind. Eng. Chem., 1912, 4, 246, and J. Amer. Cer. Soc., 1918, 1, 76.
‖ Discussion on paper by C. J. Peddle, J. Soc. Glass Tech., 1920, 4, 232.
¶ J. Phys. Chem., 1924, 28, 1167.

relationships between specific refraction, specific volume or other properties and composition may in the future be more fruitful than suggested by the single case referred to by Sosman of the $CaSiO_3$–$MgSiO_3$ system in which no break was observed in the properties of the glasses, whereas a break was found in the curve for the mixture of crystals.

Tillotson * has already explored the relationship between specific refraction and composition for a series of glasses containing binary or ternary mixtures in which a choice from the components Na_2O,

Specific volume–composition curve for the system Na_2O–SiO_2.

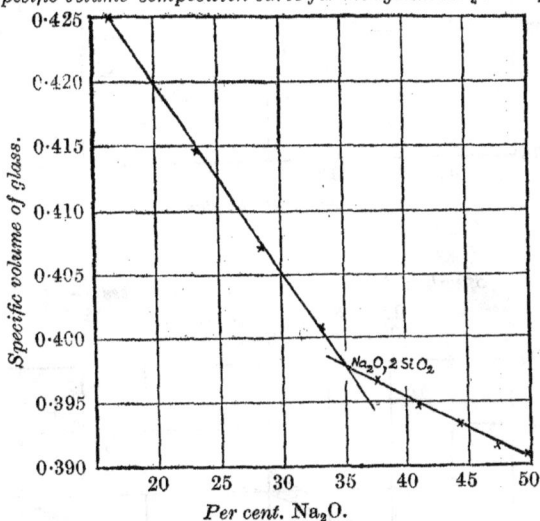

Per cent. Na_2O.

Fig. 2.

CaO, BaO, SiO_2 was made. Breaks of more or less distinctness were found in several of the series and the composition of several compounds was suggested. The study is well worth pursuing further; for although samples of thoroughly homogeneous glasses, containing only negligible amounts of impurity and of composition known by analysis should be employed in any such investigation, the case of Na_2O, $2SiO_2$ certainly justifies the basis underlying Tillotson's work.

Very few systems of glasses hitherto examined have sufficiently wide compositions for conclusions to be drawn as to whether any compounds exist or do not exist. I have had plotted the specific volume-composition curve of the Na_2O–SiO_2 system, utilising the data of C. J. Peddle,† and the curve shows a break (Fig. 2) at

* Loc. cit.   † J. Soc. Glass Tech., 1920, **4**, 1 (T).

$Na_2O,2SiO_2$ confirming the specific refraction curves plotted by Tillotson and by Stott. Further, curves have been plotted for (a) the specific volume-composition (Fig. 3), (b) the specific refraction-

Specific volume–composition of glasses :—
1. $100SiO_2,20Na_2O,xCaO.$
2. $100SiO_2,40Na_2O,xCaO.$

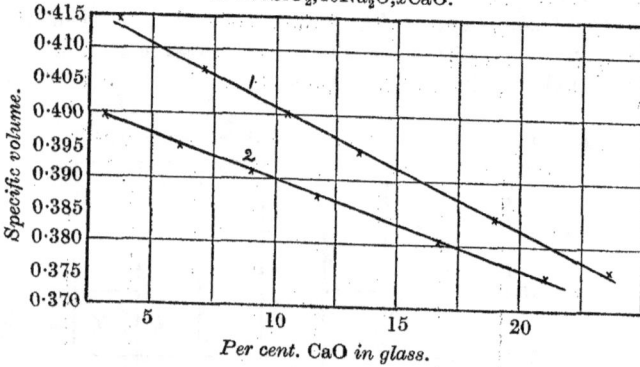

Per cent. CaO in glass.

FIG. 3.

Specific refraction–composition for glasses :—
1. $100SiO_2,40Na_2O,xCaO.$
2. $100SiO_2,20Na_2O,xCaO.$

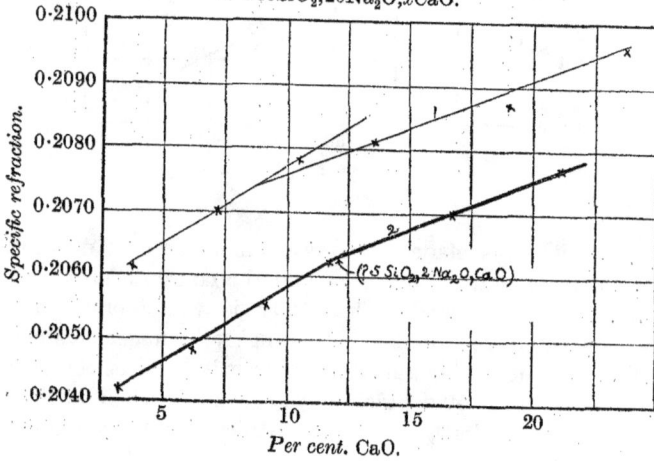

Per cent. CaO.

FIG. 4.

composition (Fig. 4) in the case of the density and refractive index data obtained by Peddle * for the series $100SiO_2$, $40Na_2O$, $xCaO$ and $100SiO_2, 20Na_2O,xCaO$; (c) for the specific refraction-composition data of J. R. Clarke and W. E. S. Turner † on the series $6SiO_2(2-x)$,

* J. Soc. Glass Tech., 1920, **4**, 20 (T).   † Ibid., 1920, **4**, 111 (T).

Na_2O, $xCaO$ (Fig. 5), and (d) the thermal expansion-volume composition data of S. English and W. E. S. Turner * for the series Na_2O-SiO_2 over a somewhat limited range of composition.

The specific refraction-composition curve for the $100SiO_2$, $40Na_2O$, $xCaO$ system of Peddle shows a possible break at the composition $2Na_2O$, CaO, $5SiO_2$, but the specific volume-composition curve for the same series of glasses is a straight line. There may possibly be a break in the specific refraction-composition curve for the $100SiO_2$, $20Na_2O$, $xCaO$ series. I have drawn it so; but a single curve may quite well fit the results both in this case and in the $100SiO_2$, $40Na_2O$, $xCaO$ series of glasses. None of the other curves, namely, Fig. 5 and the curve (not reproduced here) showing the

Specific refraction–composition of glasses $6SiO_2,(2-x)Na_2O,xCaO$.

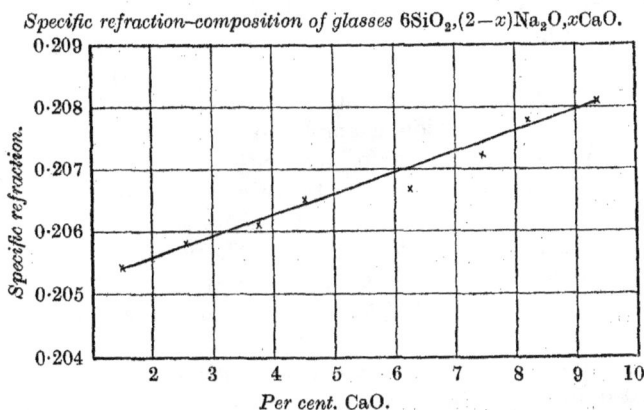

FIG. 5.

relationship between thermal expansion and volume composition for the Na_2O-SiO_2 glasses tested by Dr. English and myself, gives real indication of a break or breaks.

I would also, for the sake of completeness when dealing with this subject, refer to the curves which I have given in a paper with S. English on the physical properties of the soda–boric oxide–silica series of glasses,† and to the curves plotted by Dr. English for the viscosity of this same series of glasses at lower temperatures.‡ Changes of direction occurred in the property composition curves, but only a sharp change of direction in the case of the specific refraction curve. The occurrence of chemical combination in this series may be surmised but lacks real proof.

In reviewing this section we are bound to admit that the really helpful information which can be derived from existing information

* J. Soc. Glass Tech., 1920, **4**, 390.   † Ibid., 1923, **7**, 155.

‡ Ibid., 1924, **8**, 205.

is but small. At the same time, it seems clear that very useful work could be done in studying the composition-property diagrams of series of glasses prepared so as to cover a wide range of composition.

The Constitution of Glass.

As will be gathered from the foregoing, our available trustworthy data such as would make it possible to understand the nature of glasses are few and any structure of theory erected on them is liable to be unstable. Yet it is desirable to make a beginning, even if all that it is possible to do is to speculate, or to proceed by analogy.

The theory of the glassy state will have to provide both for single chemical substances and for mixtures of widely differing composition. In a glass consisting of fused SiO_2, B_2O_3, P_2O_5, etc., we have a supercooled liquid of very high viscosity. If, as is quite likely, these substances provide examples of dynamic allotropy, there will be present a mixture of molecules of different complexity, constituting a solution, such as is the case with $S\lambda$ and $S\mu$, the former corresponding, according to A. Smith and W. B. Holmes,* to a molecule of 8 atoms, the latter to one of 6.† The proportion of each form will be a function of the temperature and the change from the more to the less complex form with rise of temperature will involve an endothermic change such as has been observed by Tool and Valasek and by others in the case of commercial glasses. The theory propounded by Lebedeff, that the existence of a transition point in glasses in the neighbourhood of the temperature, namely, 575°, at which α and β quartz are transformed one into the other, loses its cogency when it is recalled that simple sodium borate glasses, with varying proportions of boric oxide present, show similar heat absorption to the silicate glasses although over a different temperature range. Nevertheless, despite the fact that a general and not a special interpretation must be sought, it is quite possible that the silica, the boric oxide, or other of the acidic oxides usually in excess in glass may exercise a dominant influence on glasses containing these oxides as major constituents. There is an increasing volume of evidence to warrant the view that in the absence of traces of moisture, many liquids hitherto regarded as having simple molecules are really associated ‡ and, therefore, probably con-

* Z. physikal. Chem., 1903, 42, 469.

† Smith and Carson, Z. physikal. Chem., 1911, 77, 661; Beckmann, Sitzungsber. K. Akad. Wiss., Berlin, 1913, p. 886; Preuner and Schupp, Z. physikal. Chem., 1909, 68, 129.

‡ H. B. Baker and Muriel Baker, Trans. Chem. Soc., 1912, 101, 2339; H. B. Baker, ibid., 1922, 121, 568.

tain molecules of different degrees of aggregation; whilst for a generation now it has been held that the abnormal physical properties of
certain liquids, of which water is a special example, may be explained
by the existence in them of molecules of differing degrees of complexity. Baker has shown that the melting points of elements
like sulphur and bromine and of compounds such as sulphur trioxide
and benzene can be raised in all cases by prolonged drying; and
A. Smits,[*] from a study of well-dried sulphur trioxide and phosphoric
oxide, came to the conclusion that the variation of vapour pressure
during the distillation of the former and the rise of melting point
on heating in the case of the latter were evidence of complexes in
each of these substances, so that they might possibly be regarded
as mixed crystals of forms corresponding each to a definite molecular
complexity.

It would seem as if there were a persistence of complex molecules
through a considerable range of temperature. Preuner and Schupp [†]
believed, indeed, that the μ form of liquid sulphur corresponded
to the 6-atom molecule found in sulphur vapour near or below
the boiling point of this element; whilst a proportion of complex
water molecules persists into the state of vapour. The whole of
this argument is intended to suggest that the characteristic properties of silicate and borate glasses are influenced by changes in the
molecular complexity of the free silica or boric oxide. Tammann [‡]
has pointed out that liquids which are mixtures of molecular complexes—substances like water, acetic and formic acids, sulphur
and phenol—tend to produce more than one crystal form. In view
of the existence of several forms of crystalline silica, we might
expect, if the connection suggested by Tammann holds good, and
especially if, further, complex liquid molecules, each the dissociation
product of a particular crystal form, may persist over a wide temperature range, that fused silica is a mixture of complexes. In
this connection, and in illustration of the controlling influence
which silica may exercise in silicate glasses, it may be pointed out
that Merritt [§] has found an endothermic change in fused silica in
the neighbourhood of 1200°.

One other observation concerning the forms of liquid sulphur
may be referred to at this stage. It was found by Smith and
Carson ‖ that ammonia had the power of increasing the rate at
which equilibrium was reached at any one temperature between

---

[*] A. Smits and P. Schoenmaker, *Trans. Chem. Soc.*, 1924, **125**, 2554; A.
Smits and A. J. Rutgers, *ibid.*, 1924, **125**, 2573.
[†] *Loc. cit.*
[‡] *Ber.*, 1912, **44**, 3618; *Z. physikal. Chem.*, 1913, **82**, 172.
[§] *J. Amer. Cer. Soc.*, 1924, **7**, 803.   ‖ *Loc. cit.*

the λ and μ forms, whilst sulphur dioxide retarded it. This phenomenon suggests that certain types of impurities in glass may exercise a similar influence, either stimulating or retarding the adjustment of equilibrium between different molecular complexes of silica,
of boric oxide, or, in the case of complex glasses, of other components
also. The known influence of the presence of small amounts of
sulphate in window and other glass in modifying the viscosity as
measured by practical drawing operations may be a case in point.

When we turn to the ordinary glasses of commerce we have
other factors to consider in addition to the molecular complexity
of silica, boric oxide, etc. There is first the possible formation of
chemical compounds, and secondly the problem of the molecular
complexity of these compounds. It may assist our consideration
to take as an example the widest used glasses of commerce, namely,
those of soda–lime–silica. Many of these glasses do not vary
greatly in composition from the general molecular formula $6SiO_2,(2-x)$
Na_2O, $xCaO$. We now know that sodium oxide and silica combine
to form the metasilicate Na_2SiO_3 and the disilicate, $Na_2O, 2SiO_2$.
Of these two, the latter melts at 874° and is partially decomposed
at its melting point. In view, however, of the excess of silica
present in commercial glasses the disilicate is probably present in
preponderating amount over the metasilicate. In combination
with lime, silica may form the metasilicate, $CaSiO_3$, and the orthosilicate, Ca_2SiO_4; but again, in view of the excess of silica, it is
the metasilicate $CaSiO_3$ which most likely is the prevailing form
present. Further combination may occur between any of these
different silicates, possibly in different proportions; but since
glass is formed by fusion at a high temperature and maintained
usually as glass only by more or less rapid chilling, we should expect
only the compounds stable at the high temperatures to persist
to any extent; or there may remain only the dissociation products
of compounds, possibly only the component oxides. As the rate
of reaction is modified by the viscosity of the medium, reannealing
is unlikely to favour anything but very slow formation of the compounds stable at low temperatures owing to the great viscosity.
Reheating at a somewhat higher temperature than that of annealing
may stimulate compound formation, causing deposition of crystals
and leading to devitrification.

I have already discussed the question of the existence of compounds in glasses. It must again come up for consideration in
connection with the chemical durability of glass. We have to find
some explanation for the stabilising effect on glasses of introducing
an oxide or silicate of a di-valent, tri- or tetra-valent element.
The alkaline silicates are very readily attacked by water; but the

introduction of oxides such as lime or lead oxide, even in small amount only, leads to a very great gain in chemical resistance. At first sight this fact suggests that a fairly stable compound is formed between sodium and calcium silicates. Moreover, whereas calcium metasilicate is readily attacked by hydrochloric acid, the alkali–lime–silica glasses are resistant to this reagent and the case for compound formation appears to receive support from this fact. Zulkowski argued that the chemical stability of a glass was dependent on maintaining an adequate length of time for the founding operation, since it was in the later stages that the supposed double silicates were formed. The formation of compounds does not, however, appear altogether sufficient to explain the great increase in resistance which occurs on addition of but a small amount of a dibasic oxide. Thus, in regard to resistance to boiling water, we have found * that the amount of alkali extracted was reduced in the proportion of about 36 to 1 when $0 \cdot 1$ molecule of Na_2O was substituted in $6SiO_2, 2Na_2O$ by $0 \cdot 1$ molecule of CaO. It is difficult to conceive that sufficient of a new stable compound as is repre-sented by $0 \cdot 1$ molecule of CaO is formed and can be instrumental in raising the durability so greatly.

If the formation of a stable compound is insufficient to furnish the explanation of the greater stability, some other must be sought. It is conceivable that the new oxide or silicate exercises a marked influence on the molecular complexity of the sodium silicate. I have shown † that in a solvent of low dielectric constant, like chloro-form or bromoform, two salts, each of which individually in solution is associated, appear to be still more strongly associated when present together. Since the reactivity of the amorphous and the crystalline forms of any one substance towards a reagent are often distinctly different, a difference which may be traced to complexity of structure, it is just conceivable that this cause may operate in mixed silicate glasses.

That formation of a double compound does not provide a general explanation of enhanced durability when a divalent oxide is added to the system Na_2O–SiO_2 is clear from the fact that an increase in the silica content of a glass improves the resistance to attack by water and acids. Thus, starting with a glass of composition $6SiO_2, 2Na_2O$, the durability is much enhanced by increasing the silica to the ratio $8SiO_2, 2Na_2O$, or higher, and there is no evidence of the formation of new compounds containing a proportion of silica greater than corresponds to the disilicate.

* J. D. Cauwood, J. R. Clarke, C. M. M. Muirhead, and W. E. S. Turner, J. Soc. Glass Tech., 1919, 3, 228.
† Turner and English, Trans. Chem. Soc., 1912, 101, 1786.

This fact raises the question of the influence of the residue of silica remaining in excess of any compounds which may be present. We may regard this as the solvent. As already suggested, it probably exercises a dominant influence over the characteristics of glass. It is conceivable it enters into loose combination with any silicates present in somewhat the same way as water or other liquids produce solvation of ions and of salt molecules, and leads to a reduction of ionic mobilities. The observation by Mulligan, already referred to, of the increased resistance of soda–lime–silica glass on re-annealing may be due to the fact that under stress, the contents of the silica envelope may be squeezed out to a slight extent and yield ions of greater mobility, a condition which is removed on re-annealing. We might thus account also for the production of double refraction such as is produced by mechanical deformation of solutions of gelatine as well as by strained glass.

A silicate glass might be conceived of as a sponge of silica containing silicates or their dissociation products as the filling medium; or, possibly, to use the analogy which Sosman * has employed to explain silica gels, of a mass of silica threads soaked in the silicates or their products of dissociation. The enhanced durability of silicate glasses towards the action of water and of acids which results from progressive increase in the proportion of silica is clearly due to the protective action of the latter, which is itself practically insoluble in water and in nearly all acids. The well-known protective action which silica gel, in common with various other gels, exercises over a number of colloids may conceivably have its parallel in fused silica. The improvement in resistant power to water which a glass shows after successive treatment by water and the customary treatment of window glass by dipping into hydrochloric acid, are explainable on the basis of the discovery of F. Mylius and E. Groschuff,† that a protective layer of silica gel is formed on the outer layer of the glass acting as a barrier, though not impermeable, to further action. In accordance with this explanation, we have found ‡ that whereas continuous treatment of a glass surface by water or by hydrochloric acid results in an enhanced resistance to attack, sodium hydroxide and sodium carbonate solutions maintain the severity of their corrosive action unimpaired.

The suggestion that fused silica may have properties resembling colloidal silica still requires definite proof, but the possession by fused silica and by glass of some of the properties of colloidal sub-

* " A Theory of the Structure and Polymorphism of Silica," *J. Franklin Inst.*, 1922, **194**, 74.

† *Z. physikal. Chem.*, 1907, **55**, 101.

‡ Cauwood and Turner, *J. Soc. Glass Tech.*, 1918, **2**, 235 (T).

stances may be admitted. The adsorptive power for water vapour and for gases is well known, and one of the disadvantages attending the use of fused silica is its permeability, even at comparatively low temperatures, to helium and hydrogen.

Some experiments which have been proceeding in the Department of Glass Technology at Sheffield during the past two years are suggestive in regard to the function of silica, and although they will be discussed in later papers, mention of them at this stage may be of interest. In determinations carried out by Miss V. Dimbleby on the resistance to hydrochloric acid solution of glasses of the soda–boric oxide–silica series, it was found that with the boric oxide rich glasses the curves showing the relation between corrosion and composition became nearly horizontal. This suggested to us that all soluble material had been extracted from the glasses, leaving a residue of silica. Analysis of the residues proved this to be the case. A similar result was found later in experiments by Mr. F. Winks working on Kavalier's chemical glass in which the silica had been progressively replaced by boric oxide. The glass containing about 26 per cent. of silica gave a residue of silica; indeed, the *whole* of the silica in the glass remained in the form of hard flakes containing a small amount (about 6 per cent. after drying at 110°) of water. A piece of glass tubing was treated with hydrochloric acid and it was found possible to extract all constituents save the silica, which retained the form of the tube although it proved fragile and crumbled on being dried.

Mr. Noel Heaton has also informed me that various samples of old window glass from ancient church windows which he has examined can be extracted with hydrochloric acid so completely as to leave nothing but a silica skeleton which is, to the naked eye, porcelain-like. Such action was found by him to occur with a glass of composition : SiO_2 53·85; P_2O_5 4·59; Na_2O and K_2O 15·09; CaO 12·32; MgO 7·75; Al_2O_3 3·55; Fe_2O_3 1·28; MnO 0·33. The glass was of fourteenth century make and was used in one of the windows of York Minster.

The action which has been described might conceivably be brought about by the colloidal silica precipitated from existing silicates being adsorbed by a pre-existing sponge or network of silica. It must be admitted that other explanations may conceivably be given.

Conclusion.

In preparing this communication I had no hope of being able to do anything but summarise and correlate some sections of existing knowledge. What I have probably succeeded in doing is to estab-

lish how unsatisfactory is our knowledge even of the most fundamental data which are necessary before any theory of the constitution of glass can be proposed even in a tentative manner. It seems
to me that wide fields for research lie open to the physical chemist
and the physicist in the study of the physical properties and the
molecular state of silicates, borates, aluminates, phosphates,
arsenates, zirconates, etc., of single metallic elements. When
exploratory work of this character has been accomplished, the study
of binary mixtures, and next of ternary mixtures of which one
component is silica, boric oxide, etc., may begin to lead us to a
definite conception of glasses and to enlarge our knowledge in general
of the amorphous state. Investigation at the present stage, except
for immediate industrial use, of the complex glasses entails the consideration of too many factors to permit of satisfactory fundamental
deductions. The use of fireclay for preparing the glasses to be
investigated must be discarded and platinum vessels used to obtain
specimens of that degree of purity which the chemist demands of
all other substances with which he cares to work, and the presence
of salts as impurities must be excluded. The homogeneity of the
specimen must be carefully controlled and the time and temperature
conditions to which it has been subjected should be carefully
recorded until it has been established how far these conditions
influence the ultimate physical properties of the specimen. With
investigations carried out under conditions such as these the condition of comparative ignorance in which we must confess we now
exist concerning the constitution of glass will gradually be dispelled.

DEPARTMENT OF GLASS TECHNOLOGY,
    THE UNIVERSITY, SHEFFIELD.

XIV.—*Glasses as Supercooled Liquids.*

By PROF. G. TAMMANN.

(Translated from the German by J. H. Davidson, M.Sc., F.I.C.)

*(A Contribution to the Symposium on " The Constitution of Glass,"
held at the London Meeting, May 25th and 26th, 1925.)*

FORMERLY, glasses were considered as particular modifications of
the solid state and therefore definite melting points were assigned
to them. Although Hittorf * had already shown that during the
melting of vitreous selenium no heat was absorbed, the question of
the relation of glasses to the other phases of matter remained

* *Pogg. Ann.*, 1852, **84**, 214.

indefinite until researches on the process of crystallisation indicated that glasses were supercooled liquids.

When a crystal is heated gradually, it begins to melt at a definite temperature. At this temperature it may remain in equilibrium with its melt for an indefinite period.

As a rule, a crystal, on heating, begins to melt at a definite temperature, independent of the rate of heating, and, generally speaking, a crystal cannot be superheated except in the case of certain crystals whose rate of formation is very slow, and these melt consistently. Fused crystals behave quite differently on cooling. If the melt is cooled slightly below the temperature of the melting point, crystallisation does not take place, and undercooling occurs. The melt may remain a shorter or a longer time in this undercooled state. If crystallisation follows, it begins at individual points in the supercooled liquid. A crystallisation centre is formed which may grow to a single crystal, or, as very frequently happens, from which fine needle-shaped crystals grow in all directions and form a spherolite.

The capacity of a liquid for supercooling is determined by two factors,

(1) the number of crystallisation centres formed in unit volume in unit time, and

(2) the linear rate at which the ends of the needle crystals grow into the supercooled liquid.

Both values are dependent upon the temperature of the supercooled liquid. The establishment of these relations gave the key to the relation between glasses and fusions. A glass is a non-crystallised, strongly undercooled fusion.

The Number of Crystallisation Centres.

If a substance the melting point of which is below 300° is sealed in a narrow thin-walled glass tube, melted, and the tube then brought rapidly into a bath at a known temperature, it will be observed that, according to the nature of the substance, either

(1) crystallisation begins at one or more points, from which needle crystals shoot out rapidly through the liquid,

(2) spherulitic crystals are formed more slowly, or

(3) the liquid remains completely transparent even when strongly supercooled.

In the third case it may be that invisible crystallisation centres are formed, which grow so extraordinarily slowly that their growth to visible size cannot be awaited.

To differentiate these cases, one must consider that with strong supercooling, the rate of crystallisation diminishes rapidly with increasing supercooling and attains a maximum value with 20—30° supercooling. If one brings the tube, in which no visible crystals have formed, from the low temperature into a bath the temperature of which lies between 20° and 30° below the melting point of the substance, the crystallisation centres possibly formed must so far develop in a short time that the spherolites developed around them become visible.

If crystallisation centres do not originate in this developing bath even after considerable time, numbers of spherolite crystals may be assumed to be developed. The number is equal to the number of crystallisation centres which originate with suitable supercooling.

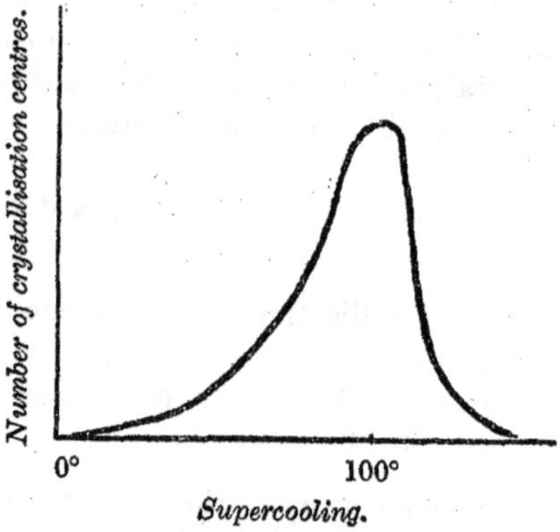

0°   100°
Supercooling.
Fig. 1.

Such numbers were collected by the author * for a series of substances. These showed that the number of crystallisation centres which formed in one minute in the same quantity of the supercooled liquid had a very definite maximum in relation to the supercooling, and that with very strong supercooling, no crystallisation centres whatever originated. Fig. 1 illustrates this relation. For easily fusible substances, the definite maximum is attained as a rule with a supercooling of about 100°.

The Rate of Crystallisation.

Apart from the number of crystallisation centres, the relation between the rate of crystallisation (signified by the term R.C.) and the supercooling is of significance for the supercooling capacity.

To measure the linear R.C. the substance is brought into a U-tube of 1—4 mm. diameter, the fused substance supercooled and one

* Z. physikal. Chem., 1898, **28**, 441.

meniscus disturbed with a crystal of the original substance. From the point of disturbance crystal threads grow out into the melt in parallel order. The ends of these threads lie on a plane perpendicular to the axis of the tube and their progression per minute can easily be measured. When the maximum value of the R.C. does not exceed 3—4 mm. per minute, its relation to the supercooling is reproduced by Fig. 2.

With small supercooling, the R.C. increases with the degree of supercooling since the released heat of crystallisaton reduces the R.C., and the more rapidly this heat can escape, the greater will be the R.C. With greater degrees of supercooling the R.C. diminishes with increased supercooling, as is the case with many rates of change

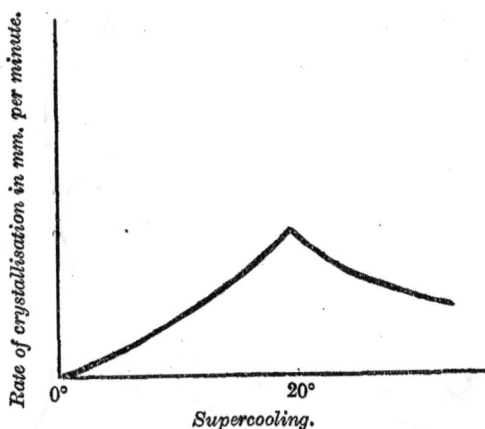

FIG. 2.

of state. There must, therefore, be a maximum R.C. The relation between the degree of supercooling and the R.C. with greater maximum values (more than 5 mm. per minute) is more complicated.*

The Viscosity of Liquids in relation to their Supercooling.

During the supercooling experiments the following observations were made in regard to the mobility of the liquid in relation to the degree of supercooling.

With the smaller degrees of supercooling, liquids in general are still mobile, with the exception of SiO_2, As_2O_3 and the acid aluminium-silicates which are still viscous at their melting points and only become mobile at higher temperatures. The temperature of the melting point has no influence on the mobility of the liquid, and the viscosity of the liquid in this region is a constant function of the

* *Aggregatzustände*, G. Tammann, Leipzig, 1923, pp. 247—271.

temperature. With about 90° supercooling the liquids become very viscous—like syrup or glycerine—and with about 110° supercooling they become hard and brittle like glass; that is to say, the viscosity increases extraordinarily in a certain temperature interval. If the liquid, supercooled to the glassy state, is gradually heated, it may be noted that the hard glass does not melt at a definite temperature, as a crystal does, but that, with increasing temperature, it gradually softens.

A glass has no melting point, but a temperature-range of softening.

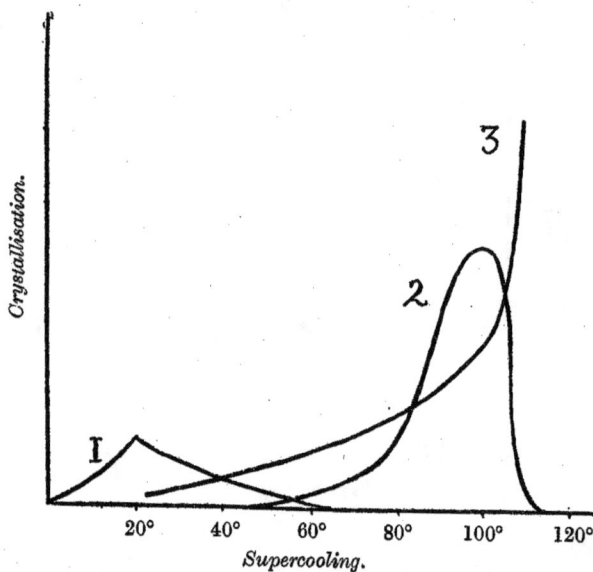

FIG. 3.

If a crystalline substance is gradually heated and a time-temperature curve taken, it is noticed that the rise of temperature is arrested and also for a longer period the temperature remains unchanged during the melting of the crystalline material because it absorbs heat during its melting. If the experiment be repeated with a glassy substance no heat absorption during the temperature range of softening can be observed. The melting of the glass cannot be identified on the heating curve. The temperature at which the thermometric apparatus in the mass becomes movable cannot be identified in either the heating or the cooling curves. Further, the volume of the glass varies according to the same law whether in glassy or liquid condition during the softening or solidifying interval.

Only the viscosity varies with the temperature in an unique manner through this temperature range.

The Conditions for the Change from Liquids to Glasses.

If we summarise our information on the relation between temperature and (a) rate of crystallisation, (b) number of crystallisation centres and (c) viscosity, in one diagram (Fig. 3), we observe that the following sequence frequently occurs. With slight supercooling we get the maximum of the rate of crystallisation (Curve 1), then follows the maximum number of crystallisation centres (Curve 2), and, finally, the extraordinary increase in viscosity (Curve 3).

To transform a substance in liquid form into a " glass " it must be cooled as rapidly as possible to not less than 110° below the melting point, and the dangerous region of from 30—100° below the melting point must be passed as rapidly as possible. The smaller the mass of the liquid, the more readily may it be converted into a glass by sudden cooling, since with diminishing mass the number of crystallisation centres is correspondingly reduced. As this number is never very large, it must be possible to choose a quantity of the liquid small enough to ensure that no crystallisation centres originate during the time of chilling. Further, it is evident that it will be advantageous in the transformation of a liquid to a glass, if the maximum number of crystallisation centres in unit mass in unit time is small, the region of greater numbers of crystallisation centres lies within a temperature range as restricted as possible and in which, apart from this, the rate of crystallisation has a very low value. Should this be the case, crystallisation will not make significant headway even should a centre form.

The Stability of Glasses.

The temperatures and pressures at which two phases are in equilibrium with each other may be shown by curves in a temperature-pressure plane (Fig. 4). There must be three such curves : (1) the vapour-pressure curve of the liquid, a, k, (2) that of the crystal o, a, and (3) the fusion curve of the crystal, a, m, u, n. These three curves cut each other in a point, the triple point a. The vapour-pressure curve ends in the critical point, k, at which the liquid and vapour present in a state of equilibrium are identical. The fusion curve cannot end in such a point since, even with equal specific volumes, crystal and melt do not become identical.

Consideration of the relations between volume changes on fusion and the heat of fusion and pressure and temperature lead to the following conclusion.* At the point m, the fusion curve reaches a

* Aggregatzustände, 1923, pp. 25—34.

maximum and at the point n it is retrogressive. The region of existence of a crystal is enclosed therefore by its fusion curve, its vapour-pressure curve and the pressure ordinate. This form of the fusion curve justifies the assertion that no end point occurs on the fusion curve analogous to the critical point k on the vapour-pressure curve of the liquid.

While the critical point k corresponds as a rule to pressures about 100 kg./cm.^{2} and the triple point a to about 0·001 kg./cm.^{2}, the highest point (m) of the fusion curve corresponds generally to very high pressures—20,000 to 80,000 kg./cm.^{2}.

For the glass phase there is no particular region of existence as

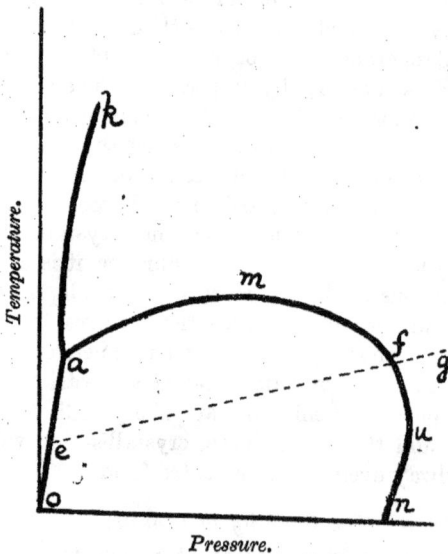

Pressure.

FIG. 4.

for vapour, liquid and crystal. Since liquids can be supercooled, they may be obtained at temperatures and pressures appertaining to the crystal field, though they are unstable under these conditions and tend to pass into the crystalline state. In order to decide at which temperature and pressure a supercooled liquid hardens, that is, assumes the viscosity of a glass, it is necessary to know the pressure-temperature lines at which the viscosity has a constant value. It is probable that the viscosity does not change when the specific volume does not change; that on the pressure-temperature lines of constant volume, the isometric lines, the viscosity will also have a constant value; and experience has confirmed this supposition, at any rate up to pressures of 2000 kg./cm.^{2}.*

* O. Faust, *Göttinger Nachrichten*, 1913, p. 489.

About 100° below the temperature of the melting point a super-cooled liquid becomes very viscous and at about 130° below it is already a brittle glass. The temperature at which the glass softens, increases with increasing pressure along the line e, f, g. At points below this line to the fusion curve f, u, n, this supercooled liquid is in the glassy state. At all such points the glass is unstable and only becomes stable at temperatures and pressures which lie to the right of the portion of the fusion curve f, u, n, and below the line f, g.

A glass can only become stable at very high pressures; with lower pressures it is always unstable, and if it appears stable to us, it is only because no crystallisation centres form in it at the particular temperature, and any crystallisation centres or spherolites possibly occurring in the glass grow so slowly that their growth cannot be established in any reasonable time.

In the case of only one chemically homogeneous substance, boric anhydride, B_2O_3, the formation of a crystal of the composition B_2O_3 from the glass has never been observed up to date. It is possible that this glass is stable even at ordinary temperatures and pressures.

Devitrification.

Knowledge of the temperature range through which devitrification proceeds more rapidly than at other temperatures is of practical value. To get a general idea of the nature of the devitrification at different temperatures the following procedure may be adopted.* If a glass rod, suitably bedded, is placed in an electric furnace and heated so that the temperature falls several hundred degrees from one end of the rod to the other, crystallisation centres will begin to form first, and in the greatest number, at that portion of the rod which is at the most favourable temperature for their development. At that point of the glass rod which is at the temperature of the maximum rate of crystallisation the spherolites will be the largest.

In Fig. 5 the length of the glass rod is represented by a, b. The curve c, d, shows the temperatures at different points along the length of the rod. The curve s, k, v, shows the relation between temperature and the property of spontaneous crystallisation, that is, indicates the number of spherolites present in unit time in unit mass in relation to temperature; and the curve I.R.C. shows the relation between temperature and linear rate of crystallisation.

Between the temperatures of the maxima of both curves, t_1 and t_2, the incidence of devitrification first becomes visible through clouding or the formation of spherolites, since between these two

* Z. anorg. Chem., 1914, 87, 248.

temperatures the conditions for the formation and growth of crystallisation centres are more favourable than at temperatures outside this range. The temperatures t_1 and t_2 exist at the points numbered 1 and 2 on the glass rod. If the temperature of the glass rod be measured at the points between which devitrification becomes apparent, this temperature must be between those of the maximum rate of crystallisation and the maximum capability for spontaneous crystallisation.

From the positions 1 and 2, crystallisation proceeds in both directions and is arrested on the hotter side at the temperature t_3, the equilibrium temperature between liquid and crystal. Towards

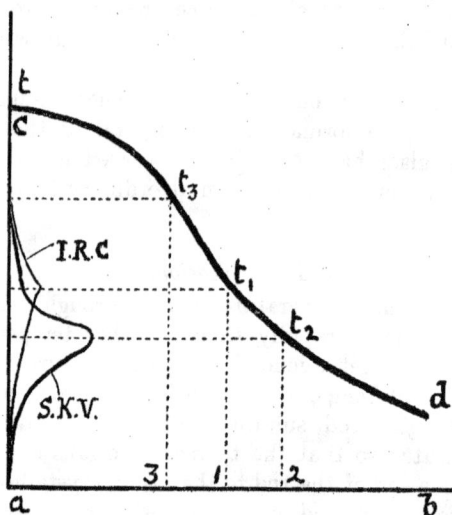

Fig. 5.

the cooler end the boundary line between clear glass and clouded devitrified material can only be sharply defined when the rate of crystallisation at the particular temperature has a significant value while the number of crystallisation centres is not very great. Should the opposite conditions hold, the boundary is not well defined. The porcelain-like layer is followed by successive layers of diminishing opacity which, on account of the smallness of the spherolites, finally passes through coloured layers into clear glass. Such colouring due to devitrification may be seen as a beautiful effect produced in resistant Jena glass due to uneven heating.

From this procedure it is possible to estimate the temperature of both maxima, and the melting point of the devitrification product. P. Ponomareff * carried out such estimations for glasses consisting

* Z. anorg. Chem., 1914, **89**, 383.

of $Na_2O,2B_2O_3$ with excess boric oxide, up to 65°. This method may be used to accentuate crystallisation in viscous liquids. For example, fused potassium silicate (K_2SiO_3) containing a little carbonate, (K_2CO_3) was brought to crystallisation in the course of an hour. Its rate of crystallisation is very small. On the other hand, sodium silicate (Na_2SiO_3) has a high rate of crystallisation, and the crystallisation is soon completed after the formation of a crystallisation centre. If fused sodium silicate, rapidly chilled so that it has solidified as a glass, be gradually heated, and the time-temperature relations mapped on a curve, heating curves are

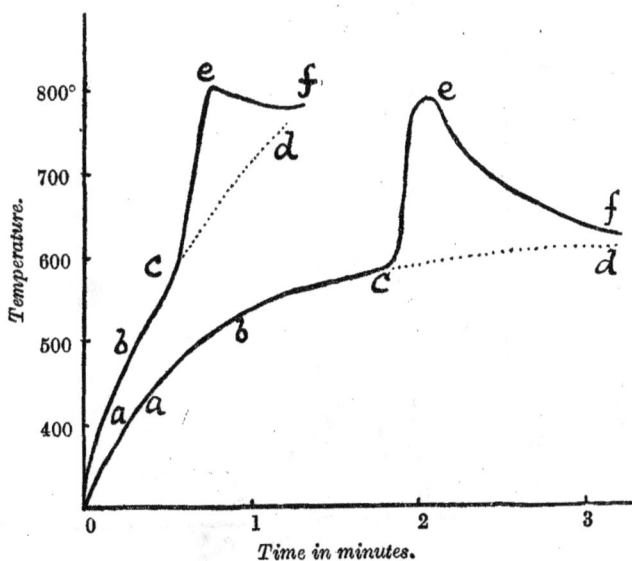

FIG. 6.

obtained as in Fig. 6. A rapid increase of temperature begins at 550° in consequence of the release of the heat of crystallisation, and the temperature rises quickly to about 800°, accompanied by glowing of the substance, to fall finally to the temperature of its surroundings.

The temperature of the beginning of devitrification is about 500° below the melting point (of the crystal) 1055°.

The Capacity for Spontaneous Crystallisation.

The number of crystallisation centres which form in unit volume in unit time—the spontaneous crystallisation property—is also termed the "nucleus number." The nucleus number has a definite

116

THE CONSTITUTION OF GLASS

maximum in its relation to the degree of supercooling, the temperature of which is only slightly influenced by foreign substances, whereas, on the other hand, the presence of foreign materials has an extraordinarily great influence on the nucleus number.

Fig. 7 illustrates this in the case of betol. The milk-white spherolites—or nuclei—which melt at 91° were counted. The melts were each supercooled for 2 minutes and the nuclei developed

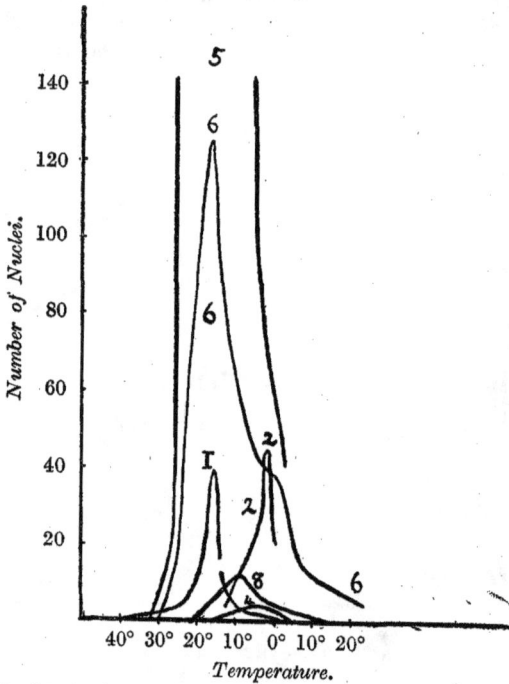

1. *Betol 3 times recrystallised.*   2. *Betol + 0·5% Naphthalene.*
4. *Betol once recrystallised.*   5. *Betol + 5% Benzamide.*
6. *Betol + 0·1% Anisic Acid.*   8. *Betol + 0·1% Perchloroethane.*

FIG. 7.

at 70°. The nucleus number of the betol thrice recrystallised from alcohol is shown in Curve 1, and that of the betol only once recrystallised in Curve 4. The curves indicate that the addition of small quantities of anisic acid and benzamide causes an extraordinary rise in the curve, whilst additions of naphthalene or perchlorethane have only slight influence. Cane sugar and salicylin to the extent of 0·2 per cent. have a contrary effect, as they entirely prevented the appearance of any nucleus even in 10 minutes.

An investigation of 153 easily fusible substances showed that, when sealed in thin-walled glass tubes, 59 of these could be converted into glasses by sudden chilling to an adequate temperature.* With still more drastic chilling the number of liquids which could be cooled into glasses would be further increased.

It may be said that the glassy state is not an exceptional condition, but that every substance may be obtained in this state if only its fusion could be chilled rapidly enough. Up-to-date metals have not successfully been transformed into the glassy condition; even by electrolytic separation at ordinary temperatures, that is, far below their melting points, the metals separate in the crystalline condition.

Of the other chemical elements, sulphur, selenium and oxygen may be obtained as glasses. Liquid oxygen is tough before crystallisation begins, with a low R.C.† So-called amorphous boron, carbon and silicon are crypto-crystalline substances. Of the oxides, B_2O_3, As_2O_3, P_2O_5 and SiO_2 may readily be obtained as glasses and V_2O_5 only through the chilling of small quantities. The property of withstanding extended supercooling is retained by the salts of those acid anhydrides, and this the more, the less basic oxide the salt contains. Acid silicates and borates can be converted into glasses even with slow cooling. The same applies to the salts of metaphosphoric acid and pyrosulphuric acid ($K_2S_2O_7$). The more basic the silicates and borates are, the more quickly must they be cooled in order to be obtained as glasses. Even in the later volcanic rocks considerable glassy residues are found frequently, which had not crystallised on account of the rate at which the temperature decreased. Of the sulphides, As_2S_3 and Sb_2S_3 have been obtained as glasses.

The fusions of hydrated salts may frequently be strongly supercooled and consequently at $-80°$ be obtained as hard glasses, for example, $CaCl_2,6H_2O$; $FeCl_36H_2O$, or $Na_2S_2O_3,5H_2O$.

The Nucleus Number and the Theory of Probability.

The number of crystallisation centres at any minute is extraordinarily small compared with the number of molecules present. The formation of a crystallisation centre requires a large number of favourable conditions to be fulfilled simultaneously. Probably eight or more neighbouring molecules with similar orientation and with velocities between definite values must be present in order that a crystallisation centre may originate. The greater the number of conditions to be fulfilled, the more improbable is this occurrence. The evidence that nucleus formation is concerned with a procedure subject to the laws of probability is evident

* Z. physikal. Chem., 1898, 25, 472.

† W. Wahl, Z. physikal. Chem., 1913, 84, 116.

(1) if the nucleus number in individual experiments varies about an average value in the same manner as other occurrences under the influence of chance, as, for example, the number of upturned spots on several dice vary about an average number;

(2) if the relative variations of the nucleus number about its average value diminish with increasing nucleus numbers as is the case with other occurrences subject to chance.

The nucleus number for supercooled piperonal (m. p. 37°) was determined, since in the case of this substance the nucleus number is independent of the number of meltings. The substance was melted at 50°, supercooled to 25° for 30 seconds and the nuclei developed at 35°.

In the following table are arranged the results of 50 such countings of nuclei. The number varies from 10 up to 37 nuclei in the same liquid during equal supercooling periods.

For comparison, six dice were thrown 50 times and the sum of the spots noted for each throw. The two experimental series are comparable with each other since the average value of the dice numbers is 19·40 and of the nucleus numbers 19·44 with each 50 trials. The highest nucleus number is 27 and the highest dice count 27; the lowest nucleus number 11 and with the dice 10. Arranging nucleus numbers and numbers obtained from the dice throws gives the following table :

Nucleus No.	No. of trials.	No. of trials yielding each 6 different nucleus numbers.	Dice counts.	No. of throws.	No. of throws yielding each 6 different counts.
10	0		10	1	
11	1		11	0	
12	1		12	1	
13	3	9	13	2	7
14	3		14	2	
15	1		15	1	
16	2		16	2	
17	5		17	9	
18	6	25	18	3	25
19	3		19	7	
20	4		20	3	
21	5		21	1	
22	5		22	5	
23	0		23	4	
24	3	16	24	3	18
25	5		25	5	
26	2		26	0	
27	1		27	1	

Half of the total trials of nucleus numbers or dice counts fall within the middle 6 values—that is, between 16 and 21. The average error $F = \pm \sqrt{\dfrac{s}{n-1}}$ (where s is the sum of the square of the differences and n the number of experiments) is practically the same for both experiments; F for the nucleus number $= \pm 4\cdot12$, for the dice numbers $\pm 3\cdot90$. The most probable variation $2/3\ F$ is, for the nucleus number $\pm 2\cdot75$. For the dice numbers it is $\pm 2\cdot6$, and between the values $+ 2\cdot6$ and $- 2\cdot4$ lie 28 per cent. of all variations. Both series of results are very similar indeed. The formation of a crystal centre is an occurrence which is subject to the laws of probability. It follows, then, that the percentage variation from the mean of a series of nucleus numbers must be the greater the less the nucleus number observed.

Nucleus Numbers with Slight Supercooling.

The nucleus number may fall to an extraordinary degree with lesser supercooling, so that the formation of a nucleus may only be occasionally observed in larger amounts of the supercooled liquid and after longer times. Even in the case of substances with very large nucleus numbers one finds, with very slight supercooling, very small nucleus numbers which become correspondingly less the slighter the degree of supercooling. In 1 g. lauric acid (m. p. $43\cdot2°$), a substance with a very large nucleus number, the formation of the first nucleus began after the times given below with the corresponding supercooling.

Supercooling ...	$3\cdot2°$	$2\cdot2°$	$1\cdot7°$	$1\cdot2°$	$0\cdot9°$	$0\cdot5°$	$0\cdot4°$
Seconds	$2\cdot5$	$3\cdot5$	$5\cdot5$	$10\cdot0$	$19\cdot5$	$120\cdot0$	1200

The equilibrium temperature of a crystal and its melt falls slightly if the size of the crystal falls below a certain limit. For a reduction in the thickness of a crystal of azobenzol from 10 to $0\cdot8\ \mu$, the equilibrium temperature was reduced by $0\cdot35°$.* Very small crystals are in equilibrium with their melts at significantly lower temperatures than larger ones. On this account a crystallisation centre must melt at the equilibrium temperature of a larger crystal. If the degree of supercooling at which nuclei no longer form be known, and also the influence of the size of the crystal on the melting point, the size of the crystallisation centres could be estimated.

The nucleus number is an extraordinarily sensitive property; it frequently diminishes with the number of melts, since by this means small quantities of materials originate which lessen the nucleus numbers.

Piperonal is a substance in which the nucleus number is inde-

* A. Meissner, Z. anorg. Chem., 1919, **110**, 169.

pendent of the number of melts. With this substance the nucleus number is significantly reduced the higher and the longer the melt be heated before the supercooling.* It follows that a part of the molecules of the melt differs from the remainder in that it forms crystallisation centres more readily, and that this property is gradually lost when the substance is kept at a temperature above its melting point for a time, and this the more rapidly, the higher the temperature lies above the melting point.

Finally another noteworthy phenomenon may be mentioned, which perhaps occurs more frequently. In the case of piperin it is not material whether the temperature at which crystallisation centres form is reached on a rising or on a falling temperature. When the piperin was brought to 0° and then gradually warmed up to the temperature at which the crystallisation centres should form, very many more crystallisation centres formed than if the piperin were cooled down gradually from its boiling point to this tempera-ture. The temperature of the maximum nucleus number—40°—did not change although the number of nuclei increased con-siderably.†

The Relation between Viscosity and Temperature.

The viscosity increases with falling temperature to an exceptional degree during the solidifying stage. In the case of easily fusible organic compounds it was shown that the temperature-viscosity curves in the solidifying range were practically parallel.‡ Similar results were obtained by S. English § for the logarithms of the viscosities of soda–lime–silica glasses of varying compositions between the temperatures of 500° and 1400°.

According to H. Le Chatelier ‖ the logarithm of the log. viscosity varies along a straight line or along two straight lines which inter-sect between 700° and 900°, according to the composition of the glass. The analytical expression of this law is as follows—

$$\log \log \eta = P - M . t,$$

where η = viscosity, t = temperature and P and M are constants.

The value of the constant M is given in the following table for three liquids during their solidifying period. The glass referred to is a soda–lime–silica glass investigated by English ¶ between 500° and 1400° (glass No. 26). These determinations gave the absolute viscosity in dynes/sq.cm. For betol and piperin the absolute viscosity values are also approximately calculated from rate of velocity of fall of small balls, according to Stokes' Law.

* Othmer, Z. anorg. Chem., 1915, **91**, 219.　† Z. physikal. Chem., 1898, **25**, 452.
‡ Ibid., 1899, **28**, 29.　　　　　　§ Trans. Soc. Glass Tech., 1924, **8**, 205.
‖ Compt. rend., 1924, **179**, 517 and 718.　　¶ P. 233 loc. cit.

	M.	Temperature range.
Glass	0·0011	500—900
Piperin	0·0091	50— 90
Betol	0·0158	15— 50

It is obvious that the constant M, which determines the relation between viscosity and temperature, is much smaller for substances whose solidifying range is at a high temperature than for substances which solidify at lower temperatures. The increase in the value of the constant M with falling values of the temperature of the solidifying range can also be anticipated from the following considerations. With solidifying ranges occurring at exceedingly low temperatures, the range from absolute zero is small, and, therefore, for any great solidifying range, no corresponding temperature difference remains available, since, in any case, solidification must be complete at the absolute zero.

For the manipulation of glasses by blowing or spinning they must be brought to temperatures corresponding to a definite viscosity range. The greater this temperature range—the smaller the value of M—the more suitable for manipulation. In the case of betol, this temperature range is in any case much smaller than with ordinary glass. It happens not infrequently that the temperature range corresponding to the higher nucleus numbers falls within that of the viscosity range most suitable for the manipulation of the glass. This occurs equally with soda–lime–silica glasses—according to the observations of S. English—as with the glasses of easily-fusible substances. With the larger nucleus numbers this circumstance may hinder or even prohibit the working of the glass.

The following table gives a comparison in the case of easily-fusible substances of the temperatures of equal mobility of the softened glass essential for its manipulation, drawn up from the velocity of a small rod moved in the fusion.

The temperatures of maximum nucleus numbers, as a rule, fall within the temperature range of the necessary workable mobility of the softened glass.

	Beginning of movement.	Velocity of 1·02 mm. in 100 secs.	Melting point.	Temperature of maximum nucleus No.
Piperin	37·0°	44·9°	127°	40°
Betol	2·1	6·4	95	20
Alphol	3·4	9·6	80·4	—
Peucedamine	−1·8	1·5	81	0 ?
Cocaine	10·0	15·6	98	—
Santonine	46·2	—	170	42
Narcotine	64·5	—	175	140
Allylthiourea	−7·5	—	74	0 u. −20
Chlorourethane	5·5	—	102	40
Cinchoninic acid	36·8	—	161	60
Papaverine	47·3	53·0	147	—
Brucine	125·2	132·2	178	—
Cane-sugar	108·6	113·3	160	—

Differences in the Properties of Glasses and Crystals.

The properties of a glass and of a crystal of the same composition differ from one another in one consideration essentially. The properties of a glass are independent of direction except in such cases where strains are existent which may readily be set up through rapid cooling, but may easily be removed by heating and subsequent slow cooling.

In a crystal, the mechanical strength has a very definite minimum value in certain directions—the cleavage planes—while in glasses no such differences are present. The fracture is conchoidal.

Further, from the chemical standpoint a glass behaves differently from a crystal. If a crystal surface be etched, hollows are produced having rectilinear contours on the etched surface, and bounded by facetted surfaces in the direction of their depth. The surface of a glass on etching shows, on the contrary, small etched pittings with curved boundaries, and, in the direction of their depth, bounded by doubly curved surfaces.

On etching, the atomic groupings of a crystal are removed from the regular structure along straight lines of definite direction, whereas, in a glass, the attack is quite irregular.

The Supercooling Capacity of Binary Melts.

A very high supercooling capacity is seldom evident with chemically homogeneous substances, and even when larger quantities of the substance have been obtained successfully in the glassy condition there is a strong tendency to devitrification. The question then arises, whether the supercooling capabilities of binary, ternary or polynary mixtures are not greater than those of chemically homogeneous substances. Regarded from the atomic standpoint, the answer is that foreign substances in general, if present in quantity will hinder the formation of crystallisation centres unless they throw mixed crystals from the mixtures.

In the case of binary metallic mixtures, when both compounds crystallise without supercooling, it is seen that eutectic crystallisation sets in without any supercooling to speak of. If, on the contrary one of the components, for example, antimony or bismuth, crystallises after slight supercooling, the crystallisation of the eutectic starts with practically the same degree of supercooling as that particular component (the one which can be supercooled). It is probable in the case of metals, that the nucleus number is so great that slight reductions in its value are not noticeable.

The influence of one substance on the supercooling capabilities of another is prominent in the case of salts. Four separate cases may be differentiated :—

(1) Both components A and B crystallise without significant supercooling, but in the mixtures containing no greatly preponderating excess of either crystallisation is strongly retarded (Fig. 8).

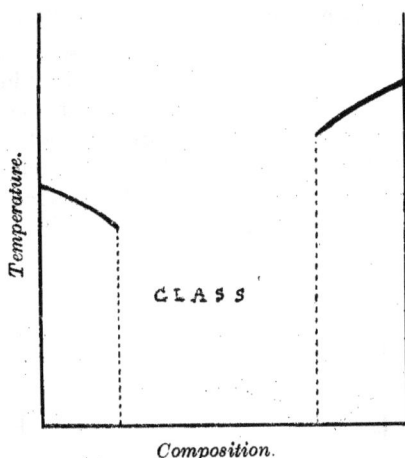

Composition.
FIG. 8.

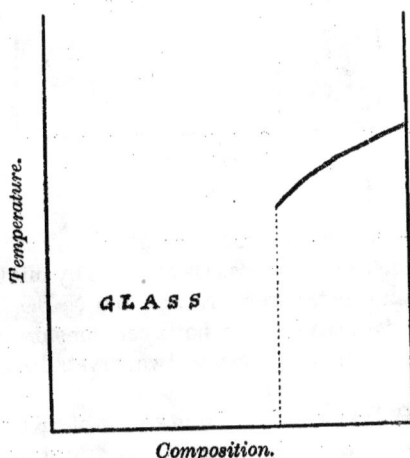

Composition.
FIG. 9.

The cooling curves of the components themselves show marked retardation of the temperature fall at their respective melting points, but the addition of small quantities of the other component causes the retardation to be less marked, and further additions cause it to disappear altogether, consequently these latter mixtures solidify as glasses. Examples of this class are mixtures of sodium and magnesium silicates (Na_2SiO_3 and $MgSiO_3$) and lithium and zinc silicates * (Li_2SiO_3 and $ZnSiO_3$).

(2) One component crystallises readily on cooling but the other does not. In this case, retardations on the cooling curve are only shown with such mixtures as are rich in the former component, whilst the others solidify as glasses (Fig. 9). Even in such mixtures as are rich in the readily crystallisable component, part of the melt solidifies as glass. Examples are furnished by mixtures of potassium and lithium silicates * (K_2SiO_3 and Li_2SiO_3).

(3) Both components crystallise readily and form a compound which also crystallises readily, but mixtures of the compound and

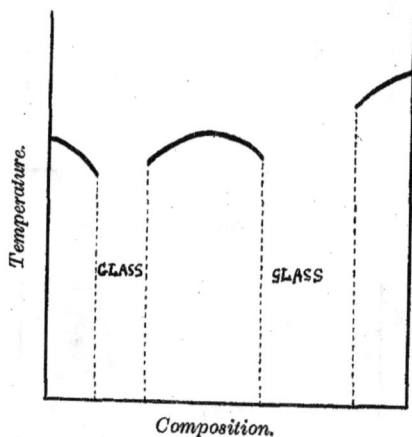

Composition.
FIG. 10.

either of the components do not crystallise but solidify as glasses (Fig. 10). An example of this case is afforded by mixtures of sodium metaphosphate and metaborate ($NaPO_3$ and $NaBO_2$).†

(4) Finally, it is possible that both components crystallise with difficulty but a compound of the two crystallises more readily (Fig. 11).

According to P. Hautefeuille,‡ from a melt of SiO_2 and P_2O_5 octahedral crystals with a melting point of 1400° separate out at 700° and monoclinic prisms at 900°. Examples of the contrary case are afforded by the formation of celluloid from nitrocellulose and camphor, and of bakelite from phenolin and formaldehyde.

The customary soda–lime–silica glasses may be regarded as

* R. C. Wallace, *Z. anorg. Chem.*, 1909, **63**, 16.
† H. S. Van Klooster, *ibid.*, 1910, **69**, 127.
‡ *Compt. rend.*, 1883, **96**, 1052; 1884, **99**, 789.

ternary mixtures of Na_2SiO_3, $CaSiO_3$, and SiO_2. The two components Na_2SiO_3 and $CaSiO_3$ crystallise readily as do practically all metasilicates, and the same applies to their mixtures, from which mixed crystals separate. The corresponding diagram has been worked out by R. C. Wallace.* With an excess of silica the nucleus number of these mixed crystals is reduced extraordinarily, so that

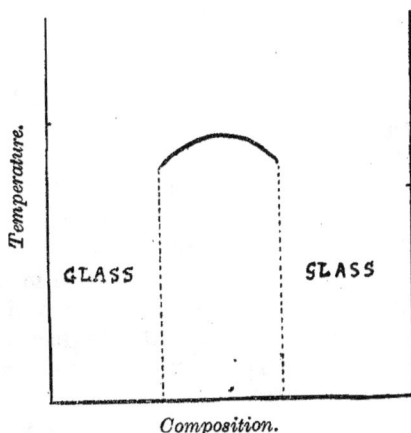

Composition.

FIG. 11.

mixtures with an excess of 8 per cent. of silica or over solidify as glasses.

Systematic researches on the crystallisation of ternary and polynary mixtures must lead to valuable information on glass batches though it is difficult to apply these without some definite point of view.

THE UNIVERSITY,
GÖTTINGEN, GERMANY.

XV.—*On the Constitution and Density of Glass.*†

By A. Q. TOOL and E. E. HILL.

*(A Contribution to the Symposium on " The Constitution of Glass "
held at the London Meeting, May 25th and 26th, 1925.)*

Introduction.

IN the past, reports have been made on certain investigations which are under way at the Bureau of Standards to determine

* *Z. anorg. Chem.*, 1909, **63**, 5.

† Published by permission of the Director of the Bureau of Standards of the U.S. Department of Commerce.

what changes are caused in glass by varying the treatments involved in its annealing. The purpose of these studies was to find whether any of the possible changes are of such a character and magnitude as to render them of either practical or theoretical importance. Finding this to be the case, the scope of the work is being extended to include the search for a better control of those changes.

Many references in regard to the occurrence of such changes in certain properties of glass are found in the literature. Statements that the refractive index and density are altered by the degree of annealing are especially frequent; however, very little is found regarding the control of such variations. From these statements and the data given to support them it is difficult, moreover, to draw any definite conclusions, since these effects have usually been observed incidentally during investigations with other primary objects in view. Furthermore, the difference in, and the uncertainty with regard to, the conditions under which the various observers have worked combined with the fact that they were seldom studying the same glasses make it impossible for others safely to co-ordinate their results and use them.

In considering the different aspects of this problem and the various changes to be expected from altering the heat treatment it was deemed advantageous to formulate some sort of tentative hypothesis regarding the constitution of glass. This focusses attention on the effects to be expected and the relations which should exist between them. Such a working hypothesis, which seems to correlate satisfactorily all the results so far obtained, resulted from a consideration of the opinions expressed from time to time by a number of investigators.

The Hypothesis Employed.

This hypothesis requires that, under proper cooling conditions, certain practically reversible processes or reactions involving some or all of the constituent molecules shall advance in such a way that other possible processes often associated with cooling, such as segregation or crystallisation, are prevented, thus preserving the vitreous character. This preventative action may result either from all the simple molecules or components becoming involved in unwieldy non-crystalline aggregates, or through such a reduction in the mobility of the mass as a whole that any of the simpler components in excess and not included in these aggregates are restrained from the diffusion necessary for crystallisation and segregation. The character of all the reactions or processes possible is not necessarily considered as similar to those occurring normally between the acidic and basic constituents to any greater degree than the latter are comparable to that of crystallisation. In many cases, for example, some of the processes appear to resemble polymerisation. Furthermore, many glasses are formed from a single constituent and are in no marked degree dissimilar to the more complex types. Then, again, the glass-forming processes are confined to certain acidic constituents or components and combinations including them. Reactions between the basic and acidic constituents may, however, greatly alter many of the other processes.

According to the view here adopted, the character of the processes and the rate at which they advance will change as the glass cools and, when it is held at any given temperature, this rate should decrease with time until a state of practical equilibrium becomes established after a period the length of which depends on the temperature. In those ranges where glass is easily deformable, this period should be relatively short and in the upper part of the annealing range will not exceed an hour, perhaps, but not far below the lower part it may approach a duration of months. In this connection the possibility must not be overlooked that some of the processes, if they become too far advanced, may interfere with the normal advancement of others. Again, certain of them may require some degree of advancement on the part of others before they become active to any marked extent.

Crystallisation and other types of segregation are considered as processes subject to the kind of interference mentioned. They also impose such interference on other processes. Even those processes normal to a perfect glass may in some cases on interfering in this way cause a variation in the period required at any given temperature for reaching a condition of practical stability. Thus they may cause the glass to have apparently two or more of these conditions for the same temperature. A marked effect of this sort, however, probably occurs very rarely with the simpler and more perfect glasses; at least in so far as the data already obtained are indicative.

The length of the period required for reaching practical equilibrium at any temperature should, of course, depend on the degree of advancement of the various processes when that temperature is reached. Thus, one sample of a glass, in equilibrium at some temperature, cooled rapidly over the interval between this and another temperature chosen for treatment, should require a different period from a second sample in the same initial condition but cooled more slowly over the same range. Reducing the interval between the original equilibrium temperature and any chosen treating point to which the glass is cooled should also change the period. In fact, the slow cooling mentioned in the case of the second sample should usually be practically equivalent to such a reduction in the interval. Whether these changes in the treatment from that given to the first sample cause an increase or decrease in the period depends, however, on the relationships between the processes involved. For instance, they may at times increase the period, since the processes reaching completion at the higher temperatures decrease the mobility and thereby cause those which must be completed at the temperature of treatment to advance more slowly than they would in the highly chilled glass. However, since these particular changes normally bring the glass nearer to equilibrium, it would seem that they should usually reduce the period necessary for reaching that condition.

Since the hypothesis adopted requires the characteristic processes to be practically reversible, a glass in a condition of equilibrium at a lower temperature should on being reheated to, and retreated at, a higher one, return after a time to the condition which is normal for the higher temperature. That is, it should return to practically the same condition it would have reached on being treated at this new temperature without being first cooled below it. Again, the period required for this retrogression should depend on the temperature of retreatment and the magnitude of the interval between this and the original effective or equilibrium temperature. In some cases it is possible that the rate of retrogression increases for a time and then after reaching an optimum decreases again to practical cessation as equilibrium is approached.

On cooling to low temperatures, for example, that of the atmosphere, certain processes may continue at an appreciable rate throughout the cooling range. In general, however, few of those processes the furtherance of which is possible through extremely long treatments within this range should reach on any normal cooling such a degree of advancement that reheating the glass over any moderate interval will cause it to show any evidence of a material retrogression. In other words, glass at too low temperatures cannot be brought into the corresponding conditions of equilibrium although the conditions corresponding with higher temperatures are easily obtained by treatments in the upper or annealing ranges. However, in these lower ranges it may be possible to obtain certain small cyclic variations * from repeatedly heating and cooling a glass ; these taking place as a result of the retrogression and advancement of any secondary processes which may be noticeably active

---

* L. Marchis, "Les Modifications Permanentes du Verre et le Deplacement du Zero des Thermometres," 1898.

ON THE CONSTITUTION AND DENSITY OF GLASS. 189

there. In addition, however, there should usually be a gradual drift, in the value obtained for the coefficient showing these changes, which would be indicative of the unstable condition of the glass.

At atmospheric temperatures certain extraordinary treatments such as exposure to powerful radiations may easily expedite normal processes or induce abnormal transformations which heating to temperatures where retrogression is rapid should often fully reverse. Observers have noted that colorations produced in this manner do disappear on heating the glass to temperatures approaching the annealing range. An example of this is found in an article by J. R. Clarke.*

If, as surmised, a drift in one direction generally occurs at low temperatures signifying instability and a trend toward equilibrium for a temperature lower than the effective temperature of treatment, then glass in these ranges must be considered as undercooled. This instability does not necessarily suggest that the glass tends toward devitrification, although certain components in excess at these temperatures might cause this if sufficiently untrammelled by the surrounding magma. It suggests rather that the glass is tending toward a condition more nearly stable at atmospheric temperatures yet fully comporting with the vitreous state. This view would not apply, however, to those regions where contamination and certain deposits induce deterioration.

The Effective Annealing Temperature.

At any low temperature where changes proceed with exceeding slowness the condition of a glass is, therefore, one of approximate equilibrium for some higher temperature. This temperature has been designated † the "effective annealing temperature." It is determined by the condition of the glass before annealing, by the treating or annealing temperature, by the period of treatment, and by the mode of cooling. In appraising the effect of cooling, allowance must naturally be made for all those effects which accompany it normally, such as the usual changes dependent on the thermal expansivity. The magnitude of such changes or effects and in fact all properties of the glass will be controlled, although presumably in varying degrees, by this approximate equilibrium condition and appear as functions of the temperature to which this condition most nearly corresponds. That is, they will be functions of the effective annealing temperature.

The determination of this temperature for a given specimen of glass requires a number of samples of the same glass to be so pre-

* J. R. Clarke, *Phil. Mag.*, 1923, **45**, 735.
† A. Q. Tool and C. G. Eichlin, *J. Opt. Soc. Amer.*, 1924, **8**, 419.

130 THE CONSTITUTION OF GLASS

pared that their effective annealing or treating temperatures are as accurately known and range over as wide an interval as possible, extending to temperatures both above and below that sought. A comparison of measurements then made on some chosen property for all these samples and for the original specimen makes it possible to estimate with more or less accuracy the unknown effective treating temperature of the original. The result, however, probably depends somewhat on the property chosen for this determination, especially when the measurements are made at atmospheric temperature. This follows from the assumption that during cooling all the processes do not advance uniformly and in consequence, reach individual limits in any continuous cooling treatment, thus causing the resultant effective temperature to have the sense of an average. Since, presumably, the different properties of the glass vary in their dependence on the various processes involved, it follows that the effective temperature as determined will also vary somewhat with the property chosen as characteristic.

As an illustration of the above, consider two samples of the same glass; one cooled rapidly from a high temperature to some treating point where it is held for a time and then cooled rapidly to atmospheric temperature; and the other, cooled at a more moderate rate through the range without holding it at any point. Within the limit of experimental accuracy two such samples might have the same density whilst their refractive indices and dispersions differed measurably. Again, their indices for a portion of the spectrum might also be practically alike so that the dispersions alone differed materially and this perhaps only when the measurements were extended over a sufficient range. It also appears possible that differences might occur between all three of these characteristics, whilst in part of the spectrum the ratio of the refractivity to the density differed little for two such samples, especially if the treatments given them varied distinctly in the upper part of the annealing range. These considerations arise both from the working hypothesis and from those observations * showing that the dispersion and absorption of light continue to increase through this range, although the indices first increase and then decrease; and also those † showing that the relationships between the density and the refractive indices for part of the spectrum at least do not change to the same extent. These facts indicate on the basis of the working hypothesis that some of the normal processes and structural units involved are not the same in the higher and lower ranges.

* J. O. Reed, *Ann. Physik*, 1898, **65**, 707.
† C. G. Peters, *J. Wash. Acad. Sci.*, 1923, **13**, 217.

After a glass has been brought into its condition of approximate equilibrium at a given temperature the difficulty of reducing it to atmospheric temperature without materially shifting the effective treating temperature from this point is probably best accomplished by chilling or, at least, by rapid cooling. By this method it should be possible to preserve to a considerable degree the condition corresponding with any treating temperature which is not too far above the annealing range. Chilling from temperatures in a wide range above this limit is not likely to produce any great change in the effective treating temperature resulting, since the processes advance so rapidly there that the final result becomes a function merely of the cooling rate. This limit corresponds with that found through the maximum strain obtainable by chilling glass from different temperatures.* One danger in treating glass above this limit, if the periods are long, is that of crystallisation or segregation. Either of these effects will alter the characteristics of the glass and may lead to erroneous conclusions.

The further the treating temperature is chosen below this rather indeterminate limit above the annealing range the more nearly it should be possible to make the actual and effective treating temperatures coincide. In a measure, this can be tested by varying the severity of the chilling. It should also be added here that treatments including chilling made well below this limit do not introduce appreciable permanent strain, especially when the samples are small. This is of importance when those properties modified by strain are being investigated. With density changes, however, such considerations are probably of little import, since in theoretical discussions covering this subject—particularly that of thermo-elastic stresses—it is assumed that the average density of such a strained block as normally determined should differ little if any from that in the unstrained condition. In the case of the refractive index, on the other hand, no such average corresponding with the true value can be obtained without using a method † which requires pulverisation of the block and this usually lacks the accuracy desirable.

Considerations such as these make it evident that discretion must be used in making a choice of the properties which are to be employed in studying the characteristic changes caused by the heat treatment. The choice should naturally be restricted to those properties which undergo the greatest and most characteristic changes and are subject to the most accurate measurements. It appears important, therefore, to discuss some of those properties which show characteristic changes.

* E. Zschimmer and H. Schulz, *Ann. Physik*, 1913, **42**, 345.
† A. A. Lebedeff, *Trans. Opt. Inst.*, *Petrograd*, 1921, **2**, 1.

Properties Showing Characteristic Changes with Heat Treatment.

Many of the processes presupposed will undoubtedly involve the liberation of heat during their advancement on cooling and absorb it while retrogressing on heating. Furthermore, they will alter by the degree of their advancement the capacity of the glass for the absorption and conduction of heat even after it has been cooled. Consequently, a study at atmospheric temperatures of the changes in the specific heat and thermal conductivity caused by varying the treatment should yield valuable information. If the measurements can be made with sufficient accuracy, chilled glass will presumably show a higher specific heat than very well annealed glass. At higher temperatures and with the specific heat methods in general use, however, the results obtained, especially in the annealing range, will undoubtedly be more or less erratic and show a dependence on the interval used and on the rate and method of changing the temperature. That is, the apparent specific heat obtained for any given glass by a method of heating will usually differ from that procured over the same interval by a method of cooling, and both results will vary with the rate employed. Only where equilibrium obtains at the beginning and the ending of the tests will the results by the two methods be equal and give the same average apparent specific heat. On the other hand, the difference between the results obtained, in one case by chilling through a given range and in the other by cooling slowly over it from a temperature where in both cases the glass was in equilibrium, may allow certain estimates to be made concerning the heat normally absorbed by the suppressed processes peculiar to the glass in this range; and also for that which may be termed the true specific heat, since it arises chiefly from processes common to all materials. In general, however, the estimate for the first will be too small whilst that for the latter will be too large. These conclusions follow, since all the processes peculiar to glass will not be suppressed by the chilling, whilst others may be suppressed in both cases unless the cooling rate is so reduced that the requisite accuracy in measurement is unattainable.

These are merely some considerations cited to show the difficulties which will be met in determining and interpreting any results on the apparent specific heat of glass. Despite these difficulties, however, it has been shown * that the heat capacity of some glasses changes rather abruptly in an intermediate range corresponding, it seems, with that where annealing could be accomplished. Above this range the heat capacities were found greater than at temperatures below it, a condition which agrees with the views expressed

* W. P. White, *Amer. J. Sci.*, 1919, (4), **47**, 44.

above. Presumably, however, slower cooling would have increased this difference and hence made the agreement more pronounced. Since, for the reasons mentioned, measurements of this sort are likely to be difficult, it seems advisable that a thorough study of the variations observable in the heating and cooling curves should first be made for qualitative purposes, especially in the case of the heating curves, since very marked differences as a result of altering the previous treatment are to be found.

These differences in the heating curves arise from variations in the magnitudes of two effects. One, an exothermic effect, becomes evident before the annealing range is reached, and the other, an endothermic effect, occurs in the upper part of this range. The first is most pronounced in chilled glasses and is almost if not wholly lacking after careful annealing, whilst the latter is common to all glasses, but is most pronounced in those which are well annealed. These marked indications of the heat effects accompanying the processes peculiar to glass are believed to be greatly accentuated by the ease with which this material may be under-cooled or super-heated. From an analysis of the curves * thus obtained under varied conditions it is possible to determine the effective annealing temperature with considerable accuracy and to learn much concerning the degree of annealing and the manner in which the processes advance or retrogress.

A study of the thermal expansivity also shows a marked difference between chilled and annealed glasses. In fact, a thorough examination of a sufficient number of thermal expansivity curves obtained from a series of differently treated samples of the same specimen of glass should yield practically the same information derivable from the heating curves. One of the advantages of the thermal expansivity method is the accuracy with which the measurements can be made. Its possibilities are evident when the results † of some of the careful and comprehensive investigations are examined. For example, expansion curves taken while heating chilled glasses show a decrease in the expansivity sometimes amounting to an actual shrinkage, which corresponds with the exothermic effect on heating. This decrease in volume persists after the glass is cooled to room temperature, provided the heating is not carried too far and the cooling is not too rapid. They also show a large increase in the expansivity which corresponds with the endothermic effect; and, when the limit of heating and the cooling rate are properly chosen, some of this increase in length will persist after room temperature is

---

* A. Q. Tool and C. G. Eichlin, *J. Opt. Soc. Amer.*, 1924, **8**, 419.

† C. G. Peters and C. H. Cragoe, Bureau of Standards Sci. Papers, No. 393, 1920.

reached. These changes in length correspond with the density changes caused by heat treatment and they must not be confused with the alteration in form produced when the expansion curves are carried into the softening or sintering range.

The electrical properties may also yield important data relative to the structure of glass and the changes caused by heat treatment. The electrical conductivity, for example, rises rapidly with increasing temperature and undergoes * a sudden although small change in the same range where the endothermic effect and abnormal increase in the expansivity are found. This agrees with the views advanced here, and, for the same reason there should also be a difference in the conductivity of chilled and well annealed glasses, a conclusion well supported by the work of Foussereau.† The fact that there still appears some question regarding the degree ‡ to which this conductivity, under certain conditions, is electrolytic in nature and the fact that it has been known for some time, on the other hand, that certain bases could be displaced and transported § through glass walls, suggest that investigations in this field should be fruitful in yielding data bearing on the nature of glass and on the cause of the changes from heat treatment. The same may be said regarding a study of the dielectric properties of glasses, since some of the changes possible from heat treatment should make them more suitable for the uses required of them in the electrical industry.||

In fact, all properties of glass on which measurements of any accuracy can be made are available in studies of this character; and some of the results attainable will doubtless be of considerable practical as well as theoretical value. This is especially true as regards the production of certain colours in glass and the elimination of certain defects associated with somewhat comparable segregations, possibly colloidal in nature; and indeed the changes in the mobility, density, refractive index, and dispersion may ultimately be carried to such proportions as to be of practical value.

Changes in the Density and Refractive Index.

That the density and refractive index may be changed by annealing has long been known, and both effects are mentioned by Hovestadt.¶ In dealing with thermometers he gives many references

* H. Schönborn, *Z. Physik*, 1924, **22**, 305.
† J. Foussereau, *J. Phys.*, 1883, **2**, 262.
‡ H. H. Poole, *Nature*, 1921, **107**, 584.
§ E. Warburg, *Ann. Physik*, 1884, **21**, 622; E. Warburg and F. Tegetmeier, *ibid.*, 1888, **35**, 455; F. Tegetmeier, *ibid.*, 1890, **41**, 23; C. A. Kraus, and E. H. Darby, *J. Amer. Chem. Soc.*, 1922, **44**, 2783.
|| J. R. Clarke, *The Glass Ind.*, 1921, **2**, 221.
¶ " Jenaer Glas," by H. Hovestadt, Jena, 1900.

ON THE CONSTITUTION AND DENSITY OF GLASS. 195

and discusses in detail some of the observed effects resulting undoubtedly from the density variations caused by the treatment given the glass. Unfortunately, in the past, these changes in the density and refractivity and even those in the expansivity have often been considered as the result of, or at least, as wholly co-existent with, the introduction and relaxation of those stresses caused by imperfect annealing. To an extent, the presence and relaxation of stresses do cause changes in most properties, more especially in the index, and, through deformation chiefly, in the apparent expansion of bulbs. Much of the change both in the index and density occurs, however, when there can be no doubt regarding the non-existence of any appreciable mechanical strains of the type generally called thermoelastic either before, during, or after the heat treatment producing it. If, therefore, this change is to be attributed to stresses they must be entirely different from the usual thermoelastic type and must arise from purely local inter-molecular conditions.

For testing these views, density experiments were performed on polished samples of medium flint and Pyrex cut from original specimens so nearly free from strain that they would normally be rated as reasonably well annealed. After the new treatments given in these tests, those samples not heated above 420° for the medium flint, and above 600° for Pyrex, were still practically free from strain. If they were strained by heating to higher temperatures they were easily reannealed by treatments somewhat below these points. The samples of medium flint were cubical and weighed about 5·8 grams. Those of Pyrex, averaging about 2·4 grams, were square plates so cut that three would make a cube for treatment as one sample. It being desirable to cool these samples rapidly, their size was the minimum consistent with reasonably accurate density determinations. For reducing the permanent strain arising from cooling, the type of sample used for Pyrex is advantageous, but it entails the disadvantages of increased surface and difficulty in handling.

In treating, the sample was placed in a small electric tube furnace held at the desired temperature, and always near or in contact with the junction of the thermo-couple used in determining the furnace temperature. With medium flint, an aluminium cylinder with a 12 cm. length, a 4 cm. diameter, and a 4 mm. wall thickness was used to equalise the temperature about the sample. Below 600°, this was occasionally used with Pyrex with only a very little difference resulting. Presumably, the temperature indicated by the thermo-couple was always somewhat lower than that attained by the glass, inasmuch as the heat flow was through the sample from its supports to the junction. However, since the temperature—when accurate control was necessary—usually varied little more than 2° from the value chosen for treatment and since the precaution of occasionally making explorations to determine the variations in temperature around the sample by means of a second couple was taken, it is improbable that this discrepancy could have reached 5° without being detected.

After treating the samples at selected temperatures for the desired periods, they were removed from the furnace, placed on a clean asbestos board, and allowed to cool in the air. Whilst this produced large temporary stresses, those permanent in character and sufficiently large to be taken into account resulted only under the conditions already mentioned. Furthermore, breakage never occurred, since to have increased the temporary stresses to that limit would have required a much more drastic treatment.

Whilst these treatments did not alter the weight in air, they did in general change the density considerably, as a study of the data represented in the two figures will show. In making the density measurements the weighings in air were accurate to 0·2 mg. and corrected for the buoyancy of the air. The volumes were determined by weighing in a kerosene oil having a carefully determined density of 0·8063 which was continually checked and showed only slight variations in the last decimal place. This oil was used because it reduces the surface tension effects and thoroughly wets the glass surfaces. The samples were suspended in it by a 0·2 mm. rough gold-plated platinum wire supporting a suitable cage for holding them. The density observations were made at 25° and the oil was maintained at this temperature by a bath surrounding its container. A more complete description of this apparatus may be found in other publications.* The weighings in oil were accurate to 0·4 mg. The deviations in the weighings in air and in oil as given were the maxima under good weighing conditions and were usually found to be the controlling factors in the errors of the density determinations. In the case of the medium flint the probable error in the density approached ± 0·0004, but was somewhat less in the values obtained for Pyrex.

For the tests on medium flint (B.S. Melt 494).† eleven samples

* N. S. Osborne, E. C. McKelvy, and H. W. Bearce, Bureau of Standards Sci. Papers, No. 197, 1913; H. W. Bearce and E. L. Peffer, Bureau of Standards Tech. Papers, No. 77, 1916.

† The composition of this glass which was made at the Bureau of Standards as an experimental melt under the direction of A. N. Finn was determined by H. G. Thomson. The analysis shows the composition to be as follows :

SiO_2 50·55	Na_2O 2·69	Al_2O_3 0·65	CaO Trace
PbO 40·14	K_2O 5·93	Fe_2O_3 0·02	MgO Not detected

free from bubbles and reasonably homogeneous were chosen. In the first series of tests, eleven treating temperatures equally spaced and ranging from 360 to 460° were employed, and at each of these one of the eleven samples was treated. The treating periods—almost doubled for each 10° decrease in temperature—varied from 1 hour at the highest to 264 hours at the lowest temperature. From previous study of the exothermic and endothermic effects and the

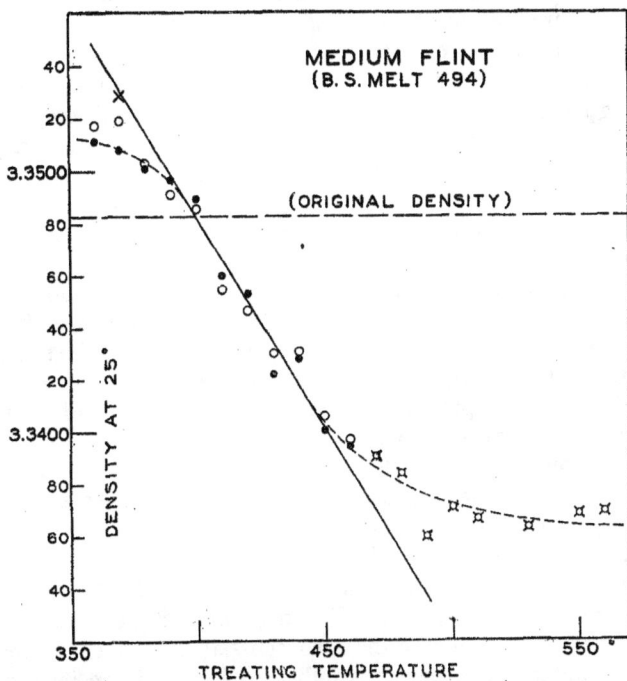

FIG. 1.

Medium Flint. Density change from heat treatment.
Dots : results after 1st treatment.
Circles : after 2nd.
Crossed circles : after 3rd, samples same as before but with different temperatures as indicated.
Cross : result at 370° for prism of same glass after a period exceeding 500-hour effective treatment.

changes produced in them by chilling and annealing, it was known that the selected periods would be insufficient at the lower temperature and probably longer than required at the higher, although a certain time is always necessary for the glass to reach temperature equilibrium. It was also known from the same study that little or no change in the density would result from treatments near 400°,

138

THE CONSTITUTION OF GLASS

since this was approximately the effective annealing temperature of the original.

The results of this series are shown by dots in Fig. 1, where the abscissæ represent the treating temperatures and the ordinates the densities as determined after treatment; whilst the broken horizontal line at 3·3483 represents the average original density of the eleven blocks. The greatest deviation from this average for any sample before treatment was 0·0005; after treatment, the greatest (− 0·0088) was exhibited by the sample treated at 460°, whilst near 400° the change became positive and rose to the value + 0·0029 for the sample treated at 360°, although, for the reasons mentioned, the treating periods were believed to be insufficient for all temperatures below 390°. The points above 390°, on considering the errors in the temperature determinations and the unavoidable fluctuations in the cooling rates, fall closely enough to a straight line to indicate that the variation in density shown is nearly a linear function of the treating temperature in this range.

In a second series, each sample was retreated at the same temperature and for the same period as before in order to test the sufficiency of the previous treatments in producing approximately the maximum deviation obtainable at each temperature. The results of this second series are represented by the circles in Fig. 1 and show that the previous treatments were all nearly sufficient except for two temperatures, 360° and 370°. In regard to these temperatures, a more accurately determined value on a prism treated more thoroughly at 370°—indicated by a cross—shows approximately the limit that may be reached there, whilst at 360° an estimate based on the study previously mentioned indicates that the treating period required to reach the limiting value there should exceed 1000 hours. The fact that the treatments at the lower temperatures are undoubtedly insufficient is indicated in the figure by a dotted branch of the curve.

A third series, to determine the deviations obtainable by treatments at still higher temperatures, was accomplished by retreating those samples previously treated at the points ranging from 390° to 460°. At each of the new temperatures between 470° and 560° to which a sample was assigned and where the results are indicated by crossed circles, the treating period was an hour except in the case of the last, where it was reduced to ½ hour because the sample treated there, and also the one at 550°, tended to adhere to the platinum foil on which they rested. The maximum decrease in density, 0·0123, from the original value was shown by the sample treated at 490°. As the temperature of treatment is increased the cooling rate becomes the chief factor in determining these density changes. Apparently this limit is reached near 480°, a fact indicated by a dotted branch of the curve in this range. Only by increasing the severity of the chilling can a further decrease in the density be expected.

The slope of the straight part of the curve corresponds with a decrease of 0·00016 in the density for each degree increase in the treating temperature. This represents a linear expansion which is about 1·7 times the linear expansivity at ordinary temperatures and almost one-half of that in the critical range where the expansion is always greater. The values of the expansivity in these ranges as determined by G. E. Merritt of this Bureau are respectively 0.9×10^{-4} and above 3×10^{-4}.

As a final test all the samples with densities now varying from 3·3520 to 3·3360 were treated at a common temperature, namely, 400° for 72 hours. Their average density after this treatment was found to be 3·3481, with a maximum variation from this value of 0·0009. The deviation of this average from the original value was only 0·0002, and corresponded with a change in the effective treating temperature of about 0·1° as derived from the slope of the curve. Hence as regards density the final condition of the glass was practically identical with the initial for all the samples.

The results on Pyrex glass which are represented in Fig. 2 are different. In view of its composition and some previous investigations * on its heating curves, this was, however, to be expected. Treating temperatures were chosen at 25° intervals starting with 450° and ending with 750° where the glass adhered slightly to the platinum foil on which it rested. In the first series, two samples were treated at 500° and one each at the other twelve temperatures. The treating periods varied from ½ hour at 750° to more than 1000 hours at 450°, being almost doubled for each 25° decrease in temperature. All samples treated at 500° and above showed a decrease in density; those treated at 475° and 450° showed an increase. As shown by a second identical treatment, that given above 525° seemed sufficient to establish density equilibrium. At 500° one sample reached this condition on the first treatment whilst the other required three treatments. Since there are excellent reasons based on previous work for the belief that a much longer time would be needed, it is practically certain that neither of those samples treated at 475° and 450° reached equilibrium. However, the opportunity has not yet been afforded for thoroughly testing this question.

As before, the broken horizontal line represents the original density, 2·2381, the greatest deviation from this by any of the

* A. Q. Tool and C. G. Eichlin, *J. Amer. Cer. Soc.*, 1925, **8**, 1.

fourteen samples being 0·0005. The dots represent the results after the first treatment and the circles those after the next one, except for the samples treated at 500°, where the results given are those obtained when apparent equilibrium had been reached after three repetitions. The curve shows that a minimum density is obtainable somewhere in the range 540° to 570°, and that treatments within certain limits both above and below this yield greater densities.

As a third test, all the samples were then treated for 72 hours at

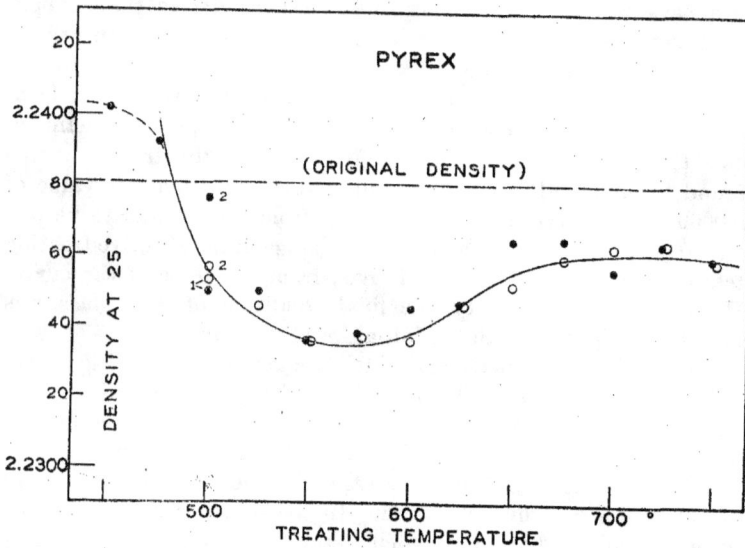

FIG. 2.

Pyrex. Density change from heat treatment.
Dots : results after 1st treatment.
Circles : after 2nd, except at 500° where they represent results after 3rd,
　Two samples were tested at this point as indicated.

560° and their resultant average density was 2·2336 with a maximum deviation of ± 0·0004 for any sample. This average corresponds reasonably well with the minimum indicated by the curve, so, presumably, this approaches the minimum attainable with this specimen of glass; at least, without altering the nature of the glass by going to temperatures above the sintering point. Other specimens, whilst yielding much the same form of curve, did show different values for this minimum density—a result, however, not surprising since the composition of Pyrex glass is also subject to variation.

After this treatment some of the samples were distributed at random between certain of the previous treating temperatures and retreated according to the schedules followed in the first tests. The same form of curve with approximately the same values for corresponding temperatures were again obtained. After this, a few of these samples were given a final treatment at 450° and whilst, as a result, they all increased in density, it appeared that those which had just previously been treated above 650° advanced much more rapidly than those which had last been treated between 500° and 600°. In fact, the former all advanced well beyond the average original density (2·2381), whilst none of the latter exceeded 2·2365. In none of these cases, however, did it appear that a sample attained its final density equilibrium for 450°. Consistent with the conclusions drawn from the working hypothesis, it may be assumed that this difference is caused by the retarding effect produced by the advancement of certain processes which treatment between 500° and 600° facilitates.

Pyrex being rich in silica and unsaturated, so to speak, the thought suggests itself that this retardation and the minimum density effect are possibly related to the inversion found near 573° on heating quartz. In view of the little that is known regarding this subject, that might well be the case, but it must be remembered that such a relationship would not of necessity entail the existence of even incipient crystallisation.

These results show that the density of a glass depends on the effective treating temperature. This, in turn, is determined by the temperature and duration of treatment and the cooling rate; and, in some cases also, by the previous history. That decreasing the rate of cooling increases the discrepancy between the effective and actual treating temperatures is known from tests not described here, but that the previous history may be effective, especially in affecting the rate at which the processes advance, is evident from the above tests on Pyrex. Although the changes which may be made in the density are quite appreciable, prediction as to the limit they may reach is precluded by the limited scope of the tests and the fact that the magnitude of the changes depends on the type of glass. A change much greater than 1 per cent. of the density is, however, undoubtedly possible in some cases, since that limit has been easily exceeded with borosilicate crown glass.* Furthermore, much of this change is attainable in strain-free glass.

At the lower treating temperatures, if the effect of a possible shift in the short wave absorption band has not become marked, one is probably justified in making the assumption that the changes in the refractivity and density are approximately proportional;

* A. Q. Tool and C. G. Eichlin, *J. Amer. Opt. Soc.*, 1924, **8**, 419.

and perhaps, also, that the relative magnitudes of these changes are practically of the same order. In that case, the above conclusions suggest the degree to which the index may be changed and yet maintain a strain-free glass without necessitating extremely long treatments. The results of the preliminary experiments on the refractive index changes for this same medium flint were in agreement with these assumptions. They have shown that by proper heat treatment its refractive index * may easily be increased or decreased and also adjusted within reasonable limits, to any value desired, with an accuracy limited chiefly by the precision of the temperature control during treatment. Also that for a change of less than 50° in the treating temperature and with the introduction of no appreciable strain, the change in the index reached the third decimal. Through long periods of treatment at the lower temperatures and by special care in the cooling from the higher ones, the range of possible effective treating temperatures consistent with the production of a strain-free product can undoubtedly be extended considerably beyond this 50° interval, with a resultant increase in the magnitude of the possible changes. Indications have also been found that the dispersion varies but little unless the treating temperatures are relatively high, and under those conditions it increases. More extended observations, however, are required before definite conclusions are possible. The results so far obtained correspond with those found by C. G. Peters † in an investigation of a somewhat different character. The magnitude of the changes observed also agree with those found by Twyman and Simeon,‡ when it is taken

---

* These conclusions are based on results obtained in collaboration with L. W. Tilton of this Bureau while investigating the changes in refractivity by altering the heat treatment and in which the probable error did not exceed a few units in the sixth decimal. The values for the index of the original medium flint at 20° relative to air at 760 mm. pressure were :

Wave-length.	Index.
6562·8	1·58574
5895·9	1·59000
4861·3	1·60069

† Loc. cit.

‡ F. Twyman and F. Simeon, this Journal, TRANS., 1923, 7, 199. At this point it is of interest to note that, in the article to which reference is made, the "full annealing value" for the index is that obtained when following a more or less arbitrarily chosen annealing schedule. On considering this schedule, it seems evident that the "full annealing value" given must in many cases be a much lower value than that often obtained by others when they deem it advisable to employ lower annealing temperatures, longer treating periods, and slower cooling rates.

In view of this, it would be desirable for comparative purposes to know what effective treating temperatures resulted from the retreatments described in their article and also what those of the original specimens were. The data

into account that their results refer apparently to a change from a certain highly strained to a certain well-annealed condition, and were obtained on glasses which yield larger differences than many others.

Conclusion.

With density and refractive index changes, no greater even than those just discussed, resulting from heat treatment, it becomes evident at once that the method employed by Winkelmann and Schott * and others for determining certain " factors " or coefficients for the various constituent oxides of glasses in order to be able to compute the density, etc., for untried compositions, can at best give no more than approximate results, especially if the compositions are greatly varied. In fact, it has been observed quite often that the consistency of the results is likely to be very indifferent, although in some cases it is practically all that could be desired. Any lack of consistency observed can probably be ascribed chiefly to those changes produced in the number, character, importance, and degree of advancement of the active processes by the differences in the heat treatments and also by any changes in the relative proportions

---

given, however, are not satisfactory for this purpose; hence any estimates would be uncertain, although it might appear that the effective treating temperatures should not have fallen below the actual treating temperatures —591° for the dense barium crown and 565° for the boro-silicate crown, it appears—by an amount greatly exceeding 20° or 30°. In the case of the dense barium crown, this view is supported by the fact that the original block had a higher index than the retreated samples. In that of the boro-silicate crown, it is supported by the definite advancement beyond the arbitrary full annealing value when a chilled sample was treated for a relatively short time at 500°.

The recorded treatments at the lower temperatures were not continued for a sufficient time to be of aid either in this regard or in determining the conditions of final stability at those temperatures, since for some of them this condition could not have been reached under months or perhaps even years. In this connection, it may be stated that to establish the sufficiency of any treatment given at a chosen temperature, for the purpose of reaching stability, requires first the fixing of the equilibrium condition for that point. This can only be done by the following method or its equivalent : Two samples of the glass, both known to be unharmed by previous treatments, are taken; one known to be in equilibrium at a lower, and the other at a higher temperature than that chosen for their simultaneous treatment. Under treatment, they will ultimately reach the same condition and then show no further change even after very long additional treatments at this same point. Should they, when relatively low treating temperatures are employed, approach apparently separate values, it merely shows that the means of test lack the precision required to indicate the drift toward the true value lying somewhere between the limits they appear to have reached.

* A. Winkelmann and O. Schott, *Ann. Physik*, 1894, **51**, 730.

of the constituents. To say that the glasses should be given equivalent treatments does not in this case wholly save the situation, as there is no means of knowing what constitutes an equivalent treatment after the composition has been changed. In fact, "equivalent treatment" and "equivalent condition" are terms that cannot be used to advantage in this sense unless the compositions of the glasses are identical.

With regard to the effect due to a difference in the proportions of the constituents, it is not surprising that this should cause changes in the active processes and introduce inconsistencies into computations based on the methods mentioned. For example, on adding small amounts of a new constituent to a "parent" glass, it may be found that the process of interpenetration assumes considerable importance. In fact, because of this effect it may appear, as the proportion of the new ingredient is increased, that for a small range little or nothing is added to a given volume of the parent glass. Furthermore, if in addition a certain affinity exists between the molecules of the former and those of the latter, an actual but small reduction in this volume may occur. On the other hand, this affinity in some circumstances may cause the formation of components or introduce effects which actually increase the resultant volume above that to be expected from the sum of the volumes before mixing. Normally, these small effects will appear to be confined to a relatively narrow range of percentages, and as the amount of the ingredient is increased above this there should be an approximate proportionality between it and the change in density. Whilst still maintaining the vitreous condition, if this ingredient is one which can be added until 100 per cent. is reached where it then becomes the parent glass, similar deviations from this proportionality may be found on approaching that limit. Under these conditions, when factors for the various constituents are determined from observations taken in the ranges of approximate proportionality and used in computing the densities of the parent glasses at 0 and 100 per cent., the results may well be too high or too low, depending on the character of the effects near these limits. Attention has been directed to these points by A. N. Finn,* who has proposed a method better adapted than any previous one for computing the relationships between the compositions and physical properties when this lack of exact proportionality exists.

These effects mentioned as occurring on the addition of small quantities of a new ingredient may at times extend to intermediate percentages of some magnitude. The appearance of critical ranges

* A. N. Finn, a paper read on February 17th, 1925, at the meeting of the Amer. Cer. Soc. at Columbus, Ohio.

ON THE CONSTITUTION AND DENSITY OF GLASS.   205

may then be observed. These can be ascribed to the fact that a type of saturation or other equilibrium condition has been reached. Such effects will often cause the appearance of two or more regions of approximate proportionality between the changes in the co-efficients of certain properties and those in the percentages of the new constituent being added. Such effects are consistent with the known behaviour of solutions. Properties which may be quite susceptible to effects of this sort are the expansivity and deformability. Examples have lately been recorded by English and Turner.*

In the case where the same constituent is added to different parent glasses, it is almost a foregone conclusion that it will be, for many constituents and with most of the properties, impossible to calculate from the effects observed in one case those to be obtained in any other, especially when the parent glasses fail to have all constituents in common. As before, fair agreement between the factors for a given constituent with different parent glasses may often be found for certain properties, but great discrepancies are likely to occur in many cases.

That these " primary effects " or variations in the factors or " partial densities " from the values predicted by the older view should occur when the composition is altered seems inevitable from the modern point of view. In other words, the older view that glass may for all practical purposes be considered as a mere mixture of a number of oxides has been to a great extent discarded.† Furthermore, it seems equally natural to most investigators that these primary effects may often be advanced or retarded by heat treatments. These " secondary effects " or changes arising from treatment are of necessity in many cases smaller than the primary ones, but they may often reach a stage of considerable importance as shown by the preceding data. For instance, in the case of the density it would be unreasonable, in view of the close packing of the molecules ‡ to expect the primary changes to exceed ordinarily more than a few per cent.; yet it has been found that the secondary changes from heat treatment, even when the temperature interval is quite restricted, may exceed 1 per cent.

Finally, although these changes from treatment mask somewhat certain effects sought when the composition is altered, yet it may be said that a concentrated study of them should so aid in clearing the field that the objectives desired may in the end be more easily

---

* S. English and W. E. S. Turner, this Journal, TRANS., 1923, **7**, 155. S. English, *ibid.*, 1924, **8**, 205.

† F. Eckert, *Jahrb. d. Rad. u. Elek.*, 1924, **20**, 266.

‡ This view is held since it would appear from those cases where comparisons can be made that usually the density of the vitreous condition is not reduced excessively below that of the crystalline state.

reached and with a better understanding than is now possible. They should give important clues regarding the realignments and processes which are involved during heat treatment and when the composition is altered by varying merely the proportions or by adding new constituents.

Summary.

In the foregoing a working hypothesis which appears to co-ordinate many of the effects observed in glass has been developed. This hypothesis requires that, under proper cooling conditions, certain processes or reactions practically reversible without drastic treatment and involving some or all of the constituent molecules of a glass shall advance in such a way that other possible and disturbing processes often associated with cooling and with treatments at temperatures above the annealing range are prevented.

A consideration of the conclusions indicates that on this basis the condition of a glass is one which is intermediate between the liquid and solid states. Also, that its condition at ordinary temperatures may be considered as under-cooled not alone with regard to the process of crystallisation, usually known as the true solidification, but also with respect to the completion of many processes normal to the vitreous condition. Among these are both chemical and physical processes including possibly certain reactions between the constituents, polymerisation, the formation of colloids, interpenetration, diffusion, and perhaps many others equally important.

It has been shown that some of these at least advance and retrogress as the heat treatment is changed and that as a consequence certain properties are modified by altering the effective treating temperature. Experiments have shown that this temperature is determined by the period and temperature of treatment and by the mode of cooling. It has also been found that the changes in certain thermal effects and in the density and refractive index, at least within limits, are practically reversible, being functions of the effective treating temperature to the degree of accuracy attainable by the methods of measurement employed.

It has also been shown from the nature of the conditions involved that these changes in the refractive index and density must be small —probably not more than one or two per cent. in the latter—but the data indicate that as the treating temperature is lowered the density will in general approach closer to that value computed from the densities of the constituents. They also show that the changes from heat treatment are of a character that may cause marked differences in the utilitarian quality of the glass.

Results of this sort are therefore of practical importance in determining the relationships between glass compositions and the resultant physical and chemical properties and they should aid materially in solving those problems met in producing all types of glass, including those used for thermometers and optical instruments.

BUREAU OF STANDARDS,
WASHINGTON, D.C.,
U.S.A.

XVI.—The Viscosity of Glass.

By VAUGHAN H. STOTT, M.Sc.

(Read at the London Meeting, May 26th, 1925.)

THE acquisition of data relating to the viscosity of glass has been very seriously impeded by the inherent difficulties of the work.

The results of a critical study of the measurements of Washburn and Shelton, and of English, have recently been published by Le Chatelier.* In the present paper these measurements will be reconsidered in conjunction with some additional determinations which have been carried out at the National Physical Laboratory. Although these determinations are by no means complete, it would appear from them that some modification of the conclusions drawn by Le Chatelier is necessary. The importance of the subject, and the slowness with which further work may be expected to proceed, justify the publication of these new considerations.

Fig. 1, reproduced from a recent paper,† shows the logarithmic viscosity of a soft laboratory glass, plotted against temperature. Fig. 2 shows a heating curve of a small button of the same glass with a rate of heating of 11° per minute. It will be observed that the logarithmic viscosity–temperature curve has a point of inflexion about 500°, and that the region of heat absorption shown in Fig. 2 begins at 534°. It has been found, as might be expected, that the temperature corresponding with the beginning of the heat absorption may be lowered by reducing the rate of heating. There can be little doubt, therefore, that the sudden increase in the slope of the logarithmic viscosity–temperature curve is due to the same change in the constitution of the glass as that which produces the heat absorption, and that when the glass is in equilibrium a point or region of transition exists in the neighbourhood of 500°. (In measuring the viscosity, the time required to eliminate both the effects of lack of temperature uniformity, and of "reactivity," is so great that equilibrium is attained, and the measured values of

* This Journal, TRANS., 1925, 9, 12.
† Stott, Irvine and Turner, Proc. Roy. Soc., 1925, A., 108, 154.

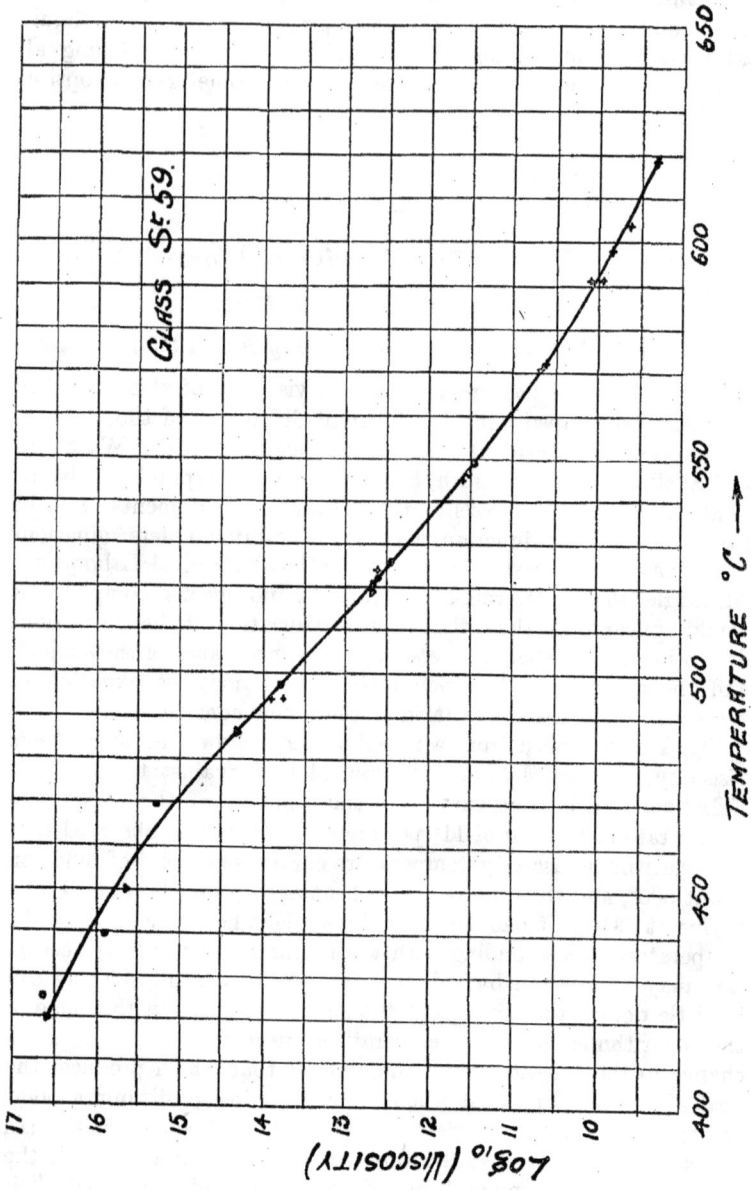

GLASS S. 59.

TEMPERATURE °C ⟶

Fig. 1.

Log₁₀ (Viscosity)

the viscosity do not depend on the order in which the measurements are made.) The viscosity of the glass at this temperature is $10^{13\cdot7}$ poises. Since Tool and Valasek * have shown that the relation between the temperature at which heat absorption begins and the " annealing temperature," † differs little from glass to glass, it may be deduced that the change occurring in the constitution of glass should happen when the viscosity is in the neighbourhood of $10^{13\cdot7}$ poises.

The measurements which have been studied by Le Chatelier do not include viscosities of this magnitude, and therefore any discontinuities which may occur in viscosity curves at points of lower

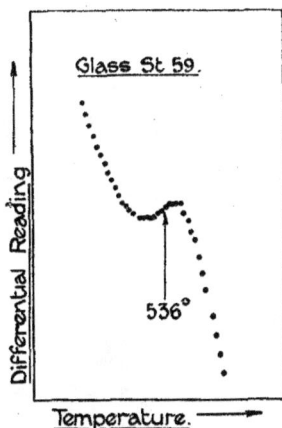

FIG. 2.

viscosity must be ascribed to causes other than that associated with the heat absorption found by Tool and Valasek.

It may be pointed out that a certain confusion may arise through the fact that English's definition of the " annealing temperature " corresponds with a considerably lower viscosity than the value adopted by Tool and Valasek. Moreover, the values of the viscosities given by English which were obtained by stretching rods are all too high, and require dividing by three, owing to an error in the calculation due to a confusion of the " coefficient of viscosity " with the " coefficient of viscous traction." The necessary corrections have been made to the log log viscosity–temperature curves

* Scientific Papers of Bureau of Standards, 1920, No. 358.

† The " annealing temperature " in the experiments of Tool and Valasek means the temperature at which the glass has a viscosity of approximately $10^{13\cdot5}$ poises.

considered later, and it will be seen that they change appreciably the general appearance of the curves.

The measurement of viscosity at low temperatures is bound up with the phenomena of "reactivity." A. A. Michelson * has shown that the application of a load to glass at room temperatures produces a displacement varying with time according to the equation $S = S_0 + A(1 - e^{-\lambda\sqrt{t}}) + Bt$, where S_0, A, λ, and B are constants.

If the load, after acting for a long time, be removed, a reverse displacement takes place which, measured from its starting point,

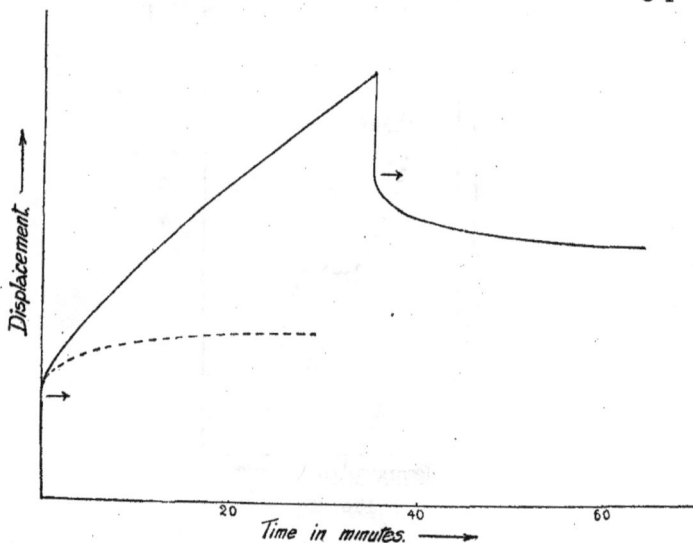

FIG. 3.

in the reverse direction, is expressed by the formula $S' = S_0 + A(1 - e^{-\lambda\sqrt{t}})$.

The displacement represented by the term $A(1 - e^{-\lambda\sqrt{t}})$ is known as the "reactivity." The viscosity is given by the value of $1/B$.

Fig. 3 shows these displacements plotted for the glass previously mentioned at a temperature of 486·5°. Fig. 4 shows the values of $\log [A - (S' - S_0)]$ plotted against $\sqrt{t}$, and Fig. 5 the values of $(S - S')$ plotted against t. It is clear from Figs. 4 and 5 that the formulæ given represent satisfactorily the experimental figures at a temperature as high as 486·5°. Measurements made at other temperatures yielded the following results (the load being always the same).

* "The Laws of Elastico-Viscous Flow," J. *Geology*, 1917, **25**, 405.

Temperature.	λ.	S_0.	A.
15° (approx.)	0·13	10·39	0·26
486·6°	0·15	10·35	8·25
529°	0·306	10·58	8·07

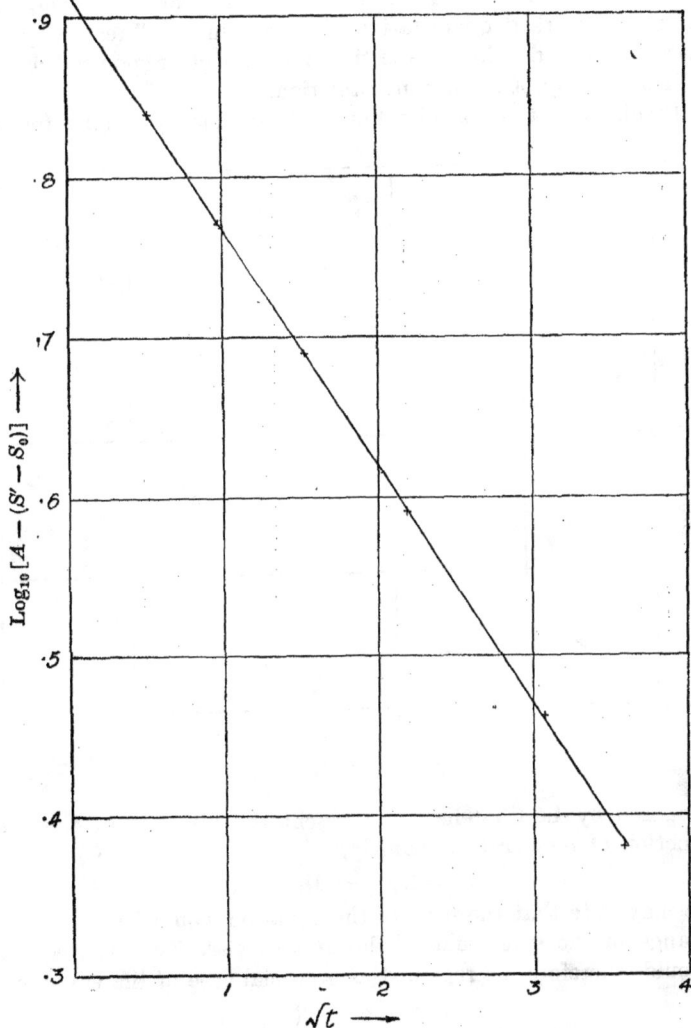

FIG. 4.

An examination of the values of λ suggests the probability of a discontinuity between the temperatures 486·6° and 529°. It may be recalled that the point of inflexion on the logarithmic viscosity–temperature curve occurs between these temperatures.

P 2

It should be noted that no special physical significance is attached to λ. Professor Filon has directed the attention of the author to other formulæ which represent the observed displacements as accurately as the one given above. It appears, however, that there is a marked quantitative difference in the "reactivity" of glass above and below the critical temperature region which may well be worthy of further investigation.

Turning now to a consideration of the double logarithmic formula

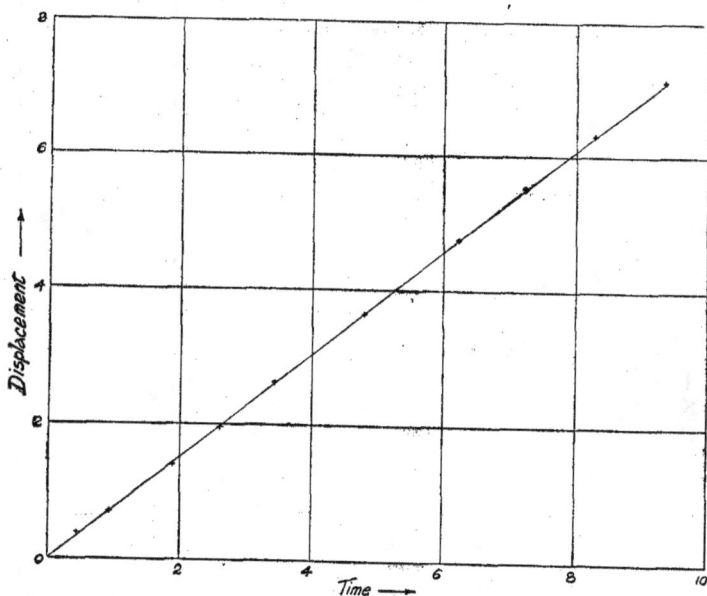

FIG. 5.

proposed by Le Chatelier to represent the viscosity of glass as a function of temperature, namely,

$$\log \log \eta = Mt + P,$$

we may note that the form of the equation would be altered by a change in the magnitude of the unit of viscosity. The equation should, therefore, be regarded as a special case of the equation

$$\log \log \eta/\eta_0 = Mt + P,$$

that is to say, η_0 happens to equal unity. This observation has been made because the original equation represents approximately the behaviour of a considerable number of glasses, and it was thought possible that by assigning suitable values to η_0 the equation might be found to be of general application. This does not appear to be the case, but it should not be overlooked that, for very viscous

THE VISCOSITY OF GLASS. 213

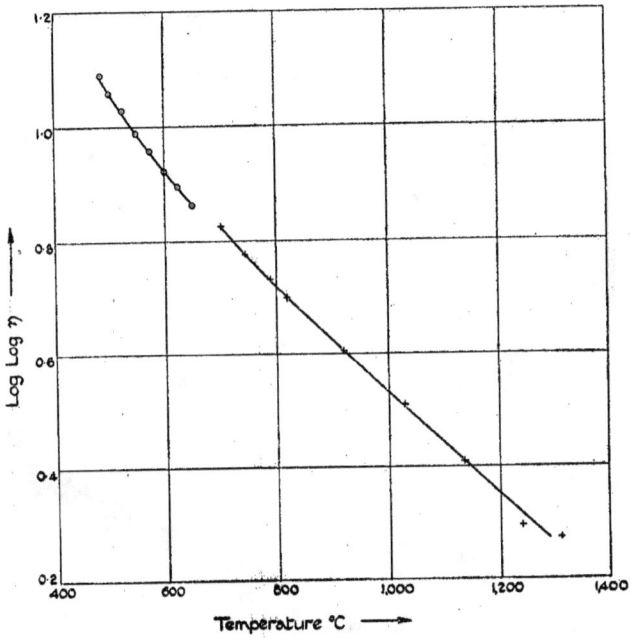

Fig. 6.

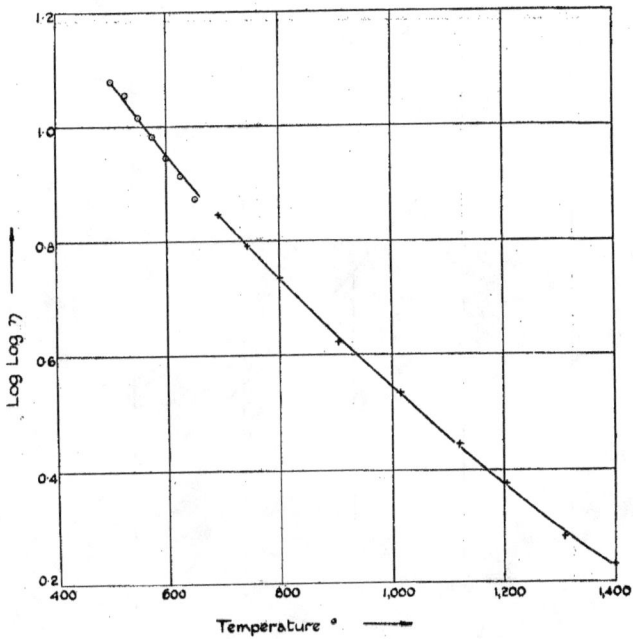

Fig. 7.

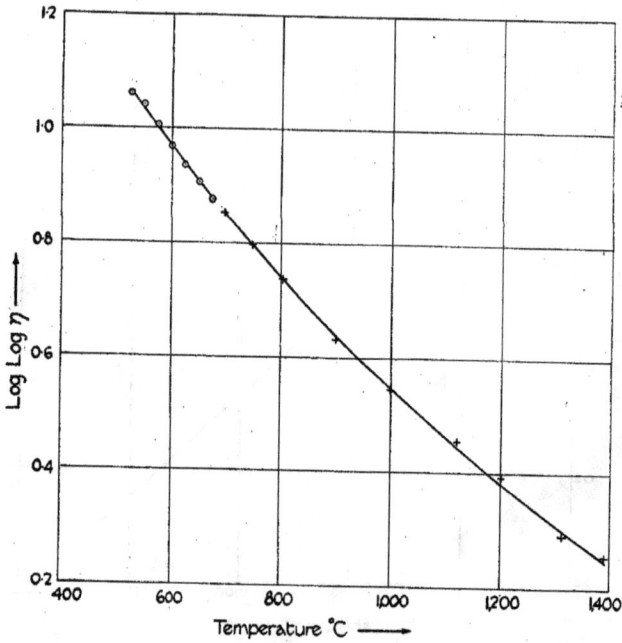

FIG. 8.

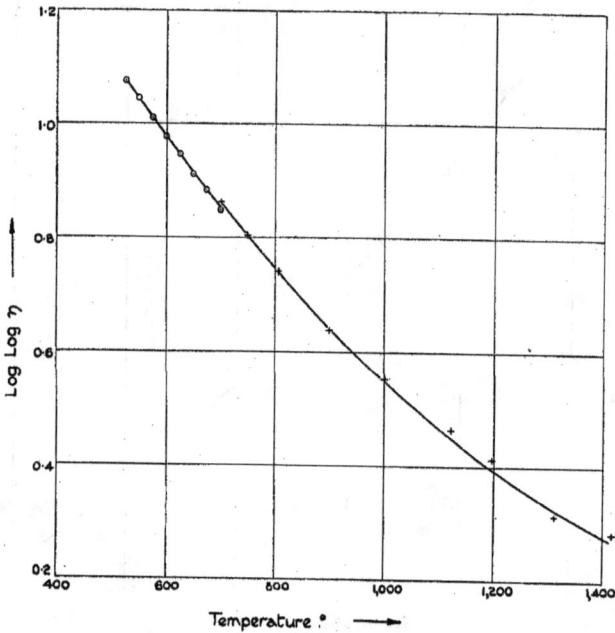

FIG. 9.

THE VISCOSITY OF GLASS.

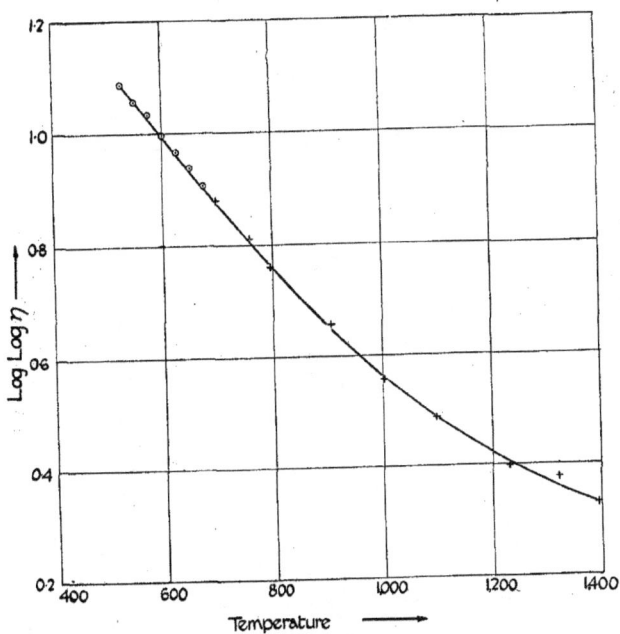

FIG. 10.

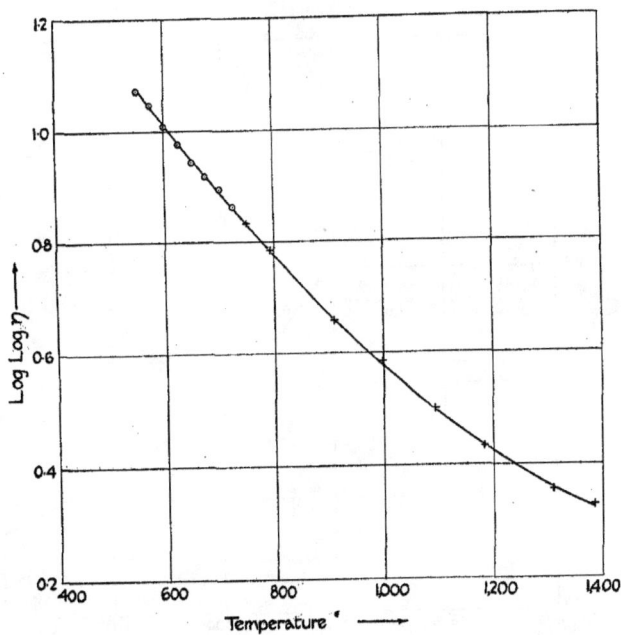

FIG. 11.

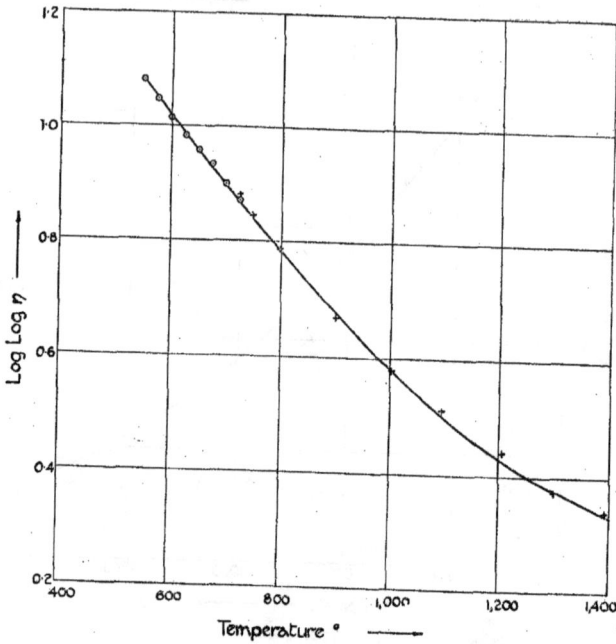

Fɪɢ. 12.

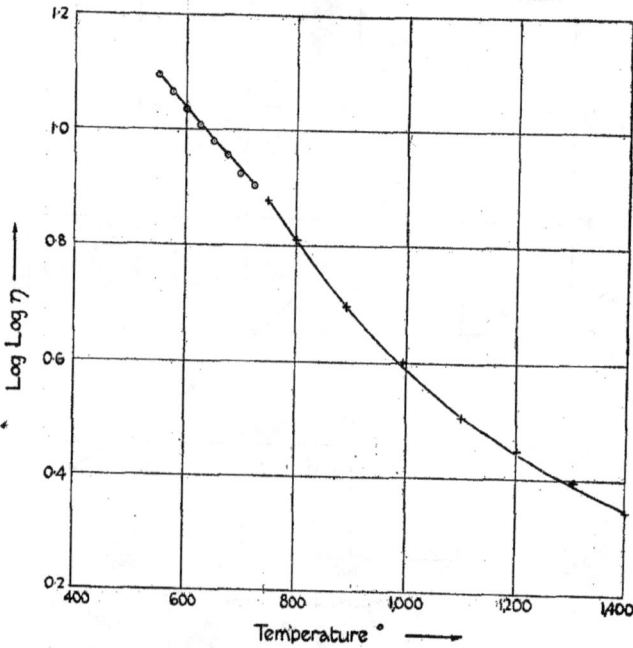

Fɪɢ. 13.

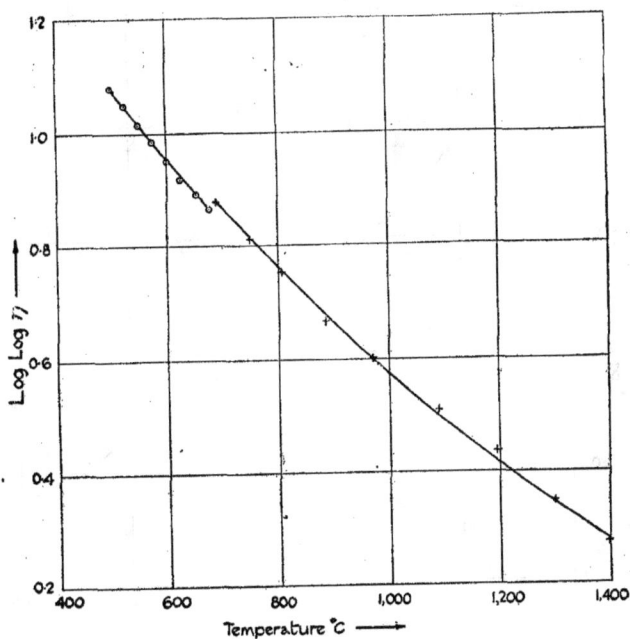

Fig. 14.

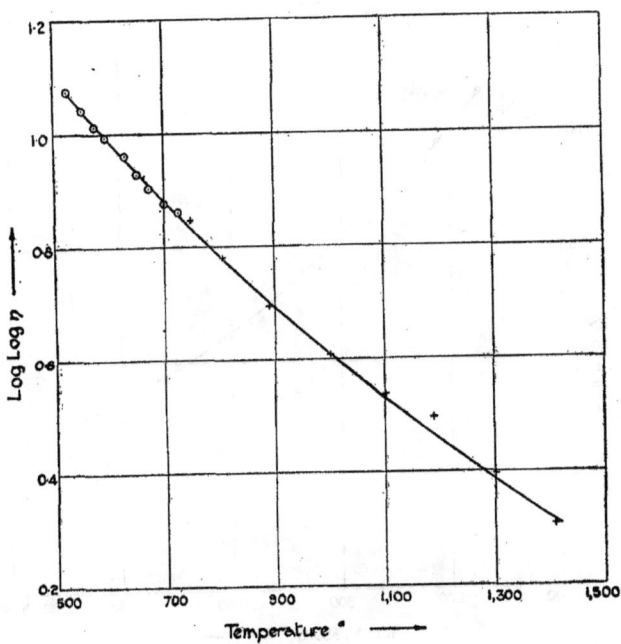

Fig. 15

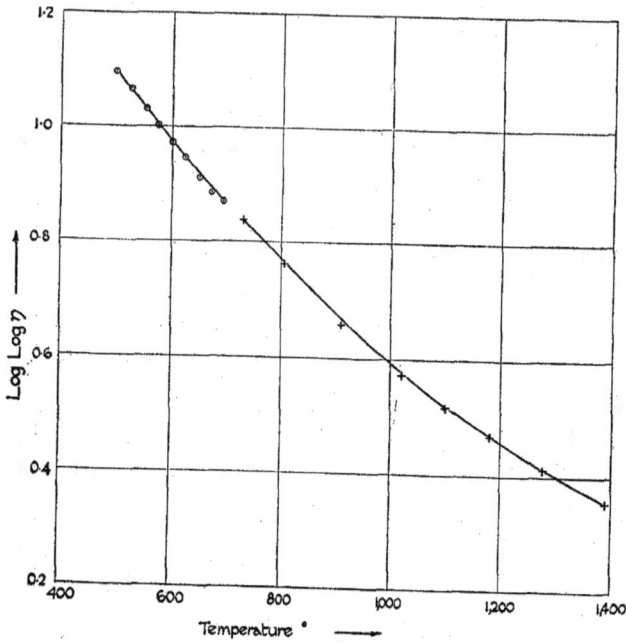

FIG. 16.

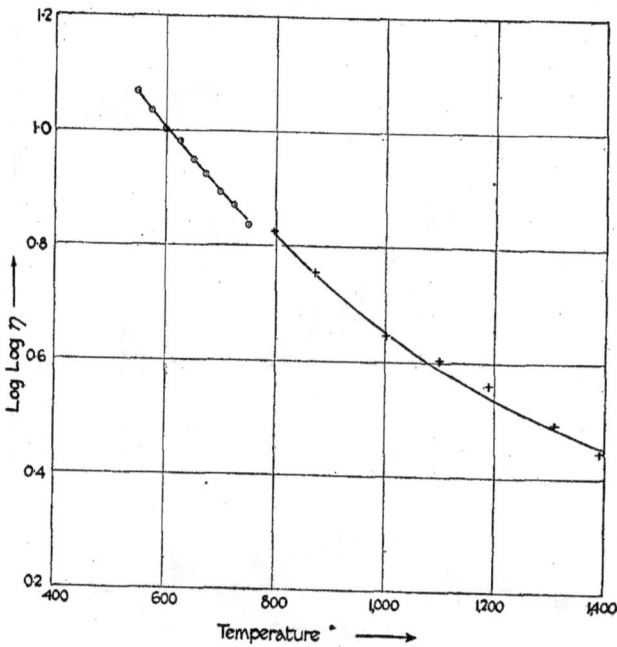

FIG. 17.

liquids other than glass, a satisfactory general formula expressing viscosity as a function of temperature has not yet been found.

Figs. 6 to 17 show the log log viscosity–temperature curves of a number of the glasses examined by English. A smooth curve can be drawn through the points * in all cases, except that of glass No. 9, if we suppose that a small variable difference may exist between the absolute values of viscosities measured by the two different types of apparatus. Consideration of the methods of measurement employed makes such an assumption very probable. The glass No. 9 was liable to devitrification, which may, perhaps, explain the discontinuity in the curve. It appears probable that, in the absence of devitrification, the viscosity of glass varies continuously with temperature except in the region of a viscosity of about $10^{13.7}$ poises and possibly, of course, at lower temperatures where few measurements have been made.

The influence of constitution on viscosity at high temperatures is shown by Washburn and Shelton's † measurements of the viscosities of the sodium silicate glasses. Fig. 18 shows the logarithmic viscosities of these glasses plotted against the percentage of silica. It is not possible to draw a smooth curve through the experimental points without at least one point of inflexion. The curves in the figure have been drawn with two points of inflexion, in such a way that the deviation of the curve from the normal form (which is a curve slightly convex to the composition axis) is a maximum at 66 per cent. silica, the composition of the compound $Na_2O,2SiO_2$. Since the existence of this compound has been completely established,‡ it is reasonable to suppose that the abnormality in the form of the viscosity curves of Fig. 18 is due to the presence of the compound in solution in the glass. It should be understood, however, that the experimental points are not sufficiently numerous to fix the exact form of the curves, although an abnormality is clearly indicated. It is interesting to note that the abnormality persists even at 1500°. Similar results are found on plotting the curves for the soda–lime–silica glasses containing 10 per cent. of lime. The persistence of the abnormality at high temperatures affords a good illustration of the sensibility of viscosity to changes of constitution, since the flatness of the melting-point curve of the system sodium metasilicate–silica at the sodium

* The points obtained by experiments on the stretching of rods are shown in the figures as circles. The other points are indicated by crosses.

† Washburn and Shelton, "The Viscosities of the Soda–Lime–Silica Glasses at High Temperatures," *Bulletin No.* 140, *Engineering Experiment Station, University of Illinois.*

‡ G. W. Morey and N. L. Bowen, "The Binary System Sodium Metasilicate–Silica," *J. Phys. Chem.*, 1924, **28**, 1167.

disilicate point shows that considerable dissociation of this compound takes place at its melting point of 874°.

Although it is very improbable that the above deductions are essentially incorrect, it must not be forgotten that most measurements of the viscosity of glass are quite unchecked by other work, and may contain unsuspected errors. In one case, however, a comparison is possible between the determinations of Washburn,

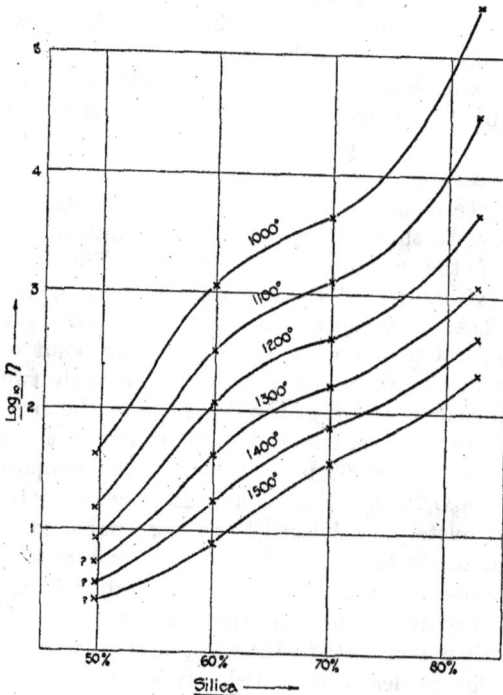

FIG. 18.

English, and the National Physical Laboratory. Fig. 19 shows the viscosities of the glass N.P.L. 15 measured in the usual way by the Laboratory.* The glass has the composition

SiO_2	....	72·22
Al_2O_3	....	0·71
Fe_2O_3	....	0·11
CaO	....	6·94
MgO	....	0·53
Na_2O	....	19·49

* The measurements of viscosity were made by Mr. D. Turner, B.Sc.Tech.

The curve marked Washburn on the figure is obtained from that investigator's diagrams. It corresponds with a composition

$$SiO_2 \dots\dots\dots 72\cdot22$$
$$Na_2O \dots\dots\dots 19\cdot49$$
$$CaO \dots\dots\dots 8\cdot29$$

The curve marked English is obtained from English's data for glass 6 after applying a correction derived from Washburn's lines of equal viscosity, or isokoms. The series of circles slightly above English's line shows the uncorrected points. As in the previous case, the compositions have been calculated from the percentages of silica and soda only.

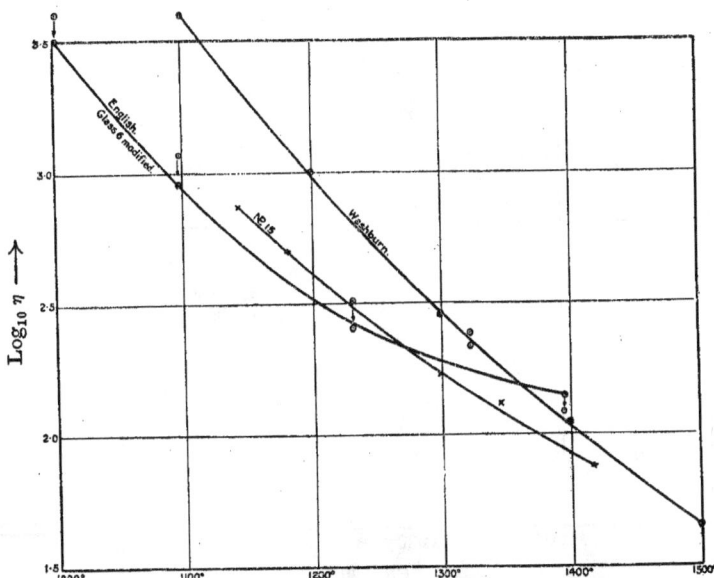

FIG. 19.

The discrepancies between the three curves are striking. The irregular nature of the last three points of English's curve suggests serious errors. The first three points of his curve, however, lie on a line nearly parallel to the N.P.L. curve. If the last two points of his curve be each shifted back a distance equivalent to 83° parallel to the temperature axis, the new curve is parallel to the N.P.L. curve. It appears, therefore, that either a considerable discrepancy exists between the two curves, or that there has probably been a large error at the two upper temperatures. Fig. 20 shows English's results for glasses 5 and 7. The middle curve is drawn by averaging the logarithmic viscosities shown on the other two curves. It may be presumed to apply to a glass having a composition midway between the compositions of glasses 5 and 7. As before, at the lower temperatures the curve is parallel to the N.P.L. curve,* but at the higher temperatures there is some divergence. Here again, however, the measured values are very uncertain. The parallelism between this curve and the N.P.L. curve is a little diminished after applying the following corrections

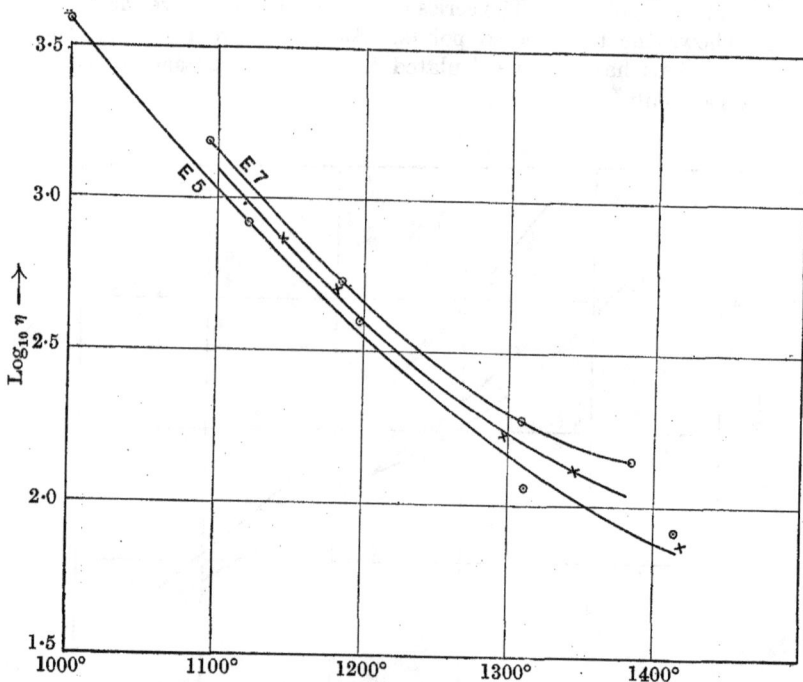

FIG. 20.

deduced from Washburn's isokoms for the small difference in composition from that of the N.P.L. glass :

At 1100° subtract 0·27
1200   „   0·23
1300   „   0·20
1400   „   0·16
1500   „   0·16

The difference between the slope of Washburn's curve, and that of English's curve or the National Physical Laboratory's curve, is so great as to throw considerable doubt on the value of Washburn's observations, since the two latter curves have approximately

* The crosses represent the experimental results for N.P.L. 15.

the same slope. It is possible that Washburn's results are due to lack of homogeneity of the glass. Variations in the quantities of the minor constituents are not of sufficient magnitude to explain the discrepancy, and a large change is necessary in the proportions of the major constituents * to alter greatly the slope of the curve. A series of measurements of the viscosities of soda–lime–silica

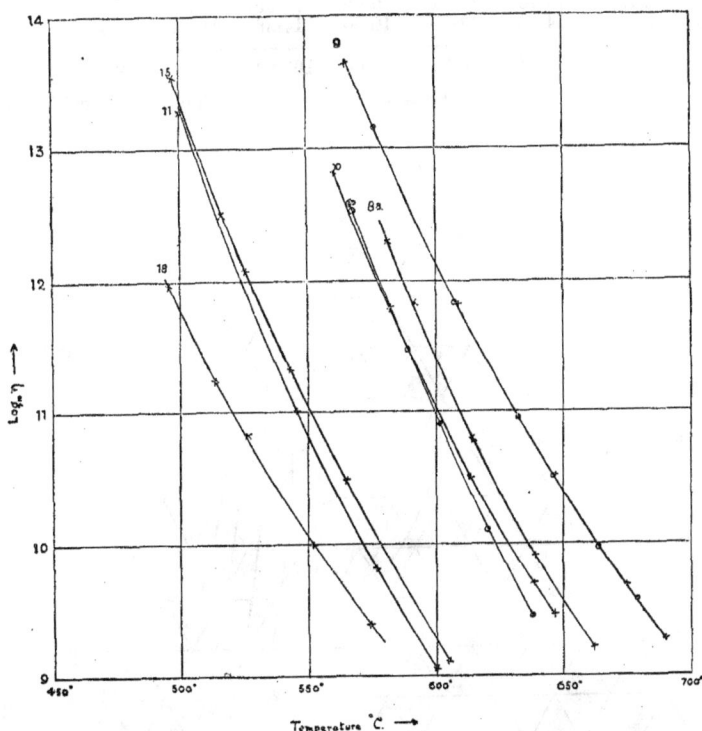

FIG. 21.

glasses has been made at comparatively low temperatures by Messrs. D. Turner, B.Sc.Tech., and H. A. Sloman, M.A., A.I.C., at the National Physical Laboratory. The method was the same as that used for previous measurements,† except that the accuracy of the determinations was increased by grinding the rods to the required dimensions on a lathe, instead of fusing together hand-

* The glass in Washburn's series which has a viscosity–temperature curve closest to that of N.P.L. 15 is his glass 3 containing 70 per cent. silica and 30 per cent. soda.

† Stott, Irvine, and Turner, *Proc. Roy. Soc.*, 1925, A., **108**, 154.

drawn rods. Fig. 21 shows, plotted against temperatures, the logarithmic viscosities of glasses of the following compositions.

No. of Glass.	8a.	8b.	9.	10.	11.	15.	18.
SiO_2	68·38	62·34	76·48	71·94	69·08	72·02	76·00
Fe_2O_3	0·22	0·22	0·30	0·16	0·29	0·16	0·26
Al_2O_3	2·22	1·48	2·54	0·54	0·59	0·60	0·38
MgO	trace	0·36	trace	0·61	0·60	0·47	0·64
CaO	14·70	18·24	9·92	12·40	8·00	6·60	1·50
Na_2O	14·62	17·70	10·80	14·38	21·18	20·16	21·04
	100·14	100·34	100·04	100·03	99·74	99·99	99·82

(The analyses were carried out by Mr. W. H. Withey, B.A., and Miss H. Millar.)

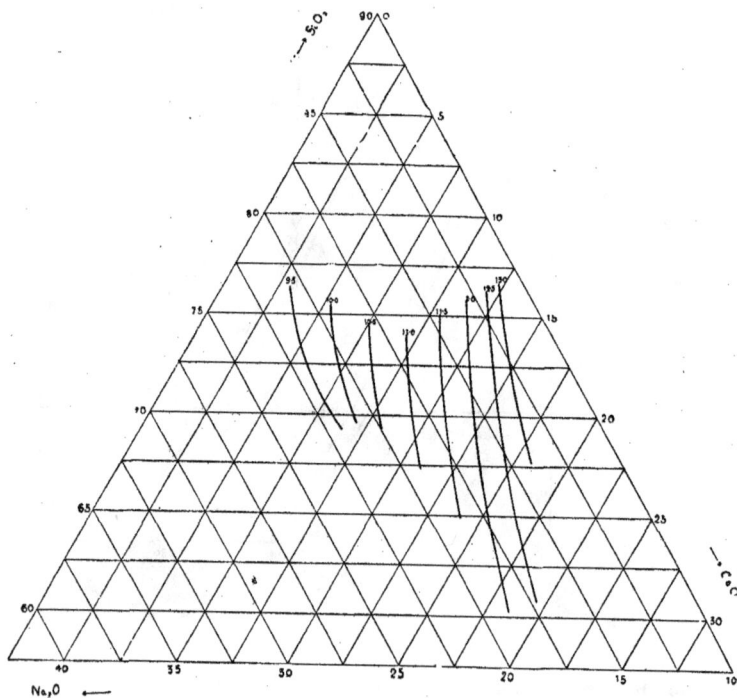

FIG. 22.

On the assumption that the viscosity is a continuous function of the composition, the curves of Fig. 21 enable lines of equal viscosity (isokoms) to be drawn over a considerable field at a temperature of 575°. Corrections should be made to the compositions of the glasses in order to allow for the effects of impurities, and also because the analytical figures do not total precisely 100 per cent. Since the viscosity values are much less sensitive to a given error (as a percentage of the total weight of glass) in the amount of silica than in that of the other constituents, the errors of analysis have been considered as errors in the silica values.

Attempts have been made to utilise some published and unpublished measurements of viscosities, very kindly communicated by Mr. English, in order to calculate the corrections which should be made to allow for the presence of magnesia and alumina. The uncertainty of the magnitudes of the corrections was so great, however, that it was decided to draw approximate contour lines without making corrections at all. The errors in the logarithmic viscosities due to neglect of the corrections are probably less than 0·4. The errors in the actual measurements of logarithmic viscosities are probably of the order of 0·05 or less, so that a recalculation of the contour lines would be desirable if, in the future, accurate values of the correction terms were available. Fig. 22 shows the curves obtained. It is desirable to emphasise the difficulty of eliminating the effects of impurities from the results of viscosity measurements. Such eliminations involve determinations of the partial differential coefficients of the viscosity with respect to variables fixing the composition. It is easy to see that these determinations must be affected by an accumulation of errors, and if it is wished to obtain final relations between viscosity and composition, in which errors due to impurities or inaccurate compositions generally, are not considerably greater than the errors of the viscosity determinations themselves, it will be necessary to prepare the glasses from materials of known purity and melt them without contamination. This procedure, which at present precludes the melting of large pieces of glass, would limit considerably the design of viscosimeters. Amongst the methods available, a modification of the Margules apparatus already used at the National Physical Laboratory might be found convenient. The necessity for extremely accurate centring of the inner cylinder (which does not arise when the shape only of the viscosity–temperature curve is to be determined) is a very considerable, although doubtless not insuperable, obstacle.

NATIONAL PHYSICAL LABORATORY,
TEDDINGTON, MIDDLESEX.

XVII.—*The Ternary System Sodium Metasilicate–Calcium Metasilicate–Silica.*

By G. W. MOREY and N. L. BOWEN.

*(A Contribution to the Symposium on " The Constitution of Glass "
held at the London Meeting, May 25th and 26th, 1925.)*

ALTHOUGH the many investigations of silicate systems carried out
in the Geophysical Laboratory have been conducted primarily
for the elucidation of geological problems, the information gained
has often been directly applicable to technologic problems. The
present investigation, however, had as its chief motive the attainment of exact knowledge in regard to a material of great technical
importance, the ordinary soda-lime glass of commerce; its application to geologic problems is not so direct as has been the case with
preceding systems, but the components are all important rock
constituents, and the system is not without interest to the geologist.

It is a matter of surprise that this system, as well as the others
the components of which are the glass-making oxides, has not been
investigated long before. The glass industry is a major one, and
the value to it of the information to be gained from a study of the
fundamental melting-point relationships of the glass-forming
oxides should long ago have inspired an exhaustive study. Such
a research has never been made, and up to the present time there
is little knowledge of the crystalline phases formed even in the most
common glass system, soda–lime–silica, and no knowledge of the
melting-point relations of these compounds. There has indeed
been much speculation as to the constitution of glass, but no systematic study of the compounds formed by the devitrification
of glass, and of the relation of this devitrification or crystallisation
to composition and temperature, information which is fundamental
to the solution of the puzzle of the constitution of glass; if, indeed,
this puzzle is capable of solution by the methods at our disposal.
It was in an attempt to supply this information that the present
study was undertaken.

The plan outlined for the work has been broad in its scope, and
has not yet been carried out in all its details; further work will be
reported in subsequent papers. Here we shall give an account of
the equilibrium relations between the crystalline and liquid phases
in the ternary system sodium metasilicate, Na_2O,SiO_2–calcium
metasilicate, CaO,SiO_2–silica, SiO_2, as well as some information
as to parts of the larger system $Na_2O–CaO–SiO_2$ containing more
Na_2O and CaO than the metasilicates. A similar study of the

THE SYSTEM SODIUM AND CALCIUM METASILICATES–SILICA.   227

potassium oxide system is partly completed, as well as a study of the effect of iron and alumina. Further work on this and other similar systems from several points of view is under way.

Previous Work.

Previous work on this system requires little discussion as little has been done, and no well-substantiated conclusions have been reached. The literature pertaining to the optical properties will be discussed in another paper; the only observation pertinent at present is that of E. W. Tillotson,* who concluded from the shape of the specific refractivity curve of a series of glasses ranging in composition from $Na_2O,3SiO_2$ to CaO,SiO_2 that the compound $2Na_2O,3CaO,9SiO_2$ was formed. Tillotson's glasses were not analysed, and their compositions were only approximately known; his refractive indices were not accurately determined; and his densities were not determined but calculated; furthermore, his work involves the assumption that deviation of a property from a straight line, when plotted against composition, is evidence of the existence of a definite molecular species. This assumption is seriously open to question. C. C. Rand † gives some qualitative observations on the devitrification tendency of some $Na_2O,3SiO_2$–$CaO,3SiO_2$–$MgO,3SiO_2$ glasses. Further qualitative observations are given by Peddle.‡ These are, of course, explicable when the melting-point diagram is known. N. V. Kultascheff § studied the melting-point curve of the binary system Na_2O,SiO_2–CaO,SiO_2, and concluded that the compounds $3Na_2SiO_3,2CaSiO_3$, or $3Na_2O,2CaO,5SiO_2$, m. p. 1090°, and $2Na_2SiO_3,3CaSiO_3$, or $2Na_2O,3CaO,5SiO_2$, m. p. 1160°, are formed. He gives no analyses of his mixtures and he drew his conclusions from crystallising temperatures determined from cooling curves, a dangerous procedure with mixtures as prone to undercooling as are the silicates. His melting point of Na_2SiO_3 is 80° low, and his results throughout are widely different from those of the present investigation. Wallace,‖ also, did some work on the binary system Na_2O,SiO_2–CaO,SiO_2, by the method of cooling curves, and reported mixed crystal formation from 0—70 per cent. $CaSiO_3$, with a maximum at the compound $2Na_2SiO_3,3CaSiO_3$, or $2Na_2O,3CaO,5SiO_2$. His results are not in accord with the facts, but are explicable when the crudity of the experimental method is considered.

* *J. Ind. Eng. Chem.*, 1912, **4**, 246.
† *Trans. Amer. Cer. Soc.*, 1915, **17**, 236.
‡ C. J. Peddle, *J. Soc. Glass Tech.*, 1920, **4**, 34.
§ *Z. anorg. Chem.*, 1903, **35**, 186.
‖ R. C. Wallace, *ibid.*, 1909, **63**, 1—48.

228. JOURNAL OF THE SOCIETY OF GLASS TECHNOLOGY.

Experimental Methods.

The method of studying the equilibrium relations between glass and crystalline phases used throughout our work was that of quenching, introduced into the study of silicate mixtures by Shepherd.* In this method a small charge of known composition is held at a constant temperature long enough for equilibrium to be attained, and the equilibrium frozen by rapid cooling. Some mixtures can be crystallised only with the greatest difficulty, and with these it is sufficient to lift the charge out of the furnace. In other cases, however, it is necessary to cool with the greatest possible speed; with these the charge, weighted down with a platinum weight, is dropped by an electrical device into cold mercury, held against the bottom of the furnace. Some silicate mixtures, for example, calcium orthosilicate, cannot be quenched even by this method, but such are not met with in the part of the system here investigated. In the greater portion of the compositions the quenching method is the only possible method of study, the crystallisation taking place far too slowly to permit of the application of the method of cooling curves.

The mixtures used were all made by melting the calculated quantities of Na_2CO_3, $CaCO_3$, and SiO_2, and all meltings were made in covered platinum crucibles to avoid contamination. The ingredients were thoroughly mixed before melting, then heated carefully over a low flame to decompose the carbonate, then heated in a gas or electric furnace until free from bubbles, or "fined." The clear glass was then powdered, to pass a 48-mesh sieve, again heated until fined and then again powdered. The heating was repeated until no inhomogeneity could be detected in the powdered glass when examined with the microscope, using an immersion liquid matching it in refractive index. By this method a variation of refractive index amounting to 1 or 2 units in the third decimal place can be detected. After this a final heating was made, and further assurance of homogeneity was given by the fact that polished prisms made from this final melting were entirely satisfactory for refractive index measurement.

In some cases it was desired to obtain mixtures of a definite composition, for example, when it was desired to establish the composition of a compound. For this purpose the amounts of the various components necessary to give a definite weight of glass were calculated. A slight excess of SiO_2 was ignited in a weighed platinum crucible, the excess removed, and the calculated quantity of $CaCO_3$ added. The crucible and contents were then again heated, after which a weighing gave the actual amount of CaO and SiO_2 present.

* E. S. Shepherd and G. A. Rankin, *Amer. J. Sci.*, 1909, **28**, 308.

Na_2CO_3 was then added, and decomposed by careful heating, as described above, followed by fining in the gas furnace, and weighing. The Na_2O content could be thus adjusted to the desired amount, and in subsequent heatings any change in composition could be followed by change in weight.

The raw materials were all of high quality. The Na_2CO_3 contained 0·009 per cent. $Fe_2O_3 + Al_2O_3$ as determined on a 10 gram portion; the SiO_2 was a good grade of quartz containing 0·07 per cent. residue after evaporating with HF and H_2SO_4. The $CaCO_3$ used was an unusually pure preparation; an analysis of which, made by Dr. E. G. Zies, gave, in percentages, SiO_2, 0·0002; $Fe_2O_3 + Al_2O_3$, 0·006; MgO, 0·003.

Analysis of the Glasses.

The information in the literature in regard to the properties of glass is not only scanty and often lacking in accuracy, but, in addition, almost without exception, suffers from the grave fault that compositions are not known or are known only by inference from the composition of the batch melted. A statement of the proportions of materials used in making the glass will give only a rough approximation to the actual glass composition because of loss of material mechanically and by volatilisation, and by gain of material from solution of the glass pot, both of which processes are highly erratic. In dealing with the properties of any silicate mixture, and especially one in which some of the ingredients are markedly volatile, it is essential that the glasses the properties of which are measured should be analysed, if there is to be any but the roughest attempt to correlate properties with composition. All of our glasses were analysed; in most cases two only of the constituents were determined, a procedure justified by the fact that the materials used were of high purity and all meltings were made in covered platinum crucibles to avoid contamination. In many of the cases, SiO_2 and CaO were determined by the usual method of fusing with Na_2CO_3, separating the SiO_2 by two evaporations with HCl, with intervening filtration, and precipitating the calcium in the filtrate by double precipitation as oxalate in the usual manner. In other cases the silicate was decomposed by evaporation with HF and $HClO_4$, CaO determined as before, and Na_2O determined in the filtrate by evaporation to dryness, driving off the ammonium salts by heating in an electric evaporator, and converting to sulphate. Some glasses low in CaO and SiO_2 were decomposed by digesting with HCl, in which case all three components were determined. No attempt was made to estimate separately the small amounts of Fe_2O_3 and Al_2O_3; special analyses made in a few cases showed that the amount of these constituents corresponded to that introduced with the raw materials, a negligible and constant amount. All fusions, evaporations, and ignitions were made in platinum.

Statement of Results of Quenchings.

The experimental results are given in Tables I and II, and graphically in Figs. 1—4. Table I gives the summarised results; the composition, in weight per cent. Na_2O, CaO, and SiO_2, the percentage

TABLE I.

Designation.	Composition by Analysis.			Temperature.	Crystalline Phases.
	Na_2O.	CaO.	SiO_2.		
2422A	20·73	5·12	(74·15) *	732°	Quartz; $Na_2O,3CaO,6SiO_2$
				810	Quartz
2423A	15·30	9·66	(75·04)	1065	Tridymite
2426A	10·08	14·97	(74·95)	1263	Tridymite
2427A	5·73	20·02	(74·25)	1475	Tridymite
2428A	39·63	9·16	(51·21)	1049	Na_2O,SiO_2;
					$2Na_2O,CaO,3SiO_2$
				1120	$2Na_2O,CaO,3SiO_2$
2438A	30·87	4·27	(64·86)	907	$Na_2O,2CaO,3SiO_2$
2438B	41·94	3·33	(54·73)	1027	$2Na_2O,CaO,3SiO_2$
2439A	50·67	2·54	(46·79)	1049	Na_2O,SiO_2
2440A	27·20	8·33	(64·47)	1014	$Na_2O,2CaO,3SiO_2$
2440B	38·32	7·42	(54·26)	1088	$Na_2O,2CaO,3SiO_2$
2442A	21·80	13·09	(65·11)	1060	$Na_2O,2CaO,3SiO_2$
2442B	33·87	10·87	(55·26)	1151	$Na_2O,2CaO,3SiO_2$
2443A	44·77	9·05	(46·18)	1024	$2Na_2O,CaO,3SiO_2$
2444A	(16·64)	17·83	65·53	1100	Pseudo-wollastonite
2444B	31·33	14·71	(53·96)	1200	$Na_2O,2CaO,3SiO_2$
2445A	(40·76)	12·32	46·92	1082	$2Na_2O,CaO,3SiO$
2446A	10·53	24·38	(65·09)	1282	Pseudo-wollastonite
2446B	(24·11)	21·26	54·63	1240	$Na_2O,2CaO,3SiO_2$
2447A	36·04	17·27	(46·69)	1107	$Na_2O,2CaO,3SiO_2$
2448A	7·44	32·17	(60·39)	1379	Pseudo-wollastonite
2448B	(16·41)	28·31	55·28	1262	Pseudo-wollastonite
2449A	20·82	27·05	(52·13)	1275	$Na_2O,2CaO,3SiO_2$
2449B	21·52	27·94	(50·54)	1277	$Na_2O,2CaO,3SiO_2$
2450B	10·61	37·63	(51·76)	1325°	Pseudo-wollastonite
2451A	16·98	35·54	(47·48)	1277	$Na_2O,2CaO,3SiO_2$
2452B	3·90	5·70	(90·40)	1611	Cristobalite
2502B	(7·98)	45·16	46·86	1358	Pseudo-wollastonite
2503A	5·18	41·50	(53·32)	1454	Pseudo-wollastonite
2505A	(11·34)	43·09	45·57	1300	$3CaO,2SiO_2$
2505x	9·88	39·03	(51·09)	1330	Pseudo-wollastonite
2524B	22·80	24·50	(52·70)	1266	$Na_2O,2CaO,3SiO_2$
2525A	(26·81)	23·57	49·62	1246	$Na_2O,2CaO,3SiO_2$
2535A	18·00	30·65	(51·35)	1280	$Na_2O,2CaO,3SiO_2$
2535B	31·87	16·08	(52·05)	1205	$Na_2O,2CaO,3SiO_2$
2703A	(12·44)	20·79	66·77	1229	Pseudo-wollastonite
2703B	(25·55)	4·61	69·84	763	$Na_2O,2SiO_2$
2714A	(21·38)	7·51	71·11	878	$Na_2O,3CaO,6SiO_2$
2717A	(9·50)	8·00	82·50	1443	Tridymite
2726B	0	37·34	62·50	1434	Pseudo-wollastonite

* The values enclosed by brackets have been obtained by difference.

TABLE I (*continued*).

Design-ation.	Composition by Analysis.			Temper-ature.	Crystalline Phases.
	Na$_2$O.	CaO.	SiO$_2$.		
2734A	(5·63)	25·73	68·64	1339	Tridymite; pseudo-wollastonite
				1341	Pseudo-wollastonite
2734B	(21·66)	6·54	71·80	850	Na$_2$O,3CaO,6SiO$_2$
2735A	(13·78)	15·66	70·56	1089	Pseudo-wollastonite
2735B	(17·23)	13·00	69·77	1006	Na$_2$O,3CaO,6SiO$_2$
2743A	(21·31)	2·84	75·85	901	Tridymite
2743B	(23·48)	3·07	73·45	745	Na$_2$O,2SiO$_2$ + quartz
				758	Na$_2$O,2SiO$_2$
2801A	(37·44)	11·68	50·88	1141	Na$_2$O,2CaO,3SiO$_2$ + 2Na$_2$O,CaO,3SiO$_2$
				1147	Na$_2$O,2CaO,3SiO$_2$
2801B	36·46	3·80	59·74	821	Eutectic Na$_2$O,SiO$_2$; Na$_2$O,2SiO$_2$; 2Na$_2$O,CaO,3SiO$_2$
				827	Na$_2$O,2SiO$_2$ + 2Na$_2$O,CaO,3SiO$_2$
				858	2Na$_2$O,CaO,3SiO$_2$ + Na$_2$O,2CaO,3SiO$_2$
				976	Na$_2$O,2CaO,3SiO$_2$
2802A	28·20	4·45	(67·35)	819	Na$_2$O,2SiO$_2$ + Na$_2$O,2CaO,3SiO$_2$
				850	Na$_2$O,2CaO,3SiO$_2$
2802B	13·39	34·76	(51·85)	1281	Na$_2$O,2CaO,3SiO$_2$ + pseudo-wollastonite
				1282	Pseudo-wollastonite
2804A	(33·47)	15·98	50·55	1212	Na$_2$O,2CaO,3SiO$_2$
2805A	(23·47)	6·06	70·47	826	Na$_2$O,3CaO,6SiO$_2$
2805B	(22·72)	7·98	69·30	901	Na$_2$O,2CaO,3SiO$_2$
2806B	(36·44)	14·17	49·39	1176	Na$_2$O,2CaO,3SiO$_2$
2811B	(19·22)	15·46	65·32	1075	Na$_2$O,2CaO,3SiO$_2$
2811C	(11·93)	15·31	72·76	1098	Tridymite; pseudo-wollastonite
				1106	Pseudo-wollastonite
2814C	(18·66)	15·97	65·37	1082	Na$_2$O,2CaO,3SiO$_2$
2818B	43·59	8·36	48·05	1086	2Na$_2$O,CaO,3SiO$_2$
2818C	(41·60)	1·82	56·58	957	Na$_2$O,SiO$_2$ + 2Na$_2$O,CaO,3SiO$_2$
				967	Na$_2$O,SiO$_2$
2819A	(39·57)	2·77	57·66	898	Na$_2$O,SiO$_2$ + 2Na$_2$O,CaO,3SiO$_2$
				965	2Na$_2$O,CaO,3SiO$_2$
2819B	(37·68)	2·32	60·00	869	2Na$_2$O,CaO,3SiO$_2$ + Na$_2$O,SiO$_2$
				914	2Na$_2$O,CaO,3SiO$_2$
2810C	(22·0)	(5·5)	(72·5)	725	Ternary eutectic : Na$_2$O,2SiO$_2$,quartz-Na$_2$O,3CaO,6SiO$_2$
2823B	(8·77)	31·70	59·53	1045	Na$_2$O,3CaO,6SiO$_2$-wollastonite
				1118	Wollastonite-pseudo-wollastonite

of the component determined by difference being inclosed in parentheses; and the equilibrium temperature for co-existence of the liquid of the given composition with the crystalline phase of the composition given in the last column of the table. In certain cases boundary curve temperatures also are given; these refer, not to the composition of the original glass, but to the liquid derived from it by crystallisation of the primary phase. This liquid must of necessity lie on the line joining the composition of the original glass with the composition of the primary phase, and its exact position can be determined either from a knowledge of the position

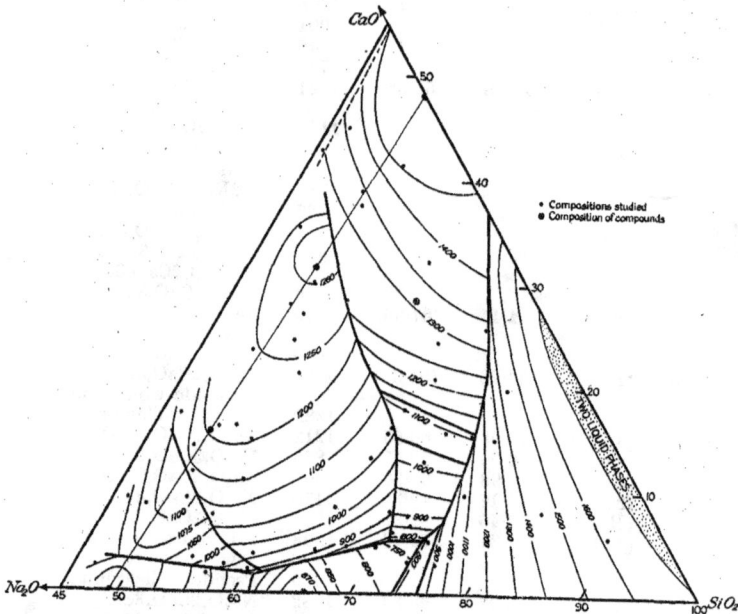

FIG. 1.

Diagram showing the composition of the mixtures studied and of the compounds formed in the ternary system $Na_2O,SiO_2-CaO \cdot SiO_2-SiO_2$. The stability fields of the various compounds are shown by the heavy lines; isotherms on the melting surfaces of the compounds, by light lines.

of the boundary curve, or from the difference between the liquidus and the boundary temperatures. Eutectic temperatures are also given for some compositions. These refer to ternary eutectic temperatures determined in a similar manner after the separation of two phases.

Table II gives enough of the experimental data for each composition to establish the conclusions summarised in Table I. Fig. 1 gives the position of each mixture studied on that portion of the triangular diagram $Na_2O-CaO-SiO_2$ which contains more than 45 per cent. SiO_2; that is, the equilateral triangle whose sides are

THE SYSTEM SODIUM AND CALCIUM METASILICATES–SILICA.   233

the Na_2O–SiO_2 and CaO–SiO_2 sides of the large triangle, and third side the line representing 45 per cent. SiO_2, all compositions being given in weight per cent. In addition, Fig. 1 gives the boundary curves between fields within which the various compounds are stable, and within each field isotherms to indicate the contours of the surface pertaining to each compound in the solid T–X model. Fig. 2 is a similar diagram, showing the boundary lines of the various fields, with the direction of falling temperature indicated by arrows,

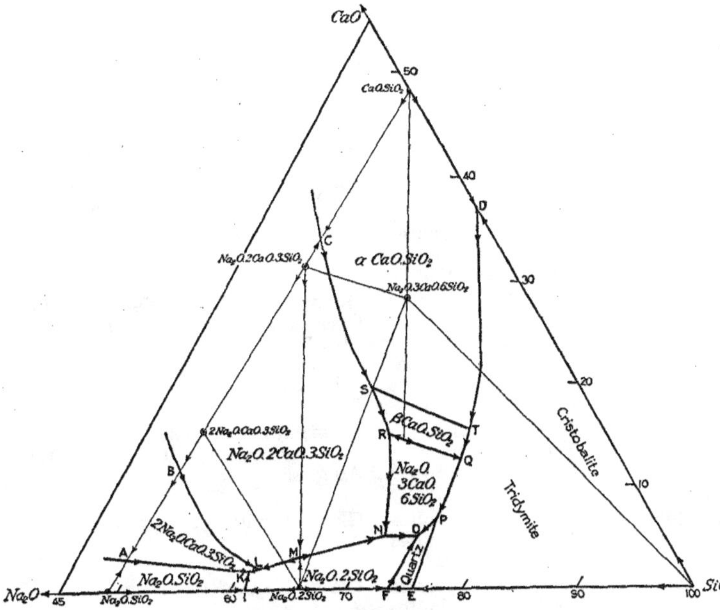

FIG. 2.

Diagram showing the compositions of the various compounds, and the fields containing the liquid compositions within which they may exist in equilibrium.

and further lines joining the composition of phases that can co-exist in equilibrium, to aid in the discussion of the course of crystallisation of the various mixtures. Fig. 3 is the T–X diagram for the binary system Na_2O,SiO_2–SiO_2, which forms one of the boundaries of the ternary system. Figs. 4 and 5 represent two binary systems within the ternary system, the system Na_2O,SiO_2–CaO,SiO_2 and the system $Na_2O,2SiO_2$–$Na_2O,2CaO,3SiO_2$.

Description of the Compounds.

In the field to which this discussion is confined, that portion of the ternary system Na_2O–CaO–SiO_2 contained within the limits

TABLE II.

Reference number.	Temperature.	Time.	Condition.
2422A	716°	60 hours	Glass; quartz; cristobalite; $Na_2O,3CaO,6SiO_2$
	752°	17 hours	Glass; quartz; cristobalite; tridymite
	803°	48 hours	Glass; quartz
	808°	18 hours	Glass; quartz
	811°	36 hours	All glass
	Boundary : $Na_2O,3CaO,6SiO_2$ + quartz 732°		
	Liquidus 810°		
2423A	1049°	1 hour	Glass; tridymite
	1063°	1 hour	Glass; very rare tridymite
	1067°	1 hour	All glass
	Liquidus 1065°		
2426A	1247°	45 minutes	Glass; little cristobalite; tridymite
	1261°	45 minutes	Glass; very little tridymite
	1264°	45 minutes	All glass
	Liquidus 1263°		
2427A	1473°	10 minutes	Glass; tridymite
	1476°	10 minutes	All glass
	Liquidus 1475°		
2428A	1046°	15 minutes	Glass; Na_2O,SiO_2; $2Na_2O,CaO,3SiO_2$
	1051°	15 minutes	Glass and $2Na_2O,CaO,3SiO_2$
	Boundary : Na_2O,SiO_2 + $2Na_2O,CaO,3SiO_2$ 1049°		
	1118°	15 minutes	Glass and few $2Na_2O,CaO,3SiO_2$
	1121°	15 minutes	All glass
	Liquidus 1120°		
2438A	903°	1 hour	Glass; $Na_2O,2CaO,3SiO_2$
	905°	1 hour	Glass; very few $Na_2O,2CaO,3SiO_2$
	907°	1 hour	All glass
	Liquidus 907°		
2438B	1025°	30 minutes	Glass; few large $2Na_2O,CaO,3SiO_2$
	1028°	30 minutes	All glass
	Liquidus 1027°		
2439A	1047°	15 minutes	Glass; Na_2O,SiO_2
	1050°	15 minutes	All glass
	Liquidus 1049°		
2440A	1013°	1 hour	Glass; very rare $Na_2O,2CaO,3SiO_2$
	1017°	1 hour	All glass
	Liquidus 1014°		
2440B	1081°	15 minutes	Glass; $Na_2O,2CaO,3SiO_2$
	1087°	15 minutes	Glass; rare $Na_2O,2CaO,3SiO_2$
	1089°	15 minutes	All glass
	Liquidus 1088°		
2442A	1044°	1 hour	Glass; much $Na_2O,2CaO,3SiO_2$
	1059°	1 hour	Glass; rare $Na_2O,2CaO,3SiO_2$
	1060°	1 hour	All glass
	Liquidus 1060°		
2442B	1147°	15 minutes	Glass; few $Na_2O,2CaO,3SiO_2$
	1150°	15 minutes	Glass; rare $Na_2O,2CaO,3SiO_2$
	1151°	15 minutes	All glass
	Liquidus 1151°		
2443A	1021°	15 minutes	Glass and numerous $2Na_2O,CaO,3SiO_2$
	1023°	15 minutes	Glass; $2Na_2O,CaO,3SiO_2$
	1025°	15 minutes	All glass
	Liquidus 1024°		

THE SYSTEM SODIUM AND CALCIUM METASILICATES—SILICA. 235

TABLE II (continued).

Reference number.	Temperature.	Time.	Condition.
2444A	1095°	1 hour	Glass and few pseudo-wollastonite
	1099°	1 hour	Glass and rare pseudo-wollastonite
	1102°	1 hour	All glass
	Liquidus 1100°		
2444B	1194°	15 minutes	Glass; $Na_2O,2CaO,3SiO_2$
	1198°	15 minutes	Glass; very rare $Na_2O,2CaO,3SiO_2$
	1201°	15 minutes	All glass
	Liquidus 1200°		
2445A	1067°	15 minutes	Glass; $2Na_2O,CaO,3SiO_2$
	1080°	15 minutes	Glass; rare $2Na_2O,CaO,3SiO_2$
	1083°	15 minutes	All glass
	Liquidus 1082°		
2446A	1078°	15 minutes	Glass; pseudo-wollastonite
	1281°	15 minutes	Glass; exceedingly rare pseudo-wollastonite
	Liquidus 1282°		
2446B	1231°	15 minutes	Glass; much $Na_2O,2CaO,3SiO_2$
	1239°	15 minutes	Glass; rare $Na_2O,2CaO,3SiO_2$
	1241°	15 minutes	All glass
	Liquidus 1240°		
2447A	1080°	15 minutes	Glass; $Na_2O,2CaO,3SiO_2$; also long crystals; inclined extinction; fair birefringence; $n = 1.63$
	1102°	15 minutes	Glass; $Na_2O,2CaO,3SiO_2$
	1106°	15 minutes	Glass; exceedingly rare $Na_2O,2CaO,3SiO_2$
	Liquidus 1107°		
2448A	1372°	10 minutes	Glass; pseudo-wollastonite
	1377°	10 minutes	Glass; rare pseudo-wollastonite
	1381°	10 minutes	All glass
	Liquidus 1379°		
2448B	1256°	15 minutes	Glass; pseudo-wollastonite
	1260°	15 minutes	Glass; rare pseudo-wollastonite
	1264°	15 minutes	All glass
	Liquidus 1262°		
2449A	1264°	15 minutes	Glass
	1274°	15 minutes	Glass; rare $Na_2O,2CaO,3SiO_2$
	1277°	15 minutes	All glass
	Liquidus 1275°		
2449B	1273°	15 minutes	Glass; much $Na_2O,2CaO,3SiO_2$
	1276°	15 minutes	Glass; many $Na_2O,2CaO,3SiO_2$
	1278°	15 minutes	All glass
	Liquidus 1277°		
2450B	1323°	10 minutes	Glass; pseudo-wollastonite
	1326°	10 minutes	All glass
	Liquidus 1325°		
2451A	1273°	15 minutes	Glass; few $Na_2O,2CaO,3SiO_2$
	1276°	15 minutes	Glass; rare $Na_2O,2CaO,3SiO_2$
	1278°	15 minutes	All glass
	Liquidus 1277°		
2452B	1604°	10 minutes	Glass; rare cristobalite
	1609°	10 minutes	Glass; rare cristobalite
	1613°	10 minutes	All glass
	Liquidus 1611°		
2502B	1352°	10 minutes	Glass; few pseudo-wollastonite
	1356°	10 minutes	Glass; rare pseudo-wollastonite
	1361°	10 minutes	All glass
	Liquidus 1358°		

TABLE II (*continued*).

Reference number.	Temperature.	Time.	Condition.
2503A	1450°	10 minutes	Glass; pseudo-wollastonite
	1452°	10 minutes	Glass; few pseudo-wollastonite
	1455°	10 minutes	All glass
	Liquidus 1454°		
2505A	1287°	10 minutes	Glass; $3CaO,2SiO_2$
	1297°	10 minutes	Glass; $3CaO,2SiO_2$
	1302°	10 minutes	All glass
	Liquidus 1300°.		
2505x	1326°	10 minutes	Glass; many pseudo-wollastonite
	1329°	10 minutes	Glass; rare pseudo-wollastonite
	1332°	20 minutes	All glass
	Liquidus 1330°		
2524B	1264°	15 minutes	Glass; $Na_2O \cdot 2CaO,3SiO_2$
	1268°	15 minutes	All glass
	Liquidus 1266°		
2525A	1244°	15 minutes	Glass; $Na_2O,2CaO,3SiO_2$
	1248°	15 minutes	All glass
	Liquidus 1246°		
2535A	1275°	15 minutes	Very little glass; about all $Na_2O,2CaO,3SiO_2$
	1277°	15 minutes	Slightly more glass; chiefly $Na_2O,2CaO,3SiO_2$
	1279°	15 minutes	Glass, about 90%; 10% $Na_2O,2CaO,3SiO_2$
	1281°	15 minutes	All glass
	Liquidus 1280°		
2535B	1195°	20 minutes	Glass; little $Na_2O,2CaO,3SiO_2$
	1204°	15 minutes	Glass; rare $Na_2O,2CaO,3SiO_2$
	1206°	15 minutes	All glass
	Liquidus 1205°		
2703A	1217°	30 minutes	Glass; pseudo-wollastonite
	1228°	15 minutes	Glass; extremely rare pseudo-wollastonite
	1231°	15 minutes	All glass
	Liquidus 1229°		
2703B	610°	75 hours	Glass; $Na_2O,2SiO_2$
	737°	60 hours	Glass; $Na_2O,2SiO_2$
	752°	17 hours	Glass; $Na_2O,2SiO_2$
	756°	7 hours	Glass; $Na_2O,2SiO_2$
	767°	3 hours	All glass
	Liquidus 763°		
2714A	719°	17 hours	Glass; $Na_2O,3CaO,6SiO_2$; quartz
	868°	17 hours	Glass; $Na_2O,3CaO,6SiO_2$
	876°	16 hours	Glass; rare $Na_2O,3CaO$ $6SiO_2$
	879°	3 hours	All glass
	Liquidus 878°		
2717A	1438°	10 minutes	Glass; fair amount tridymite
	1442°	10 minutes	Glass; rare tridymite
	1444°	10 minutes	All glass
	Liquidus 1443°		
2726B	1427°	10 minutes	Glass; pseudo-wollastonite
	1433°	10 minutes	Glass; exceedingly rare pseudo-wollastonite
	Liquidus 1434°		
2734A	1331°	10 minutes	Glass; pseudo-wollastonite; cristobalite, inside pseudo-wollastonite crystals

THE SYSTEM SODIUM AND CALCIUM METASILICATES–SILICA.   237

TABLE II (continued).

Reference number.	Temperature.	Time.	Condition.
2734A	1336°	15 minutes	Glass; pseudo-wollastonite; cristobalite, in larger crystals than before
	1339°	15 minutes	Glass; pseudo-wollastonite in vanishingly small amount; cristobalite
	1342°	15 minutes	All glass

Boundary : Pseudo-wollastonite-tridymite 1339°
Liquidus 1341°

2734B	715°	60 hours	Glass; some $Na_2O,3CaO,6SiO_2$; rare cristobalite
	750°	18 hours	Glass; rare $Na_2O,3CaO,6SiO_2$
	847°	60 hours	Glass; very rare $Na_2O,3CaO,6SiO_2$
	852°	3 hours	All glass

Liquidus 850°

2735A	1082°	15 minutes	Glass; pseudo-wollastonite
	1088°	15 minutes	Glass; exceedingly rare pseudo-wollastonite
	1089°	15 minutes	All glass

Liquidus 1089°

2735B	977°	15 minutes	Glass; excellent $Na_2O,3CaO,6SiO_2$
	1003°	15 minutes	Glass; rare $Na_2O,3CaO,6SiO_2$
	1007°	15 minutes	All glass

Liquidus 1006°

2743A	775°	6 hours	Glass; tridymite
	808°	6 hours	Glass; quartz; tridymite; cristobalite
	899°	30 minutes	Glass; little tridymite
	903°	30 minutes	All glass

Liquidus 901°

2743B	717°	72 hours	Glass; quartz; large crystals $Na_2O,2SiO_2$
	734°	18 hours	Glass; quartz; $Na_2O,2SiO_2$
	747°	16 hours	Glass; $Na_2O,2SiO_2$

Boundary : $Na_2O,2SiO_2$ + quartz 745°.

| | 754° | 3 hours | Glass; little $Na_2O,2SiO_2$ |
| | 762° | 1 hour | All glass |

Liquidus 758°

| 2801A | 1139° | 15 minutes | Glass; $2Na_2O,CaO,3SiO_2$; in cubes, octahedra, rhombic dodecahedra, and pentagonal dodecahedra |
| | 1143° | 15 minutes | Glass; $Na_2O,2CaO,3SiO_2$ |

Boundary $2Na_2O,CaO$ $3SiO_2$ + $Na_2O,2CaO,3SiO_2$ 1141°

| | 1150° | 15 minutes | All glass |

Liquidus 1147°

| 2801B | 817° | 30 minutes | No glass; Na_2O,SiO_2; $Na_2O,2SiO_2$; $2Na_2O,CaO,3SiO_2$ |
| | 826° | 2 hours | Glass; rare $Na_2O,2SiO_2$; $2Na_2O,CaO,3SiO_2$ |

Eutectic Na_2O,SiO_2 - $Na_2O,2SiO_2$ + $2Na_2O,CaO,3SiO_2$ 821°
Boundary Na_2O,SiO_2 – $2Na_2O,CaO,3SiO_2$ 827°

| | 855° | 1 hour | Glass; $2Na_2O$ $CaO,3SiO_2$ |
| | 861° | 30 minutes | Glass; $2Na_2O,CaO,3SiO_2$; little $Na_2O,2CaO,3SiO_2$ |

Boundary : $2Na_2O,CaO,3SiO_2$ + $Na_2O,2CaO,3SiO_2$; on tie-line $(2Na_2O,CaO,3SiO_2)$ – (2801B) 858°

| | 975° | 30 minutes | Glass; rare $Na_2O,2CaO,3SiO_2$ |
| | 977° | 30 minutes | All glass |

Liquidus 976°

TABLE II (*continued*).

Reference number.	Temperature.	Time.	Condition.
2802A	722°	20 hours	Glass; $Na_2O,2SiO_2$; $Na_2O,2CaO,3SiO_2$
	818°	30 minutes	Glass; $Na_2O,2SiO_2$; $Na_2O,2CaO,3SiO_2$; both crystalline phases in small amount
	820°	20 minutes	Glass; $Na_2O,2CaO,3SiO_2$
	Boundary : $Na_2O,2SiO_2$ + $Na_2O,2CaO,3SiO_2$ 819°		
	848°	20 minutes	Glass; rare $Na_2O,2CaO,3SiO_2$
	852°	20 minut	All glass
	Liquidus 850°		
2802B	1279°	15 minutes	Glass; $Na_2O,2CaO,3SiO_2$; pseudo-wollastonite
	1281°	15 minutes	Glass; pseudo-wollastonite
	Boundary : $Na_2O,2CaO,3SiO_2$ + pseudo-wollastonite 1280°		
	1283°	15 minutes	All glass
	Liquidus 1282°		
2804A	1150°	30 minutes	Glass in small amount; $Na_2O,2CaO,3SiO_2$; rare $2Na_2O,CaO,3SiO_2$
	1206°	30 minutes	Glass; $Na_2O,2CaO,3SiO_2$
	1210°	30 minutes	Glass; very little $Na_2O,2CaO,3SiO_2$
	1215°	30 minutes	All glass
	Liquidus 1212°		
2805A	760°	67 hours	Glass; rare needles $Na_2O,3CaO,6SiO_2$
	825°	2 hours	Glass; rare needles $Na_2O,3CaO,6SiO_2$
	827°	30 minutes	All glass
	Liquidus 826°		
2805B	862°	20 minutes	Glass; $Na_2O,2CaO,3SiO_2$
	899°	20 minutes	Glass; rare $Na_2O,2CaO,3SiO_2$
	903°	20 minutes	All glass
	Liquidus 901°		
2806B	1140°	15 minutes	Glass, 10—15%; 2—3% $Na_2O,2CaO,3SiO_2$; rest, $2Na_2O,CaO,3SiO_2$
	1158°	15 minutes	Glass; $Na_2O,2CaO,3SiO_2$
	1175°	15 minutes	Glass; rare $Na_2O,2CaO,3SiO_2$
	1177°	15 minutes	All glass
	Liquidus 1176°		
2811B	1072°	20 minutes	Glass; $Na_2O,2CaO,3SiO_2$
	1074°	20 minutes	Glass; rare $Na_2O,2CaO,3SiO_2$
	1076°	20 minutes	All glass
	Liquidus 1075°		
2811C	1075°	2 hours	Glass; tridymite; pseudo-wollastonite
	1097°	20 minutes	Glass; extremely rare tridymite; pseudo-wollastonite
	Boundary : tridymite + pseudo-wollastonite 1098°		
	1105°	20 minutes	Glass; rare pseudo-wollastonite
	1107°	20 minutes	All glass
	Liquidus 1106°		
2814C	1054°	18 hours	Glass; $Na_2O,2CaO,3SiO_2$
	1081°	20 minutes	Glass; exceedingly rare $Na_2O,2CaO,3SiO_2$
	1083°	20 minutes	All glass
	Liquidus 1082°		
2818B	1085°	20 minutes	Glass; $2Na_2O·CaO·3SiO_2$
	1088°	20 minutes	All glass
	Liquidus 1086°		

THE SYSTEM SODIUM AND CALCIUM METASILICATES–SILICA. 239

TABLE II (continued).

Reference number.	Temperature.	Time.	Condition.
2818c	954°	20 minutes	Glass; Na$_2$O,SiO$_2$; 2Na$_2$O,CaO,3SiO$_2$
	956°	20 minutes	Glass; Na$_2$O,SiO$_2$; exceedingly rare 2Na$_2$O,CaO,3SiO$_2$
	Boundary : Na$_2$O,SiO$_2$ − 2Na$_2$O,CaO,3SiO$_2$ 957°		
	965°	20 minutes	Glass; Na$_2$O,SiO$_2$
	968°	20 minutes	All glass
	Liquidus 967°		
2819A	897°	20 minutes	Glass; Na$_2$O,SiO$_2$; 2Na$_2$O,CaO,3SiO$_2$
	899°	20 minutes	Glass; 2Na$_2$O,CaO,3SiO$_2$
	Boundary : Na$_2$O,SiO$_2$ − 2Na$_2$O,CaO,3SiO$_2$ 898°		
	972°	20 minutes	Glass; rare 2Na$_2$O,CaO,3SiO$_2$
	967°	20 minutes	All glass
	Liquidus 965°		
2819B	864°	20 minutes	Glass; 2Na$_2$O,CaO,3SiO$_2$; Na$_2$O,SiO$_2$
	870°	20 minutes	Glass; 2Na$_2$O,CaO,3SiO$_2$; extremely rare Na$_2$O,SiO$_2$
	Boundary : 2Na$_2$O,CaO,3SiO$_2$ − Na$_2$O,SiO$_2$ 869°		
	911°	20 minutes	Glass; very rare 2Na$_2$O,CaO,3SiO$_2$
	916°	20 minutes	All glass
	Liquidus 914°		
2810c	Mixture initially wholly crystalline		
	724°	25 hours	Unaltered
	734°	18 hours	Much glass; remnants of all three crystalline phases
	Ternary eutectic : Na$_2$O,2SiO$_2$ − Na$_2$O,3CaO,6SiO$_2$ − quartz 725°		
2823B	1042°	20 minutes	Na$_2$O,3CaO,6SiO$_2$; wollastonite; no glass present
	1047°	20 minutes	Glass; wollastonite; Na$_2$O,3CaO,6SiO$_2$, in small amount.
	Boundary : Na$_2$O,3CaO,6SiO$_2$ − wollastonite 1045°		
	1116°	20 minutes	Glass; wollastonite
	1119°	20 minutes	Glass; pseudo-wollastonite
	Boundary : wollastonite-pseudo-wollastonite 1118°		

Na$_2$O,SiO$_2$–CaO,SiO$_2$–SiO$_2$, there are seven compounds; one of these, SiO$_2$, coexists with liquids in the system in three different crystalline modifications, quartz, tridymite, and cristobalite; and a second compound, CaO,SiO$_2$, is found in two modifications. All these compounds are too well known to need further description here. The crystalline modifications of silica and their transformations have been described by Fenner [*] and by Ferguson and Merwin,[†] while forms of CaO,SiO$_2$ and its transformation have been studied by E. T. Allen and W. P. White.[‡] The two other binary compounds, Na$_2$O,SiO$_2$ and Na$_2$O,2SiO$_2$, have recently been described by us.[§]

In addition to the above previously known compounds, three new compounds have been encountered in the present work. Two

* C. N. Fenner, Amer. J. Sci., 1913, 36, 331.
† J. B. Ferguson and H. E. Merwin, Amer. J. Sci., 1918, 46, 417.
‡ Amer. J. Sci., 1906, (4), 21, 89.
§ J. Phys. Chem., 1924, 28, 1167.

of these have the metasilicate ratio, namely, $2Na_2O,CaO,3SiO_2$ and $Na_2O,2CaO,3SiO_2$, and the third has the composition
$$Na_2O,3CaO,6SiO_2.$$

The Compound $2Na_2O,CaO,3SiO_2$.

The compound $2Na_2O,CaO,3SiO_2$, which may also be written $2Na_2SiO_3,CaSiO_3$, occurs in crystals of the pyritohedral group of

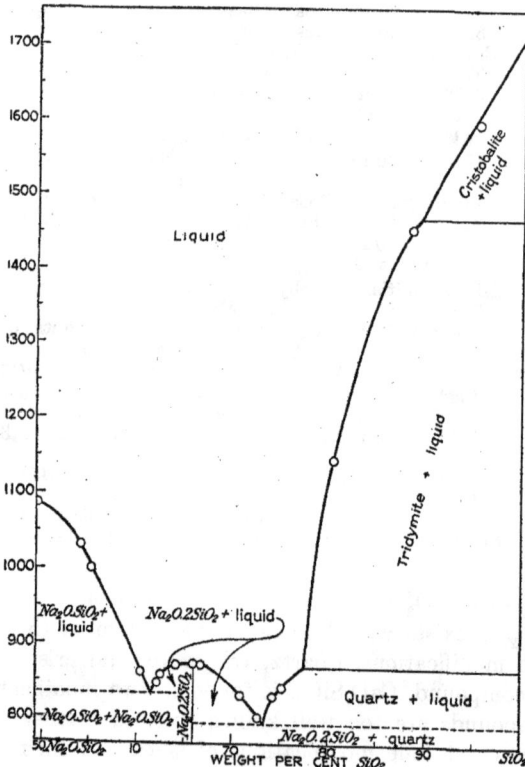

Fig. 3.

Melting point diagram of the binary system Na_2O,SiO_2–SiO_2. Composition is expressed as weight per cent. SiO_2.

the isometric system. It may occur as simple octahedra but these are usually highly modified by the presence of the cube, rhombic dodecahedron, and pentagonal dodecahedron. The refractive index is $1·571 \pm 0·002$.

This crystalline compound has an incongruent melting point at 1141°; that is, at 1141° it decomposes, forming crystals of $Na_2O,2CaO,3SiO_2$, and a liquid containing 11·5 per cent. CaO.

The composition of the compound was established by preparing a mixture of exactly the percentage composition calculated for this compound; Na_2O, 34·42, CaO, 15·54, SiO_2, 50·04, and crystallising it below the temperature of decomposition. The resulting crystalline mass was entirely homogeneous. The peculiar shape of its field, and the similarity between the properties of this compound and of the adjacent $Na_2O,2CaO,3SiO_2$, made the differentiation of the two difficult until their compositions and approximate fields had been established. The field of $2Na_2O,CaO,3SiO_2$ (Figs. 1 and 2) is a small one, bounded by the field for Na_2O,SiO_2, extending past the ternary eutectic (quadruple (P) point)*

$$Na_2O,SiO_2-Na_2O,2SiO_2-2Na_2O,CaO,3SiO_2-liquid,$$

to the quadruple (P) point

$$Na_2O,2SiO_2-2Na_2O,CaO,3SiO_2-Na_2O,2CaO,3SiO_2-liquid.$$

It will be noted that this point differs from the preceding quadruple (P) point in not being a eutectic; this difference will be more fully discussed subsequently. From this invariant point the field is bounded by the field of the other metasilicate compound,

$$Na_2O,2CaO,3SiO_2.$$

The Compound $Na_2O,2CaO,3SiO_2$.

The compound $Na_2O,2CaO,3SiO_2$, which may be written

$$Na_2SiO_3,2CaSiO_3,$$

occupies the largest field of any of the ten solid phases, extending to over 70 per cent. SiO_2. It can exist in contact with five different crystalline phases, the only fields not coming into contact with the field of $Na_2O,2CaO,3SiO_2$ being those of Na_2O,SiO_2 and of SiO_2 in its various forms.

Crystals of this compound, though usually well formed, are of indeterminate symmetry on account of the universal presence of twinning. They have, however, a common pseudo-cubic aspect and in their simplest form show square sections which are bisected diagonally by the trace of the twinning plane, to which the extinction is highly inclined. Usually the twinning is polysynthetic and then more complicated. On account of the twinning and the very low birefringence, a satisfactory interference figure was not obtained, so it is not known whether the crystals are uniaxial or biaxial. The refractive indices are $\gamma = 1·599$ and $\alpha = 1·596$. The crystals are readily distinguished from all other solid phases

---

* We have adopted the convention proposed by Schreinemakers for designating the invariant point in a condensed three component system as a quadruple (P) point; the symbol (P) indicating that the pressure variable is regarded as arbitrarily held constant, hence invariance being produced by the co-existence of four phases, in this case, liquid and three solids.

in the system by their very weak birefringence and twinning, though when exceedingly small they may be confused with the truly isotropic crystals of the compound $2Na_2O,CaO,3SiO_2$.

The composition of the compound $Na_2O,2CaO,3SiO_2$ was rendered almost certain from the results of the quenching experiments. For example, mixture 2535 A (see Table II), close to this compound in composition, remained almost entirely crystalline when heated at 1275°; two degrees higher it was still largely crystalline, and two degrees higher, at 1279°, it still contained a considerable proportion of crystals, which disappeared on further increase of two degrees in temperature. This behaviour is characteristic of an approximately pure compound. The composition was definitely established by crystallising a mixture of the calculated composition, Na_2O, 17·50; CaO, 31·61; SiO_2, 50·89 per cent. which was found to consist wholly of one crystalline phase. It has a congruent melting point at 1284°, and the upper portion of its fusion surface is characterised by an unusual flatness. It resembles in this respect $Na_2O,2SiO_2$ in the binary system, which also is characterised by a congruent melting point with a flat maximum. In the ternary system, as in the binary,* this marked flattening is to be ascribed to dissociation of the compound in the liquid phase, and it is evident that both these compounds are highly dissociated.

The stability field of $Na_2O,2CaO,3SiO_2$ is, as we have noted, a large one. Starting at the temperature at which it is formed in the binary system Na_2O,SiO_2–CaO,SiO_2 by the dissociation of $2Na_2O,CaO,3SiO_2$, 1141°, its field is bounded by that of the latter compound until the quadruple (P) point

$$Na_2O,2SiO_2-2Na_2O,CaO,3SiO_2-Na_2O,2CaO,3SiO_2-liquid$$

is reached. With further increase in the SiO_2 content, $Na_2O,2SiO_2$ becomes the adjoining phase; this boundary curve is cut by the tie-line $Na_2O,2SiO_2$–$Na_2O,2CaO,3SiO_2$, at which point there is a maximum temperature, and these two phases form a binary system (see Fig. 4). On further increase in the SiO_2 content the quadruple (P) point $Na_2O,2SiO_2$–$Na_2O,2CaO,3SiO_2$–$Na_2O,3CaO,6SiO_2$–liquid is reached, at the highest SiO_2 content of any glass that can co-exist with this compound. The field under discussion then becomes bounded successively by those of the compound $Na_2O,3CaO,6SiO_2$, followed by the two forms of CaO,SiO_2, wollastonite and pseudo-wollastonite. The latter is the stable phase which co-exists with $Na_2O,2CaO,3SiO_2$ at the eutectic in the binary system

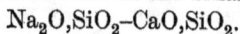

$$Na_2O,SiO_2-CaO,SiO_2.$$

It should be noted in connection with the stability field of this compound that, although at higher temperatures the curvature

* See the discussion of this subject by Morey and Bowen, *op. cit.*, p. 1177.

THE SYSTEM SODIUM AND CALCIUM METASILICATES–SILICA.   243

of the fusion surface is slight, at the lower temperatures, and especially in the region of the quadruple (P) point,

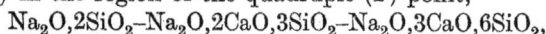

$$Na_2O,2SiO_2–Na_2O,2CaO,3SiO_2–Na_2O,3CaO,6SiO_2,$$

the curvature becomes much greater, the isotherms become crowded closer together, and temperature falls more rapidly as the liquid approaches the boundaries.

The Compound $Na_2O,3CaO,6SiO_2$.

The compound $Na_2O,3CaO,6SiO_2$ is one of the greatest importance in glass technology, being the primary phase to separate in a large proportion of commercial glasses, yet its stability field is small, and far from its own composition. It is observed in crystals of orthorhombic symmetry with the form of four-, six-, or eight-sided prisms. The elongation is γ. The refractive indices are $\gamma = 1.579$ and $\alpha = 1.564$, and the optic axial angle 2V is about 75°.

Crystals having the above properties were first identified a few years ago as a devitrification product of commercial glasses by Professor A. B. Peck, then of the U.S. Bureau of Standards. He has published no record of them, but he showed the material containing them to one of us, by whom they were mentioned in a brief note discussing the composition of rivaite.* More recently they were encountered and described in similar glasses by Insley,† who gives properties that agree entirely with those stated above within the limits of error of measurement on such crystals. Since the compound occurred in the above instances as minute crystals embedded in glass, it was not possible to determine its composition and Insley mentions it simply as " sodium–calcium silicate."

Fixing the composition of these crystals was difficult. The area of glass compositions which may be in equilibrium with them is small, and it is a low temperature area. Experiment showed that above approximately 1000° the crystals were decomposed, wollastonite or pseudo-wollastonite being formed; hence to obtain the pure compound it was necessary to crystallise the glasses below 1000°. A number of glasses crystallised at such low temperatures gave fibrous aggregates so fine-grained that individual components could not be distinguished. Under these conditions it was difficult to determine whether the material was within a few per cent. of being homogeneous or whether the admixed material was large in amount. For example, it was at first thought that the crystals had the composition $Na_2O,2CaO,6SiO_2$, a disilicate; this material, however, when crystallised for only a few hours appeared as described above, but low in index; crystallised for a longer time

---

* N. L. Bowen, *Amer. Mineralogist*, 1922, **7**, 64.

† H. Insley, *J. Amer. Cer. Soc.*, 1924, **7**, 17.

it became evident that there were distinct patches of glass. Another disilicate mixture was tried, $Na_2O,3CaO,8SiO_2$, but this proved equally unsatisfactory; it was rather less well crystallised, and hence appeared more homogeneous, but the mass effect was of a mixture of lower index than the crystals whose composition was sought, and further gave a brownish effect characteristic of fine-grained inhomogeneous material. A mixture of the composition $Na_2O,2CaO,5SiO_2$ proved indistinguishable from the preceding. By such laborious trial and error methods it was found that a mixture of the percentage composition Na_2O, 7·44; CaO, 32·17; SiO_2, 60·39 (glass 2448A in Tables I and II) gave a crystalline mass more nearly homogeneous than any other that had been tested, and, furthermore, contained crystals, in addition to those sought, which could be identified as wollastonite. A mixture of the composition $Na_2O,3CaO,6SiO_2$, containing Na_2O, 10·50; CaO, 28·45; SiO_2, 61·05 per cent., was next prepared, and after crystallising for 16 hours at 950° was found to be homogeneous and to have the same optical properties as the well-formed crystals that grow freely in contact with liquids lying in its field of stability. Further experiments on the same mixture confirmed the conclusion that the above formula is correct.

The field of $Na_2O,3CaO,6SiO_2$ is small, and situated far from the composition of the compound. Heated to 1045°, the compound breaks up into wollastonite and liquid; the decomposition takes place readily, but the reverse combination of wollastonite and liquid takes place very slowly indeed. In this connection it should be mentioned that the primary phase separating from a glass of this composition is pseudo-wollastonite, and that this initial crystallisation takes place with such rapidity that in order to obtain the glass necessary for the preparation of crystalline $Na_2O,3CaO,6SiO_2$, it must be quenched in mercury. Slower cooling results in formation of some pseudo-wollastonite, whose reaction with glass to form the desired compound is too slow a process to be of practical value as a method of preparation.

The field for $Na_2O,3CaO,6SiO_2$ is bounded by the fields of wollastonite, tridymite, quartz, $Na_2O,2CaO,3SiO_2$ and $Na_2O,2SiO_2$. This compound is one of the phases at the ternary eutectic $Na_2O,2SiO_2$–$Na_2O,3CaO,6SiO_2$–quartz, at 725°, the lowest melting mixture in the system. It is also the primary phase to separate in all soda-lime glasses ranging in percentage composition from approximately 66—74 SiO_2, 5—17 CaO, a range which, exclusive of the iron and alumina content, represents much of the glass of commerce.

THE SYSTEM SODIUM AND CALCIUM METASILICATES–SILICA. 245

The Boundary Curves and Quadruple (P) Points.

The discussion of the boundary curves and quadruple (P) points is facilitated by reference to Fig. 2. We shall first consider the boundary curve AK, along which Na_2O,SiO_2 and $2Na_2O,CaO,3SiO_2$ are co-existing phases, and IK, along which Na_2O,SiO_2 and $Na_2O,2SiO_2$ co-exist. Along each of these boundary curves the temperature falls, reaching a minimum at the quadruple (P) point K. This point lies within the triangle formed by joining the compositions of the reacting phases, and it is accordingly an eutectic. Along each of these curves the reaction taking place with rising temperature is of the type $S_1 + S_2 = L$, and the reaction taking place at the ternary eutectic is of the type $S_1 + S_2 + S_3 = L$. The temperature also falls as we pass from L to K, along which $2Na_2O,CaO,3SiO_2$ and $Na_2O,2SiO_2$ co-exist, and along which the reaction $S_1 + S_2 = L$ takes place on adding heat.

The boundary curve, BL, which outlines the region of stability of the compound $2Na_2O,CaO,3SiO_2$, is of a different type. Along this boundary we have a reaction of the type $S_1 = S_2 + L$ taking place; a decomposition of a solid phase into another solid phase and liquid. On it also as we pass from B to L the temperature falls, but the quadruple (P) point L differs markedly from the preceding. Here the composition of the liquid which exists in equilibrium with the three solid phases lies outside the triangle formed by joining the compositions of the three solid phases. The phase reaction taking place on addition of heat is of the type $S_1 + S_2 = S_3 + L$; the temperature is a minimum only for one of the phases, $Na_2O,2CaO,3SiO2$; and one of the three intersecting boundary curves goes to lower temperatures, the other two to higher temperatures.

The boundary curve CSRN, along which $Na_2O,2CaO,3SiO_2$ co-exists successively with pseudo-wollastonite, wollastonite, and $Na_2O,3CaO,6SiO_2$, is one of constantly falling temperature. At S we have the quadruple (P) point $Na_2O,2CaO,3SiO_2$–pseudo-wollastonite–wollastonite–liquid S, and at R the quadruple (P) point $Na_2O,2CaO,3SiO_2$–wollastonite–$Na_2O,3CaO,6SiO_2$–liquid R. The reaction taking place at S, and along the boundary curve ST, is primarily the inversion of pseudo-wollastonite to wollastonite, a reaction which would take place at the constant temperature of the inversion in the unary system CaO,SiO_2 if both phases had the same composition. Since the temperature is found to be lower in ternary mixtures it follows that there is some solid solution in the high-temperature phase. The reaction taking place at R is essentially that of wollastonite and liquid combining to form the 1 : 3 : 6 compound, the reaction which characterises the boundary curve RQ. The temperature will rise from R along RQ to the intersection of RQ with the prolongation of the straight line joining the composition of CaO,SiO_2 with $Na_2O,3CaO,6SiO_2$. Of the curves intersecting to form the point R, two go to higher temperatures, and one, RN, to lower temperatures, as was the case at L. Here also the composition of the liquid, R, lies outside of the triangle formed by joining the compositions of the reacting solids.

With further decrease in temperature the curve, RN, meets the two curves LMN and NO at N. Here again the composition of the liquid lies outside the solid triangle, and the point, N, is a minimum temperature for the phase $Na_2O,2CaO,3SiO_2$.

The curve, LMN, marks the co-existence of the phases $Na_2O,2SiO_2$ and $Na_2O,2CaO,3SiO_2$, and along it the phase reaction is of the type $S_1 + S_2 = L$. The temperature will rise along it from both L and N to a maximum at M, the eutectic in the binary system $Na_2O,2SiO_2$–$Na_2O,2CaO,3SiO_2$. This system is discussed subsequently.

The curve, DTQPO, gives the composition of the liquids which can co-exist with silica and its adjoining phases, and along its entire course the temperature falls. From D to T the phases are pseudo-wollastonite and tridymite; from T to Q, wollastonite and tridymite; from Q to P, $Na_2O,3CaO,6SiO_2$ and tridymite; and from P to O, $Na_2O,3CaO,6SiO_2$ and quartz.

The third boundary curve intersecting at O is the boundary curve $Na_2O,2SiO_2$–quartz. The three solid phases at O are $Na_2O,2SiO_2$, $Na_2O,3CaO,6SiO_2$ and quartz; the composition of the liquid, O, is evidently inside the triangle formed by joining the compositions of the solid phases, and this quadruple (P) point is a eutectic. The composition and temperature of this eutectic was arrived at with the greatest difficulty. It proved impossible to crystallise completely the glasses near it in composition, even when held for several months slightly below the eutectic temperature. As can be seen by reference to Table II, the primary phase could be obtained in some glasses not far from the eutectic, but in no case was more than a small percentage of the material crystallised. Three other mixtures, all of them still closer to the eutectic than any included in Table II, were prepared, and attempts made to crystallise them without success. It finally proved necessary to obtain the eutectic composition by extrapolation, but the composition so deduced cannot be in error by more than 1 per cent. A mixture of the percentage composition : Na_2O, 22; CaO, 5·5; SiO_2, 72·5, was prepared by mixing the crystallised solids in finely powdered form, and held for periods of from 16 to 20 hours at various temperatures. At 724° it remained unaltered; at 733° a con-

THE SYSTEM SODIUM AND CALCIUM METASILICATES–SILICA. 247

siderable amount of glass had formed; so the eutectic temperature may safely be put at 725°.

The boundary curve, PE, represents the transition temperature of the high temperature form of quartz to tridymite. It will be observed that there is a noticeable flattening of the silica field when quartz becomes the solid phase, just as was observed in the binary system $Na_2O,2SiO_2$–SiO_2.* A further boundary curve, that between tridymite and cristobalite, is not shown; the inversion takes place at 1470°, and the boundary curve will follow closely parallel to the 1500° isotherm, which is indicated.

A further boundary curve is that outlining the region of immiscibility in the liquid phase. It has been shown by J. W. Greig † that mixtures of CaO and SiO_2 containing more than 72 per cent. SiO_2 separate into two liquid layers, the silica-rich one containing but little CaO, and inasmuch as no such limited miscibility has been found in the system Na_2O–SiO_2, it follows that the region of immiscibility in the ternary system must have some such shape as outlined. We have made no experiments in this region, but the necessary shape of the isotherms found in the regions of high SiO_2 content make such an interpretation highly probable.

Binary Systems.

Fig. 3 gives the T—X diagrams for the binary system sodium metasilicate Na_2O,SiO_2–silica SiO_2, previously published by us,‡ and Figs. 4 and 5 the further binary systems sodium metasilicate Na_2O,SiO_2–calcium metasilicate CaO,SiO_2 and sodium disilicate, $Na_2O,2SiO_2$–$Na_2O,2CaO,3SiO_2$. In the binary system sodium metasilicate–calcium metasilicate, starting from Na_2O,SiO_2, on addition of CaO,SiO_2 temperature falls to the binary eutectic at 3·0 per cent. CaO, 1060°. On further addition of CaO,SiO_2 the compound $2Na_2O,CaO,3SiO_2$ is formed, and the temperature rises to the decomposition temperature of the latter, 1141° at 11·5 per cent. CaO. On further increase in the CaO,SiO_2 content the compound $Na_2O,2CaO,3SiO_2$ is formed, and the temperature rises to a flat maximum at 1284°, the melting point of this compound. On further addition of CaO,SiO_2 the temperature falls to 1282°, the binary eutectic between $Na_2O,2CaO,3SiO_2$ and pseudo-wollastonite, then rises rapidly to the melting point of wollastonite at 1540°.

In the binary system sodium disilicate–$Na_2O,2CaO,3SiO_2$, on

---

* G. W. Morey and N. L. Bowen, J. Phys. Chem., 1924, 28, 1167.

† A complete account of this work will appear in a forthcoming paper by Greig. A preliminary discussion was presented at the Christmas, 1923, meeting of the Geological Society of America.

‡ J. Phys. Chem., 1924, 28, 1167.

addition of sodium disilicate to the latter compound, the melting
point falls at first very slowly indeed; the two compounds in question are both characterised by extreme dissociation in the liquid
phase; both probably consist in large part not of molecules of the
composition of the compounds, but of the various possible dissociation products of these compounds, and addition of sodium disilicate probably adds no new molecular species, but merely causes

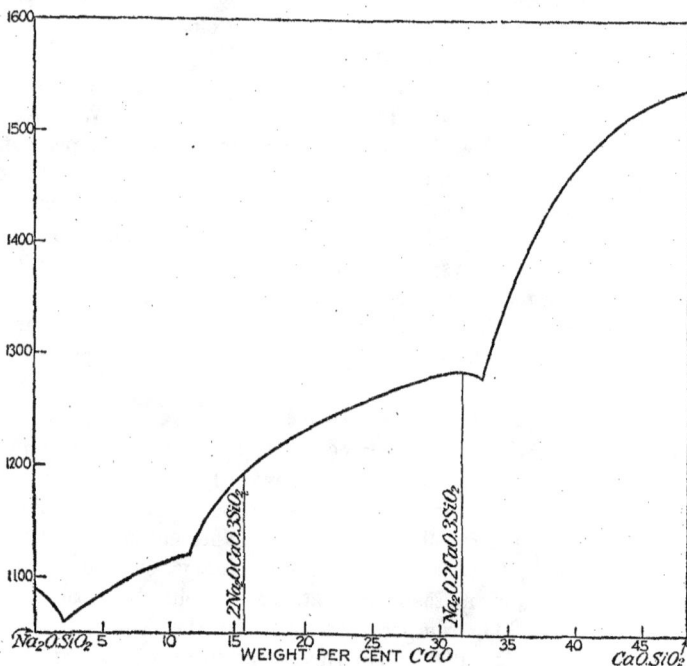

FIG. 4.

Melting-point diagram of the binary system Na₂O,SiO₂–CaO,SiO₂.
Composition is expressed as weight per cent. CaO.

a slight alteration in the proportions of those already present.
$Na_2O,2CaO,3SiO_2$ remains the solid phase until the binary eutectic
at 862°, 3 per cent. CaO is reached; the melting-point curve then
rises to $Na_2O,2SiO_2$ at 874°.

Crystallisation of the Ternary Mixtures.

The course of crystallisation of any mixture in the ternary diagram
can be deduced from Fig. 2, in which the boundary curves are drawn,
as well as the lines joining the composition of the various solid
phases,

THE SYSTEM SODIUM AND CALCIUM METASILICATES–SILICA.   249

Consider first a mixture in the area Na_2O,SiO_2–AKI. Any composition within the area deposits Na_2O,SiO_2 as primary phase; the liquid follows a crystallisation path lying on the straight line joining the liquid composition with that of Na_2O,SiO_2, until either the boundary AK or IK is reached. Here, either $2Na_2O,CaO,3SiO_2$

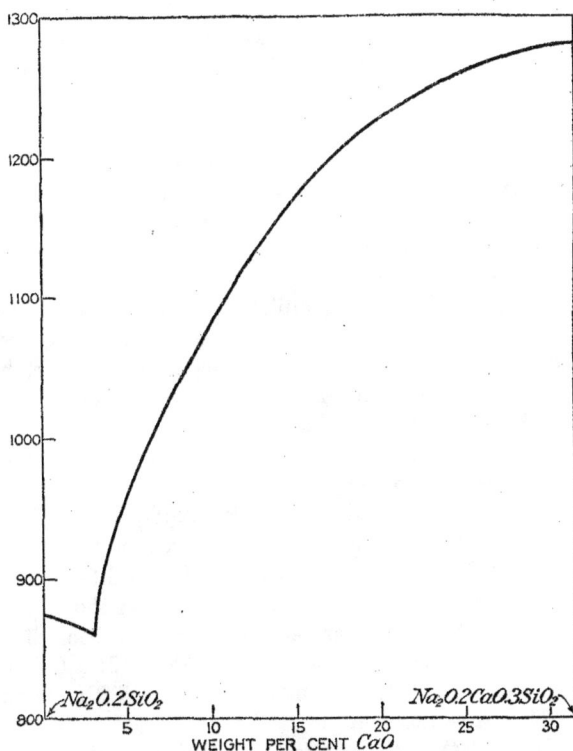

FIG. 5.

Melting-point diagram of the binary system $Na_2O,2SiO_2$–$Na_2O,2CaO,3SiO_2$. Composition is expressed in weight per cent. CaO.

or $Na_2O,2SiO_2$, respectively, appears, and the liquid follows the boundary curve down to the ternary eutectic at K, where it solidifies completely.

A composition within the area, ABLK, follows an equally simple course, depositing first $2Na_2O,CaO,3SiO_2$, then, in all cases except those compositions lying within the small area between L and the line joining K with $2Na_2O,CaO,3SiO_2$, depositing Na_2O,SiO_2 as secondary phase. Within the small area mentioned, $Na_2O,2SiO_2$ is the second phase to separate; in both cases, the mixture crystallises completely at the ternary eutectic K.

Crystallisation of mixtures lying within the region where $Na_2O,2CaO,3SiO_2$ is the primary phase is usually a less simple matter. Consider first a composition within the triangle Na_2O,SiO_2–$Na_2O,2SiO_2$–$2Na_2O,CaO,3SiO_2$, bounded by the curve, BL, and the straight line joining L with $Na_2O,2CaO,3SiO_2$. The primary phase is $Na_2O,2CaO,3SiO_2$; the liquid follows a crystallisation path passing through the composition of the primary phase and of the original mixture, until this path cuts the boundary curve BL. At this intersection $2Na_2O,CaO,3SiO_2$ crystallises; the liquid then follows the boundary curve, the reaction $Na_2O,2CaO,3SiO_2 +$ liquid $= 2Na_2O,CaO,3SiO_2$ takes place, and the proportion of the former compound decreases, while that of the latter increases, until at L, where $Na_2O,2SiO_2$ appears, the compound $Na_2O,2CaO,3SiO_2$ has been decomposed entirely. The liquid then follows the boundary curve, LK, finally solidifying completely at the ternary eutectic, K, to a mixture of Na_2O,SiO_2, $Na_2O,2SiO_2$, and $2Na_2O,CaO,3SiO_2$. If the composition lies within the same triangle, but to the right of the join L–$Na_2O,2CaO,3SiO_2$, the end products will be the same, but the secondary phase will be $Na_2O,2SiO_2$, on the boundary ML; at L, reaction takes place resulting in the formation of $2Na_2O,CaO,3SiO_2$ and the disappearance of $Na_2O,2CaO,3SiO_2$.

Any mixture in the triangle $Na_2O,2CaO,3SiO_2$–$2Na_2O,CaO,3SiO_2$–$Na_2O,2SiO_2$ will solidify completely to a mixture of the above phases and $Na_2O,2SiO_2$, but the crystallisation path will differ according to the side of the line L–$Na_2O,2CaO,3SiO_2$ on which the original composition lies. If the point representing the composition of the glass is to the left of the above line, the secondary phase is $2Na_2O,CaO,3SiO_2$, and the liquid follows the boundary, BL, until the point, L, is reached. Here the reaction

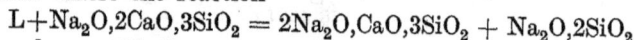

$$L+Na_2O,2CaO,3SiO_2 = 2Na_2O,CaO,3SiO_2 + Na_2O,2SiO_2$$

takes place until the mixture has solidified completely. If the original mixture lies to the right of the line L–$Na_2O,2CaO,3SiO_2$, the same three phases are finally formed, but $Na_2O,2SiO_2$ is the second crystalline phase to separate, and at L the liquid of composition L reacts with the primary phase to form $Na_2O,2SiO_2$ and $2Na_2O,CaO,3SiO_2$.

Glasses whose compositions lie within the field of $Na_2O,2CaO,3SiO_2$, and also in the triangle

$$Na_2O,2SiO_2-Na_2O,2CaO,3SiO_2-Na_2O,3CaO,6SiO_2$$

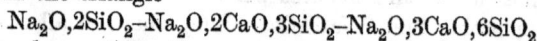

have a simple crystallisation sequence, namely, initial separation of $Na_2O,2CaO,3SiO_2$, crystallisation of $Na_2O,2SiO_2$ on the boundary MN, and final solidification at N as the result of a reaction

THE SYSTEM SODIUM AND CALCIUM METASILICATES–SILICA. 251

$Na_2O,2CaO,3SiO_2$ + liquid   N = $Na_2O,2SiO_2$ + $Na_2O,3CaO,6SiO_2$. This quadruple (P) point is thus a minimum temperature for $Na_2O,2CaO,3SiO_2$, and both boundary curves, NM and NR, go to higher temperatures. At M there is a maximum temperature on the curve, NML, and any mixture exactly on the join

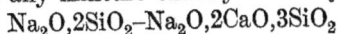

$$Na_2O,2SiO_2–Na_2O,2CaO,3SiO_2$$

will crystallise completely at M to a mixture of these two phases.

Mixtures in the above field, but lying between the boundaries, MN and NRS, and the tie-line $Na_2O,2SiO_2–Na_2O,3CaO,6SiO_2$ run a less simple course. If the composition is to the left of the tie-line N–$Na_2O,2CaO,3SiO_2$ the second phase to separate is again $Na_2O,2SiO_2$; if to the right of this line, $Na_2O,3CaO,6SiO_2$; but in either case the primary phase, $Na_2O,2CaO,3SiO_2$ must disappear at the reaction point N, as it is not possible for a mixture having the composition under discussion to solidify completely to a mixture of three phases whose composition triangle does not include the original mixture. All of these mixtures must solidify as conglomerates of quartz, $Na_2O,2SiO_2$, and $Na_2O,3CaO,6SiO_2$, and at the reaction point, N, all of the $Na_2O,2CaO,3SiO_2$ will be consumed. The liquid will then follow the boundary curve NO, until at O quartz appears, and the mixture will solidify completely. This is, of course, the sequence when equilibrium is attained; in the mixtures under discussion, and in all mixtures whose crystallisation gives rise to a liquid near the ternary eutectic, O, in composition, the attainment of equilibrium is an extremely slow process; one that we have not been able to carry out satisfactorily under the most favourable conditions in experiments extending over weeks of time.

Glasses in which pseudo-wollastonite is primary phase may follow several different crystallisation paths, depending on the composition, but in all cases the reactions are complicated. Mixtures within the triangle $Na_2O,2CaO,3SiO_2–Na_2O,3CaO,6SiO_2–$ CaO,SiO_2 will solidify ultimately as a conglomerate of these three phases, but the path followed is not a simple one. Initial crystallisation of pseudo-wollastonite is followed by a secondary appearance of $Na_2O,2CaO,3SiO_2$, along the boundary curve, CSR, entirely outside of the triangle under consideration. The liquid follows this curve until the quadruple (P) point, S, is reached, at which temperature the pseudo-wollastonite is inverted to wollastonite. At the still lower temperature, quadruple (P) point R, the reaction CaO,SiO_2 + liquid R = $Na_2O,2CaO,3SiO_2$ + $Na_2O,3CaO,6SiO_2$ takes place, going on until the liquid is entirely consumed. If the initial composition of the mixture is within the triangle $CaO,SiO_2–SiO_2–Na_2O,3CaO,6SiO_2$, but in the pseudo-wollastonite field, the result will be similar, except that the second phase to separate is tridymite, on the boundary, DTQP, and reaction takes place at point Q, CaO,SiO_2 + liquid Q = $Na_2O,3CaO,6SiO_2$ + SiO_2, the same type of reaction as that taking place at the quadruple (P) point, R, but with the compound $Na_2O,2CaO,3SiO_2$ replaced by SiO_2.

All mixtures lying within the triangle

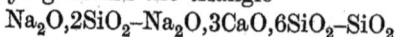

$$Na_2O,2SiO_2–Na_2O,3CaO,6SiO_2–SiO_2$$

will solidify completely as these crystalline compounds at the ternary eutectic point, O, if equilibrium is attained, the only effect of change in composition being to change the crystallisation path by which these phases are formed. If the mixture lies within the area to the left of the $CaO,SiO_2–Na_2O,3CaO,6SiO_2$ join, but within the wollastonite or pseudo-wollastonite fields, crystallisation will result either in intersection of the boundary curve NRSC, with consequent separation of $Na_2O,2CaO,3SiO_2$, which must later disappear at the reaction point N; or the boundary curve, RQ, is intersected by the crystallisation path; $Na_2O,3CaO,6SiO_2$ is formed directly, followed by the crystallisation of $Na_2O,2SiO_2$ along the boundary NO, followed by complete solidification at O. If the mixture lies to the right of the $CaO,SiO_2–Na_2O,3CaO,6SiO_2$ join, the initial separation of wollastonite will in all cases be followed by reaction at the line RQ, with formation of $Na_2O,3CaO,6SiO_2$ and disappearance of CaO,SiO_2. Depending on the composition, the next phase to separate will be $Na_2O,2SiO_2$, or tridymite, or quartz; in all cases the liquid runs down to the ternary eutectic at O, where crystallisation becomes complete.

Crystallisation paths in the various silica fields all run down to the boundary DPO or FO. In the former case either pseudo-wollastonite, wollastonite, or $Na_2O,3CaO,6SiO_2$ is the second phase to separate, and the liquid traces the path, DPO, solidifying at O to a mixture of $Na_2O,3CaO,6SiO_2$, quartz, and $Na_2O,2SiO_2$. The same final mixture of crystalline phases will result if the crystallisation path cuts the boundary FO; in this case $Na_2O,2SiO_2$ is the second phase to crystallise, and $Na_2O,3CaO,6SiO_2$ first appears at the ternary eutectic.

The Lower Ternary Eutectic, and its Relation to Glass Compositions.

It has been seen that any mixture whose composition lies within the triangle formed by the three phases

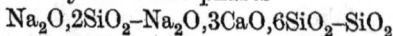

$$Na_2O,2SiO_2–Na_2O,3CaO,6SiO_2–SiO_2$$

should solidify at the ternary eutectic the percentage composition of which is : CaO, 5·2; SiO_2, 73·5, and the temperature 725°. We have seen also that mixtures near this composition are extra-

THE SYSTEM SODIUM AND CALCIUM METASILICATES–SILICA. 253

ordinarily difficult to crystallise. Inasmuch as these same compounds at higher temperatures crystallise with comparative ease, it is probable that this difference in behaviour is to be ascribed to the greatly increased viscosity of the glass at the lower temperature.

This eutectic is of great importance to the glass-maker. The area described above includes all glass compositions which are commercially practicable; mixtures higher in Na_2O are too rapidly attacked by water; mixtures higher in CaO, or, indeed, as high in CaO as the upper limit of this area, devitrify too readily to be of value; and the SiO_2-rich portion of the field is practically inaccessible because of the high temperature required, and is further impracticable because of the tendency toward devitrification. Indeed, almost the only compositions free from too great a tendency toward devitrification are those near the ternary eutectic, or those the crystallisation of which would bring them quickly to this eutectic. Thus narrowed down, the field is still seen to contain all commercial glass compositions.

All of the glasses within this restricted area are above their melting temperature throughout their working range; that is, they are above the temperature at which there can be any tendency toward crystallisation. Glasses containing 15 per cent. CaO would tend to show devitrification a little below 1050° if they were pure soda-lime glasses; glasses containing 10 per cent. CaO would have no tendency toward devitrification until cooled below 1000°; and glasses containing 5 per cent. CaO would not devitrify above about 750°. In other words, the low temperature of this eutectic and of the fields which slope down to it makes it possible for glasses to be maintained and manipulated at temperatures where their viscosity is high, without danger of devitrification. It is further to be noted that slight departure in any direction results in prompt crystallisation—crystallisation of silica, wollastonite, or $Na_2O,2CaO,3SiO_2$ —although the field of this latter compound lies at lower SiO_2-contents than are apt to be met with in soda-lime glasses.

From the point of view of devitrification tendency, a glass containing from 5 to 6 per cent. CaO, 71 to 73 per cent. SiO_2 would be best, but unfortunately this composition is impracticable because of the rapid attack of such a glass by water. Higher CaO contents are accordingly necessary, and this brings the glass into a region within which crystallisation should take place at blowpipe temperatures. Indeed, it is well known that pure soda-lime glasses are not suitable for blowpipe work because of their tendency toward devitrification, and the usual remedy is the addition of alumina. It is a fortunate circumstance that the raw materials always contain Fe_2O_3 and Al_2O_3, oftentimes MgO and K_2O, and that some or all these ingredients are introduced by pot corrosion. We have to do, not with a ternary system, but with one of far greater complexity, and we can only speculate on the effect of these added components. One effect, however, of addition of any of these substances in small amounts, or of addition of several of them, each in small amounts, can be predicted with safety. Such addition will of necessity lower the temperature of this ternary eutectic, and will also displace it to a lower SiO_2 content. The lowering of the eutectic temperature will also cause a lowering of all the adjoining surfaces, the intersection of which forms the eutectic, and hence a lowering of the devitrification temperature of the resulting glass. In the case of K_2O, we know from the work of G. W. Morey and C. N. Fenner * that the temperature will be lowered at least 200°, and the SiO_2 content lowered at least 2 per cent., both of which alterations are highly significant. In a mixed Na_2O–K_2O glass, as a result, the initial temperature of crystallisation will be greatly lowered; this is in harmony with common knowledge. Addition of Fe_2O_3 and Al_2O_3 will have a similar effect. Experiments are at present under way to ascertain the exact effect of such additions on the melting surface, additions which from practical experience are expected to be of great importance.

The composition and temperature of this ternary eutectic are thus seen to be of fundamental importance to the glass technologist. Indeed, it may be said that commercial glassware as a whole consists essentially of this ternary eutectic mixture, modified sufficiently to increase its chemical stability without sacrificing the major requirement of freedom from any tendency towards devitrification inside the working range.

Mixtures Containing Less SiO_2 than the Metasilicate Ratio.

While the system under discussion is limited to those compositions containing an amount of SiO_2 equal to or greater than the metasilicate ratio, much experimental work has been carried out on mixtures containing less SiO_2 than this ratio. Some of these are included in the tables and figures; others are outside the limits of the diagram, and are not included in the tables. The region approaching the orthosilicate ratio offers great experimental difficulty. The melting temperatures are higher than in most of the SiO_2-rich mixtures, and the rapid volatilisation of Na_2O from the mixtures approaching $2Na_2O,SiO_2$ in composition introduces a still further difficulty. Moreover, mixtures near sodium

* *J. Amer. Chem. Soc.*, 1917, **39**, 1176.

THE SYSTEM SODIUM AND CALCIUM METASILICATES–SILICA. 255

orthosilicate in composition are difficult to prepare free from CO_2; indeed, when such mixtures are heated in a CO_2-free atmosphere, Na_2O is lost more rapidly than CO_2, just as was previously found to be the case with potassium metasilicate mixtures.* The existence of sodium orthosilicate has, however, been confirmed, and a new compound, Na_2O,CaO,SiO_2, has been prepared. It forms isotropic crystals, of refractive index a little below 1·60, and the melting point is probably high. In addition to this compound, in one preparation, 2447 A (of Table II), at 1080°, an unknown phase was found in addition to $Na_2O,2CaO,3SiO_2$; this mixture is evidently not far from the boundary line between the metasilicate compound and this unknown phase. Furthermore, in the mixture 2505 A, containing 45·6 per cent. SiO_2, the primary phase was found to be $3CaO,2SiO_2$, the compound in the binary system $CaO–SiO_2$ adjoining the metasilicate, and the field of which must adjoin that of wollastonite in this system, as it does in the ternary system $CaO–Al_2O_3–SiO_2$.† The probable boundary in this system has been indicated by a broken line. This portion of the ternary system will be discussed in a subsequent publication.

Correlation of the Melting Point Relations with the Properties of the Glasses.

A considerable amount of information is available as to the properties of glasses included within the composition limits of this system, and it will be of interest to examine the variation of these properties with composition with reference to equilibrium relationships which have been brought out in this paper.

The optical properties of glasses are most important, and several investigations of the variation of optical properties with composition have been made. The optical properties of glasses covering the entire range of composition dealt with in this paper have been determined; an account of this work was given at the February, 1925, meeting of the American Optical Society by G. W. Morey and H. E. Merwin, and will be published soon. We may dismiss the subject here with the statement that no correlation is evident between the melting-point diagram and either the refractive index or the dispersion diagram.

Recent work by E. W. Washburn, G. R. Shelton, and E. E. Libman ‡ and by S. English § has given us an excellent knowledge of the viscosity of soda–lime glasses, and the variation of viscosity with

---

* G. W. Morey and C. N. Fenner, *loc. cit.*
† G. A. Rankin and F. E. Wright, *Amer. J. Sci.*, 1915, **39**, 1.
‡ No. 140, Bull., University of Illinois, April 14, 1924.
§ *J. Soc. Glass Tech.*, 1924, **8**, 205.

composition and with temperature. The results of Washburn, Shelton, and Libman are presented in a series of triangular diagrams, on which are plotted the "log isokoms," that is, the curves connecting those glass compositions which have the same value of log of viscosity η, at the given temperature. One of these, that for 1100°, is reproduced in Fig. 6; comparison with Figs. 1 and 2 shows that there is no possibility of correlation between the two diagrams.

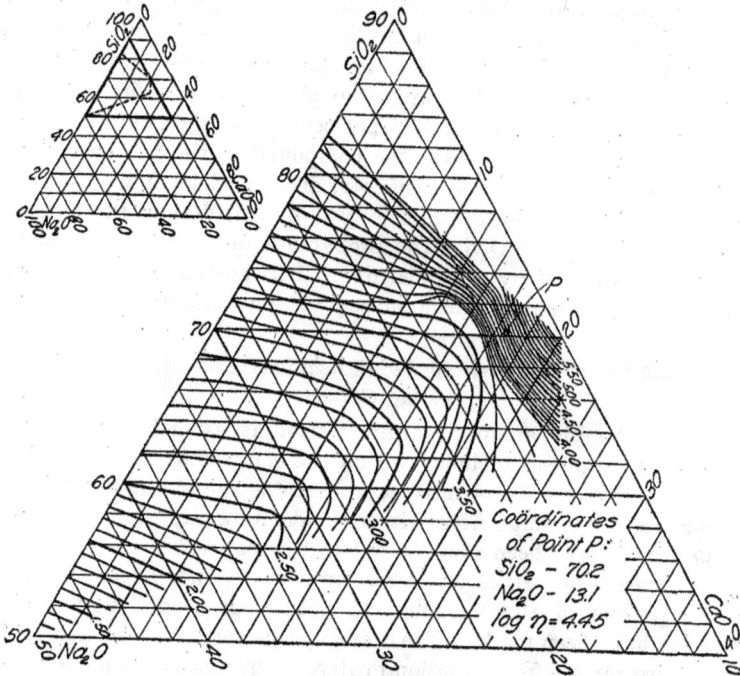

Coördinates
of Point P:
$SiO_2 - 70.2$
$Na_2O - 13.1$
$log \eta = 4.45$

FIG. 6.

Diagram showing log isokoms at 1100° C. After Washburn, Shelton and Libman.

The temperature of this isotherm is above that of the melting surface over the greater portion of its area, but a similar lack of concordance is to be found in all the viscosity diagrams. On Fig. 6 there is indicated the composition of an arbitrary point "P," the composition of which is given; this point is used as an illustration of the manner in which the log isokom diagrams may be used to obtain the viscosity-temperature curve of any mixture, and the curve so obtained is given, and we have reproduced it as Fig. 7. The course of crystallisation of this particular mixture will prove of interest. At 1110°, wollastonite will begin to crystallise; this point is indicated by arrow 1 in Fig. 7, and it is evident that there is no discontinuity at this point. As cooling continues, the amount of wollastonite increases, until the intersection of the liquid with the boundary curve, RQ, at 1045°; this point is indicated by arrow 2 in Fig. 7. The mixture at this point consists

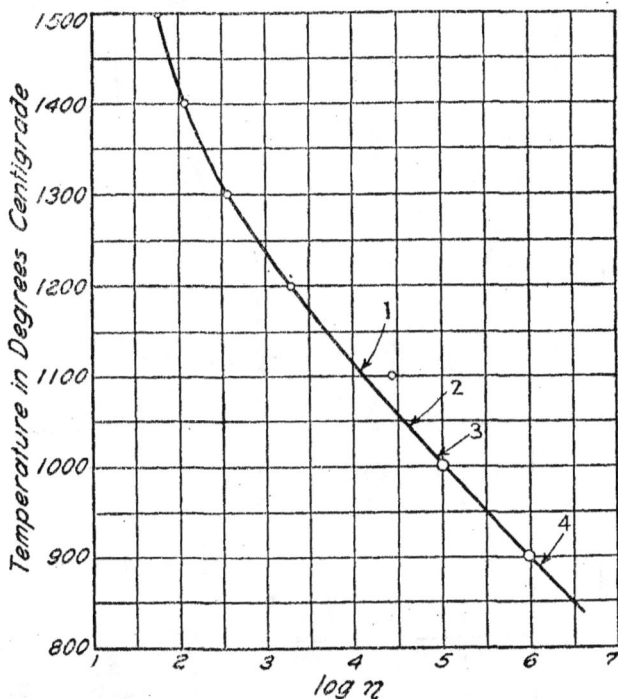

FIG. 7.

Diagram showing the change in viscosity with temperature of a glass. After Washburn, Shelton and Libman. The arrows refer to temperatures at which discontinuities would take place in the crystallisation of this glass if equilibrium were constantly attained.

of 15 per cent. wollastonite, 85 per cent. of glass of the approximate percentage composition : CaO, 12·5 ; SiO₂, 72. At this intersection reaction takes place between wollastonite and glass, with formation of $Na_2O,3CaO,6SiO_2$; this reaction continues along curve RQ until the point determined by the intersection of the line $Na_2O,3CaO,6SiO_2$–P is reached; at this point, only a few degrees cooler, the wollastonite disappears, and the mixture consists of 22 per cent. $Na_2O,3CaO,6SiO_2$, 78 per cent. glass, the com-

THE CONSTITUTION OF GLASS

position of which differs but little from the preceding glass. On further cooling the liquid crosses the $Na_2O,3CaO,6SiO_2$ field, and at 1010° meets the tridymite boundary; this point is indicated in

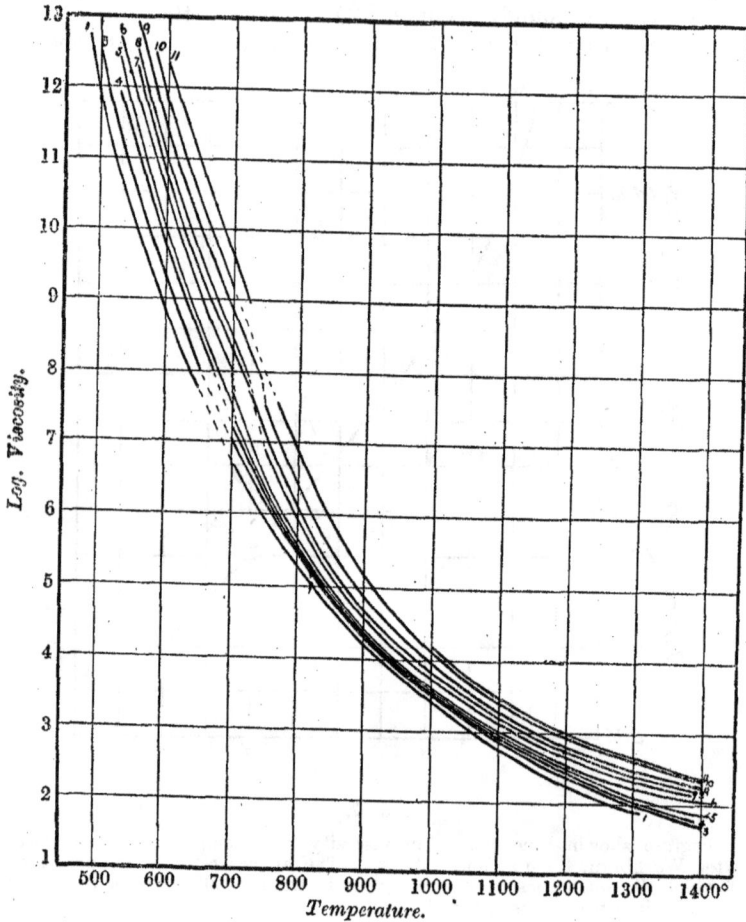

FIG. 8.

Diagram showing the change in viscosity with temperature of a series of glasses. After English. Table V gives the composition of each of these glasses, the temperature of primary crystallisation, and the primary phase.

Fig. 7 by the arrow 3. On further cooling, the liquid follows the tridymite–$Na_2O,3CaO,6SiO_2$ boundary, until at 870°, shown by arrow 4 in Fig. 7, the tridymite inverts to quartz. At this point the mixture consists of 9 per cent. tridymite, 48 per cent.

THE SYSTEM SODIUM AND CALCIUM METASILICATES–SILICA. 259

$Na_2O,3CaO,6SiO_2$, and 43 per cent. liquid, of composition that of the quadruple (P) point quartz–tridymite–$Na_2O,3CaO,6SiO_2$–liquid. The mixture becomes entirely crystalline at the ternary eutectic at 725°, off the scale of this diagram. In spite of the complicated course of crystallisation followed by this mixture, its viscosity curve is a smooth curve. There is no correlation between viscosity and crystallisation temperatures.

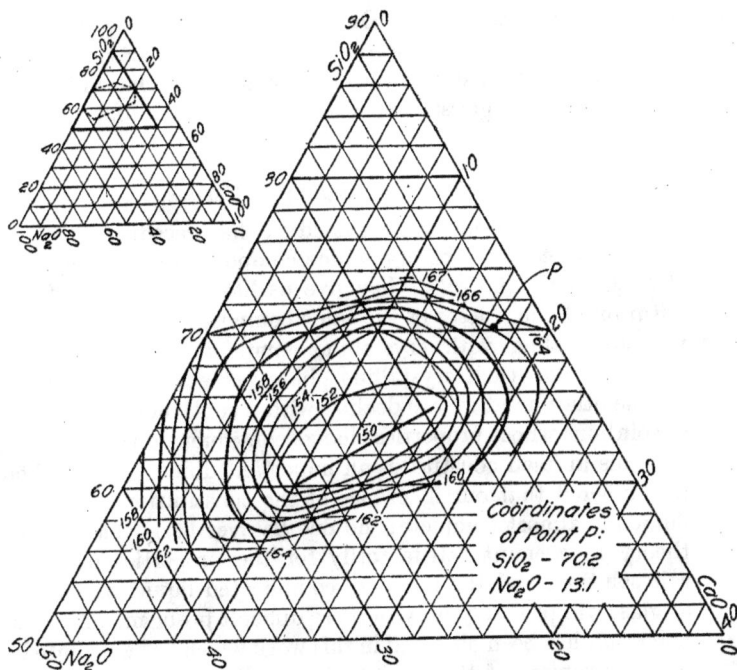

FIG. 9.

Diagram showing isoepitatic line at 1206° C. After Washburn, Shelton and Libman.

The results of English on the temperature–viscosity relations of several glasses are given by him in curves, reproduced in Fig. 8. The compositions of the various glasses, as determined by analysis, are given in the original, and include small amounts of Fe_2O_3, Al_2O_3, and MgO. We have assumed them to be pure soda-lime glasses, locating their composition by means of the Na_2O and SiO_2 contents, and in Table V are given the percentage of these constituents in the glasses, and their temperatures of primary crystallisation, as well as the primary phase. It will be observed that in none of the glasses is there a relation between the crystallisation temperature and the viscosity curve.

The work of Washburn, Shelton, and Libman also included a determination of the surface tension of glass mixtures of the same composition range as the viscosity determinations, and they also are presented by means of triangular diagrams, on which curves are drawn connecting glass compositions having the same values of surface tension at a constant temperature; these curves are called "isoepitatic lines." One of these, that giving the isoepitatic lines at 1206°, is reproduced in Fig. 9. Comparison with Fig. 1 or 2 will show that in this case also there is no relation between the properties of the glasses and the liquid–solid equilibrium relations.

The Constitution of Glasses.

While the experiments discussed in the preceding pages give us exact knowledge of the crystalline compounds formed by mixtures of Na_2O, CaO, and SiO_2, they give us little knowledge of the constitution of the glasses of the same composition. On the one hand we are dealing with crystalline compounds, possessing definite structure, and each crystalline phase possesses unique properties. On the other hand we are dealing with a continuous series of liquids or solutions whose properties change continuously with composition, and this in itself constitutes an important difference. It should be remembered that our ignorance of the constitution of the simplest liquids is almost complete; no one has yet been able to suggest a theory to account satisfactorily for the properties of a common salt solution, and the far more complicated liquid which is known as glass will probably prove a more difficult problem. Nevertheless, one result has been obtained in this work which has a definite bearing on the problem of the constitution of glass. There are four compounds which exist in or near the composition of all soda-lime glasses of commerce, the compounds $Na_2O,2SiO_2$, $Na_2O,2CaO,3SiO_2$, $Na_2O,3CaO,6SiO_2$, and SiO_2. We have found all of these, except silica, to possess in common, and to a marked degree, the property of being largely dissociated in the liquid phase. This property is characteristic of all the ternary soda–lime–silica compounds studied, as well as of the binary sodium silicates. The compound $2Na_2O,CaO,3SiO_2$ is decomposed before its melting point, and that portion of its field in which it is stable shows a marked tendency toward dissociation; in this connection note the flatness of the melting curve of this compound in Fig. 4. The other ternary metasilicate, $Na_2O,2CaO,3SiO_2$, is stable up to its melting point, but, as has been emphasised before, it is characterised by a very

THE SYSTEM SODIUM AND CALCIUM METASILICATES—SILICA. 261

flat melting surface, indicating great dissociation. The other ternary compound, $Na_2O,3CaO,6SiO_2$, is decomposed long before its melting point; the field of its stability in contact with liquid, shown in Fig. 2, lies far from the composition of the compound. All the compounds which are to be considered as possible molecular species in soda-lime glasses are highly dissociated, even in the pure state, and a liquid of the composition of any one of them will probably consist not of molecular groupings of one species, or of two, but of several; and it is probable that the same molecular groupings exist over a large range in composition. What these groupings are it is impossible to say; but it is probable that not only the molecular groupings characteristic of the compounds met with in this study are to be found, but also groupings corresponding to the ortho-silicate molecules, probably the most stable molecular species likely to be formed. However that may be, there can be no doubt that the molecular constitution of a glass is a complex one, involving equilibrium between several molecular species. The relative proportions of these species will, of course, be determined by the usual thermodynamic considerations; these proportions will change with temperature in accord with the usual van't Hoff principle of mobile equilibrium, and any such change will be continuous, the system remaining homogeneous. Such being the case, it is difficult to conceive of any change in the system, so long as it remains homogeneous, which would give rise to a discontinuous heat effect, such as has been observed by some investigators.

Glass has by some been considered to consist of different "phases"; in this connection it may be well to quote the original definition of a phase, in the words of Gibbs, whose concept it is. In his famous "Equilibrium of Heterogeneous Substances" * he defines :

> "We may call such bodies as differ in composition or state different *phases* of the matter considered, regarding all bodies which differ only in quantity and form as different examples of the same phase. Phases which can exist together, the dividing surfaces being plane, in an equilibrium which does not depend upon passive resistances to change, we shall call *coexistent.*"

When this definition is considered, it will be seen that any attempt to explain a stress pattern in a glass by assuming the glass to consist of more than one phase, as has been done by Filon and Harris,† results merely in a confusion in terms.

---

* "The Scientific Papers of J. Willard Gibbs," Vol. I., p. 96.
† L. N. G. Filon and F. C. Harris, *Proc. Roy. Soc.,* London, 1923, A, **103**, 561.

The problem of the constitution of glass is one whose solution is probably still far distant, and it will long offer a fascinating field of research. It is believed, however, that the relationships which have been outlined in this paper are fundamental to any such solution, and that the marked tendency towards dissociation of the compounds is a factor of which account must be taken in any theory of glass constitution.

Summary.

The work discussed in this paper consists of an experimental study of the ternary system Na_2O,SiO_2–CaO,SiO_2–SiO_2, by the method of quenching. The following new compounds have been found and their properties determined : the compound $2Na_2O,CaO,3SiO_2$, which melts incongruently, forming a liquid richer in Na_2SiO_3, and $Na_2O,2CaO,3SiO_2$; the compound $Na_2O,2CaO,3SiO_2$, which has a congruent melting point at 1284° ; and the compound $Na_2O,3CaO,6SiO_2$, which melts incongruently at 1045°, forming a mixture of wollastonite and a glass containing approximately 15 per cent. CaO, 67 per cent. SiO_2. These compounds are all characterised by a large amount of dissociation in the liquid phase. The melting-point surface of the various unary, binary, and ternary compounds existing as solid phases have been determined, and the results are given in tables and curves. The relation between the surfaces giving the solid–liquid equilibrium as a function of temperature and the properties of the liquids as determined by other investigators, is discussed, as is the relation of facts brought out in this work to speculations on the constitution of glass.

GEOPHYSICAL LABORATORY,
   CARNEGIE INSTITUTION OF WASHINGTON.   *May,* 1925.

Appendix.

TABLE III.

Crystallographic and Optical Properties of the Solid Phases.

Compound.	Crystal system.	Habit.	2V.	Elong-ation.	Opt. Sign.	Refractive Indices γ	β.	α.	Remarks.
SiO_2 : Quartz	Hexagonal	Bipyramids	0°	—	pos.	1·553	1·544	1·544	Prism absent β-form
SiO_2 : Tridymite	Pseudo-hexagonal	Plates	—	—	—	1·473	—	1·469	
SiO_2 : Cristobalite	Pseudo-cubic	Octahedra	—	—	—	1·487	—	1·484	
β-CaO,SiO_2 : wollastonite	Monoclinic	Needles	40	β	neg.	1·631	1·629	1·616	
α-CaO,SiO_2 : Pseudo-wollastonite	Pseudo-hexagonal	Hexagonal plates	0	—	pos.	1·654	—	1·610	No twinning observed
$Na_2O,2SiO_2$	Ortho-rhombic	Plates	50	—	neg.	1·518	1·514	1·504	
Na_2O,SiO_2	Ortho-rhombic	Needles	80	γ	neg.	1·528	1·520	1·513	
$2Na_2O,CaO,3SiO_2$	Isometric	Octahedra	—	—	—	1·571	—	1·571	
$Na_2O,2CaO,3SiO_2$	?	Equant	—	—	—	1·598	—	1·595	Twinning characteristic
$Na_2O,3CaO,6SiO_2$	Ortho-rhombic	Prisms	75	γ	pos.	1·579	1·570	1·564	

TABLE IV.

Invariant Points.

Compounds.

Phases.	Type.	Temperature.
SiO_2 : β-quartz	Inversion	870°
SiO_2 : tridymite	Inversion	1470
SiO_2 : cristobalite	Melting	1710
β-CaO,SiO_2 : wollastonite	Inversion	1180
a-CaO,SiO_2 : pseudo-wollastonite	Melting	1540
$Na_2O,2SiO_2$	Melting	874
Na_2O,SiO_2	Melting	1088
$2Na_2O,CaO,3SiO_2$	Decomposition into $Na_2O,2CaO,3SiO_2$ and liquid, 11·5% CaO.	1141
$Na_2O,2CaO,3SiO_2$	Melting	1284
$Na_2O,3CaO,6SiO_2$	Decomposition into β-CaO,SiO_2 and liquid, 15% CaO, 67% SiO_2	1047

Binary Invariant Points.

System Na_2O,SiO_2–CaO,SiO_2.

Phases.	Type.	Composition.	Temperature.
$Na_2O,SiO_2 + 2Na_2O,CaO,3SiO_2$	Eutectic	3·0% CaO	1060°
$2Na_2O,CaO,3SiO_2 + Na_2O,2CaO,3SiO_2$	Decomposition	11·5% CaO	1141
$Na_2O,2CaO,3SiO_2$	Melting point	31·61% CaO	1284
$Na_2O,2CaO,3SiO_2 + CaO,SiO_2$	Eutectic	33·0% CaO	1280

System $Na_2O,2SiO_2$–$Na_2O,2CaO,3SiO_2$.

$Na_2O,2SiO_2 + Na_2O,2CaO,3SiO_2$	Eutectic	3·0% CaO	862

Ternary Invariant Points.

Phases.	Type.	Composition.	Temperature.
$Na_2O,SiO_2 + Na_2O,2SiO_2$ + $2Na_2O,CaO,3SiO_2$	Eutectic	1·8% CaO : 60·7% SiO_2	821
$Na_2O,2SiO_2 + 2Na_2O,CaO,3SiO_2$ + $Na_2O,2CaO,3SiO_2$	Reaction point	2·0% CaO : 61·4% SiO_2	827
$Na_2O,2SiO_2 + Na_2O,2CaO,3SiO_2$ + $Na_2O,3CaO,6SiO_2$	Reaction point	5·2% CaO : 70·7% SiO_2	740
$Na_2O,2SiO_2 + Na_2O,3CaO,6SiO_2$ + quartz	Eutectic	5·2% CaO : 73·5% SiO_2	725
$Na_2O,2CaO,3SiO_2 + Na_2O,3CaO,6SiO_2$ + wollastonite	Reaction point	14·5% CaO : 66·5% SiO_2	1030

TABLE V.

Temperatures of Primary Crystallisation of the Glasses of Fig. 8.

Glass No.	Composition		Temperature.	Primary phase.
	Na_2O.	SiO_2.		
1	25·3	74·1	825°	Quartz
3	23·0	74·1	825	Quartz
4	21·5	70·1	825	Quartz
5	20·8	73·8	800	Quartz
6	19·4	73·2	880	$Na_2O,3CaO,6SiO_2$
7	17·2	74·4	930	Tridymite
8	·16·0	75·0	1040	Tridymite
9	14·9	75·0	1080	Tridymite
10	14·2	74·6	1095	Tridymite
11	13·0	74·9	1150	Tridymite

XVIII.—X-Ray Diffraction Measurements on Some Soda–Lime–Silica Glasses.

(A PRELIMINARY NOTE.)

By RALPH W. G. WYCKOFF and GEORGE W. MOREY.

*(A Contribution to the Symposium on " The Constitution of Glass,"
held at the London Meeting, May 25th and 26th, 1925.)*

EARLY in the study of X-ray diffraction effects it became known that glasses give rise to very broad powder bands. No serious attempt has yet been made to establish the origin and cause of such X-ray patterns from glasses, but considerable discussion has arisen over the nature of the supposedly similar patterns from liquids. Existing " explanations " of these fall under three heads. In the first place it has been assumed * that they arise from constant distances between atoms in chemical molecules, or secondly † from the average distances apart of the molecules themselves. Lastly,‡ it has been thought that these bands are due to more or less transitory groupings of atoms or molecules into crystal-like aggregates. The first of these possibilities is eliminated as the predominating one by the fact that liquid argon gives the same type of pattern as do liquids composed of poly-atomic molecules. At the present time there do not exist the experimental data necessary for a satisfactory and quantitative explanation of liquid patterns in terms of a diffraction between molecular groupings.

The following experiments were undertaken as a preliminary survey of an extensive system of glasses to see whether much can now be learned from such a study concerning the fundamental nature of glasses themselves and the mechanism of a quantitative explanation of their X-ray patterns. Our observed results would not have been predicted from previous experience. In some instances the broad bands thought to be characteristic of glasses have been found. In others, however, narrow bands, or lines, have been obtained which are as sharp as the lines produced by crystals of colloidal dimensions. Sometimes only one such broad line is observed, in other cases the pattern consists of several such lines. In still other instances the photograph from a single glass is a com-

---

* P. Debye and P. Scherrer, *Nachr. Gesell. Wiss.*, Göttingen, 1916, p. 16.

† This idea was amplified by P. Debye in a recent paper read before the National Academy of Sciences in Washington.

‡ W. H. Keesom and J. De Smedt, *Proc. Roy. Acad. Sci.*, Amsterdam, 1922, **25**, 118; 1923, **26**, 112. C. V. Raman, *Nature*, 1923, **111**, 185; *Proc. Indian Assoc. Cultivation of Science*, 1923, **8**, 127.

posite of lines and broad bands. All of the glasses studied, no matter what the character of their X-ray diffraction, showed no signs of devitrification under the microscope and layers of considerable thicknesses exhibited no milkiness. Both the lines and the bands seem to differ considerably amongst themselves in width. The positions of the lines are sometimes different from glass to glass, though frequently several glasses agree in having lines in the same positions. The available data are not yet sufficient to show definitely whether or not the diffraction lines and bands have a

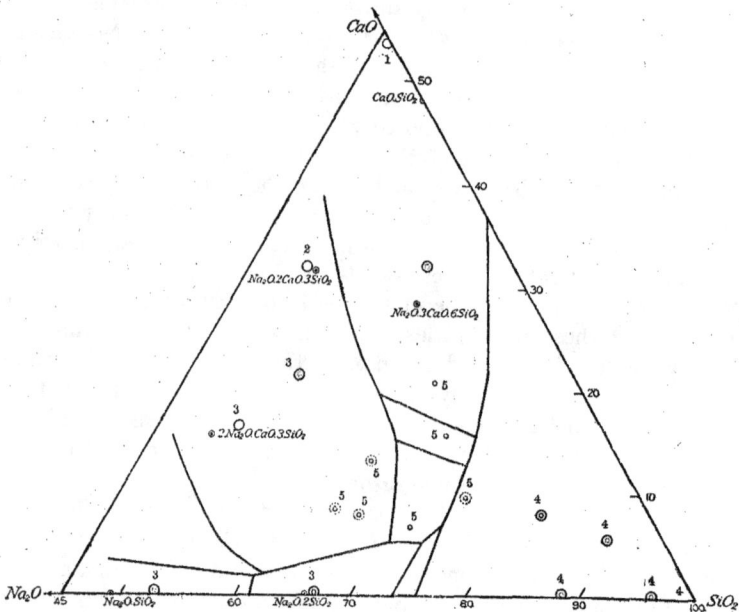

FIG. 1.

similar origin and thus can grade imperceptibly into one another. Neither do they prove, though they strongly indicate it, that entirely clear and apparently homogeneous glasses may contain large amounts of crystals of colloidal dimensions.

The results of the present experiments are shown with the aid of Fig. 1. In it the results of the X-ray examination are plotted as large and small circles upon a partial reproduction of Fig. 2 of the paper by G. W. Morey and N. L. Bowen on "The Ternary System : Sodium Metasilicate–Calcium Metasilicate–Silica." Large circles signify the presence of line, small circles of broad band patterns. Where both are produced by a single glass, the predominant pattern is shown by full circles, the one present only faintly by dotted circles. Measurements of the positions of the lines and bands indicate that within the limits of experimental error, these patterns fall into five groups. The numbers beside each plotted point of Fig. 1 show which of these patterns, numbered arbitrarily from 1 to 5, is predominant in each of the glasses studied. Thus, for instance, the chief feature of each of the glasses in the silica corner of the figure is a broad line (pattern 5), the position of which is essentially the same in each glass and the same as that of silica glass.

A close relation exists between the diffraction patterns of these glasses and certain aspects of the melting-point diagram. Those glasses which crystallise with greatest difficulty are the ones which give the broad bands that were previously thought to be the distinctive property of all glasses. Line reflections are observed with more readily crystallisable glasses when the products of ready crystallisation—either primary or final—form a considerable part by weight of the glass.

A number of explanations of these phenomena have been considered; but we believe it more profitable to defer extended discussion of the present results until we have examined carefully many more glasses differing from one another and from those already studied both in chemical composition and in mode of preparation. The X-ray patterns and the structures of some of the compounds of this ternary system have already been investigated; measurement of the others, together with a comparative study of the potash–lime–silica system of glasses, is now in progress.

GEOPHYSICAL LABORATORY,
    CARNEGIE INSTITUTION OF WASHINGTON, U.S.A.   *May*, 1925.

XIX.—*Some Remarks on the Constitution of Glass.*

By F. ECKERT, D.Phil.

(Translated from the German by Arnold Cousen, M.Sc.)

THE problem of the constitution of glass has often been obscured because the question of the state of aggregation has been confused with that of constitution. From its state of aggregation, glass is undoubtedly a supercooled liquid for it exhibits a series of thermodynamic characteristics of such a condition. Since, however, at the same time, glasses show physical and chemical properties such as are usually exhibited quantitatively by solid bodies only, the vitreous state possesses a certain right to be regarded as a special condition. A similar case occurs in defining the colloidal state since colloids differ quantitatively only in properties from mixtures or solutions, and not, in a strict sense, qualitatively.

The vitreous state arises always when the conditions for crystallisation in a cooled solution, namely viscosity, power of spontaneous crystallisation, and velocity of crystallisation are within certain fixed limits.* The same causes which bring about the vitreous state exercise a retarding effect on all possible reactions.

This view permits us, in its wider application, to draw a picture of the constitution of glass, which allows an explanation of several facts, some of them recent observation, others hitherto unexplained phenomena. In general, we obtain glasses when the molecular structure of the molten liquid is such as to produce a relatively high viscosity at temperatures at which there is still a small possibility of crystallisation, and only few molecules gather to form crystal nuclei. Under the same conditions as those which produce the high viscosity, the possibility of reaction of the molecules amongst themselves is retarded. Indeed, with melts of glasses of complex composition a series of reactions proceeds with change of temperature (*i.e.*, with increase of temperature). Compounds are, for example, in part decomposed into simpler compounds, in part dissociated, as the temperature rises. But above all, oxides into which the compounds split up, may give off oxygen at higher temperatures, producing lower oxides, as is the case with the multivalent heavy metal oxides. Now it must not be imagined that reactions such as, say, the last named, set in spontaneously at a certain temperature and proceed, according to the law of mass action, to the production of the corresponding equilibrium. Rather is it that at relatively low temperatures a certain proportion of the molecules concerned proceed to react, although the greater part do not do so— in a manner somewhat analogous to the law of the kinetic gas theory. Since, further, the reaction can proceed to the production of the equilibrium which corresponds to the temperature, it is dependent, therefore, on whether or not the molten liquid remains for a sufficiently long time at this temperature. In further developing this explanation it is obvious that the duration of the complete carrying out of the reaction and the production of chemical equilibrium is dependent upon the various periods of time during which the molten liquid has previously been kept at various temperatures. It can be said, in short, that in this viscous, slowly-reacting liquid the state of combination or constitution of the molecule is *dependent upon the total past heat treatment* (Wärmevergangenheit), that is upon the various temperatures through which

* G. Tammann " Kristallisieren und Schmelzen," 1903.

SOME REMARKS ON THE CONSTITUTION OF GLASS.   269

it has passed and the time during which it was maintained at any temperature.

The condition for the formation of the vitreous state is a fixed minimum velocity of cooling. We must hurry through the dangerous temperature range in the neighbourhood of the greatest velocity of crystallisation and the maximum formation of spontaneous nuclei, with a minimum velocity of cooling, in order to obtain as a final product a structure which contains so few molecules in the crystalline state that it still may be spoken of as glass (amorphous). This rapidity of cooling—quenching—can so far be increased as the heat conductivity of the substance and its shape allow. The more rapidly we can bring about this cooling, the more is the initial state of the liquid preserved. We see also that the constitution of the final product is dependent not only on the initial state but also on the length of time during which the material has had to adjust itself to the temperature range. We have thus a distinct influence of the " past heat treatment."

We see now how this conception of the dependence of glass upon its heat treatment simplifies the explanation of its properties. A number of cases is known in which the composition of the glass as measured by the proportions of oxides which it contains must be regarded (within a large range of accuracy) as a constant. But the *properties* of the glasses, under differing conditions, vary by a considerable amount. Thus, for example, a series of optical glasses is known in which the optical constants, measured with very great accuracy, have been shown to vary, although the glasses came from the same melt, according as to whether the glass was rapidly or slowly cooled, further reheated, etc. The influence of strain in this case had been considered and eliminated. These (as yet unpublished) investigations show that a far-reaching reversible action is exercised by this influence of the past heat treatment on the magnitude of the refractive index, depending in the first degree upon the final heat treatment, that is upon the method of final cooling from the fluid into the quasi-solid state. I must refrain from quoting here from the large number of indications in literature which furnish evidence of this tendency. I have compiled a portion which came to my notice up to a year ago, in another place.*

The density of glass is also similarly influenced by the past heat treatment. In observations in this direction the influence of strain caused by cooling must be very carefully eliminated.

But above all I hope to show the influence of the past heat treatment on the *mechanical-elastic* properties. Factory experience over

* F. Eckert, *Jahrb. Rad. und Elektro.*, 1923, **20**, 116.

many years at the most varied places, supplemented by a series of careful chemical analyses, has proved that glass, when repeatedly melted, or when kept fluid at one temperature for a long period, frequently shows changes in its elastic mechanical properties in the solid state, as it does of its viscosity in the plastic condition. The familiar observation of the glass-maker that remelted cullet is more viscous in working than corresponding glass made from batch materials, and that with such cullet glasses breakages more frequently occur, is pertinent in this direction. Analyses have proved that the relative change of glass composition due to the remelting of the cullet is small as compared with the relative alteration of viscosity and of elastic properties. There is certainly an addition to the proportions of some of the constituents present on repeated remelting, particularly in the case of silica and alumina, as well as a volatilisation of some of the more fluid constituents, but these changes can be considered and allowed for in preparing from batch glass of identical composition to glass which has been often remelted. Even when this is done the glasses will have pronouncedly varying properties. It can be stated, on the other hand, that the changes of physical properties effected by *small* changes of composition are of strikingly smaller importance than the above.

Even in itself, this fact, which cannot be now denied, proves that the original view that glass is purely a mixture of oxides is now untenable, not only since this view is a restricted one, since the properties of oxide mixtures are purely additive, or practically so, but because such a point of view gives no clear indication as to the effect of the past heat treatment.

As we know, from the mode of representation of modern physics of the atom, the various physical properties have their origin partly in the inner, partly in the outer sphere of the atom; that is, they are divided into nuclear and peripheral properties. Nuclear properties play the chief rôle in the region of the short electro-magnetic waves—the Röntgen rays. They are not influenced by the union of atoms, and in that case we see that the absorption phenomena, in the case of Röntgen rays, are, for glasses, purely additive, being derived from those of the elements. On the other hand, properties caused by the peripheral electron ring of the atom, as, for instance, molecular volume, elasticity and cohesion, optical absorption, melting point, etc., show more or less strongly the influence of past heat treatment, especially those properties which we are accustomed to call constitutive. In the case of glass the properties of cohesion and viscosity appear to be constitutive to a special degree. If, indeed, in pure mixtures, deviations from the additive relationships are obtained, such deviations, particularly in general those in the case of the mechanical-elastic properties and also the molecular bonding, point to large molecular compounds. One can conclude not only from the great constitutivity of the properties, but also from influence of the past heat treatment, that complex molecular compounds exist in glass (in the liquid as well as in the solid state).

I believe that the existence of complexes in glass is no longer seriously denied; this is the hypothesis which we have assumed for the above explanation. Since the object of this paper is solely to deepen and strengthen this point of view, but not to give experimental evidence, I shall conclude with some brief references.

In several recent papers, particular those of A. Q. Tool, J. Valasek, C. G. Eichlin, H. Jackson, A. A. Lebedeff, P. Lafon,* etc., molecular rearrangements in glass at relatively low temperatures are demonstrated. The work of Weber and Wiebe on the mechanical and thermal afterworking of glass falls under the same heading, these being obviously traceable to the changes of constitution of the K_2O and Na_2O compounds. As indicated in the above examples, those properties of every complex which may be termed " heat resistance " are particularly affected by the heat treatment.

The influence of the heat treatment on the chemical condition of the glass was conjectured at a relatively early date by G. Keppeler, † and in a certain sense by R. L. Frink. As yet, the chemical properties have not been examined in this respect. The practical interest with regard to glass, its application as chemical ware, etc., ensures that interest is concentrated, in a chemical sense, on the behaviour of the surface of the material, which, in consequence of the absorption processes there occurring, is of considerable complexity. How far the chemical properties of the interior of the material are constitutive and dependent upon the heat treatment has not yet been fixed.

Some recent results,‡ however, in the case of the " hardening " of the interior of the glass by very rapid cooling, indicate at least an influence of the past heat treatment on the elimination of alkali.

According to our conception, then, glass is constitutively a concentrated solution of simple and complex silicates, aluminates, borates, etc., in silica principally. It must be noted, in this respect, that even pure molten silica as quartz glass cannot be regarded as a chemically homogeneous fluid, independent of any heat treatment. But glass is no "ideal" homogeneous solution, since indications are wanting of the complete molecular distribution of the individual

* F. Eckert, loc. cit.
† R. Dralle, Glasfabrikation, 1911, Vol. I., 225.
‡ I thank Prof. Keppeler for a brief note on the result of this research.

constituents, the solute as well as the solvent. One finds, as distinct from the condition of a pure solution, all variations up to a more or less ordered aggregation of simple molecules. This fact causes many to regard glass as a super-cooled liquid existing in a colloidal condition, but in a strict sense this may prove correct only in special cases.

In which form the compounds are retained in the molten liquid when it is more or less rapidly cooled and how the constitutional properties are altered thereby depend upon the total and above all on the immediately preceding heat treatment. It remains to be added that the influence of the past heat treatment can extend as far back as the formation of the batch materials from which the molten glass is prepared.

GLASWERKE RUHR,
    ESSEN, GERMANY.   *May 13th, 1925.*

XX.—*The Structure of Quartz.*

By SIR WILLIAM H. BRAGG, K.B.E., D.Sc., F.R.S.

(Read at the London Meeting, May 26th, 1925.)

THE problem of the structure of quartz was one of the first to be attacked by the new methods of X-ray crystal analysis : and a paper on the subject was communicated to the Royal Society.* The results then published carried the investigation to the end of the first stage of the X-ray analysis; that is to say, to the point at which it was possible to state the general arrangement of the SiO$_2$ groups of which the crystal is composed.

It is convenient to remember that observations of the external form of a crystal showing what degrees of symmetry it possesses determine its " class." Quartz has a trigonal axis, and three digonal axes at right angles to the trigonal axis. It has no other symmetry. It belongs therefore to Class 18.†

Within this class there are seven different modes of arranging the constituent atoms and molecules, which modes can with certain exceptions be distinguished from one another by the X-ray analysis carried to the end of the first stage. In the work already referred to it was shown that the unit cell of the crystal contained the substance of three molecules of silicon dioxide, so arranged in a spiral fashion that each could be derived from its neighbour by a

* *Proc. Roy. Soc.*, 1914, **A, 89,** 576.

† See A. E. H. Tutton, " Crystallography," 2nd ed., Vol. I., p. 350. London, Macmillan, 1922.

THE STRUCTURE OF QUARTZ. 273

rotation of 120° round the crystal axis accompanied by a translation of 1·79 Ångström Units along the axis. The disposition is illustrated in the accompanying figures which are taken from " X-Rays and Crystal Structure," pp. 261, 262.* In Fig. 1, the black spots represent silicon atoms lying in a plane perpendicular to the axis, the white spots are the projections, upon the plane, of silicon atoms lying 1·79 Å.U. below the plane, and the shaded spots represent the projections of the silicon atoms at the same distance above. The point marked T is the intersection of the trigonal axis with the plane : and the digonal axes are marked aa, bb, cc.

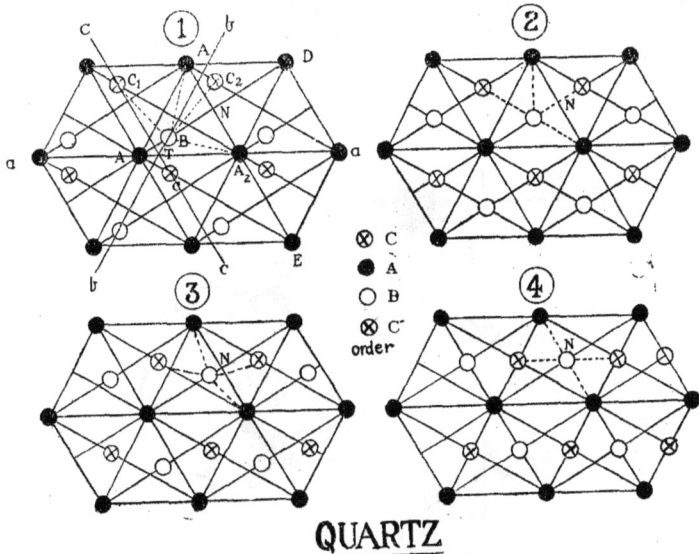

QUARTZ

FIG. 1.

This arrangement is known in mathematical crystallography as either D_3^4, or D_3^6, the difference between these two being no more than that one is right-handed to the other's left. D_3^4 and D_3^6 are two of the seven modes of arrangement in class 18. The X-rays cannot distinguish between the two : being unable to distinguish between any arrangement and its reflection in a plane mirror; and this is the difference between a right-handed and a left-handed screw.

The length of the side of the equilateral triangle ABC as projected on the plane is unknown so far, and four separate figures show possible variations in its value.

* W. H. Bragg and W. L. Bragg, " X-Rays and Crystal Structure," 4th ed. London, Bell, 1924.

The other figure (Fig. 2) shows the arrangement of the atoms in the spiral round the trigonal axis; in one part of the figure the axis lies in the plane of the paper, in the other it is at right angles to it. The black balls in this figure represent silicon atoms and the white balls oxygen. The positions of the oxygens with respect to the silicons require three parameters for their definition. They may be taken to be :

(1) the distance between a silicon centre and either of the neighbouring oxygen centres;

(2) the angle between the line joining the silicon to either of these oxygen centres and the digonal axis;

(3) the angle between the plane of the paper and the plane

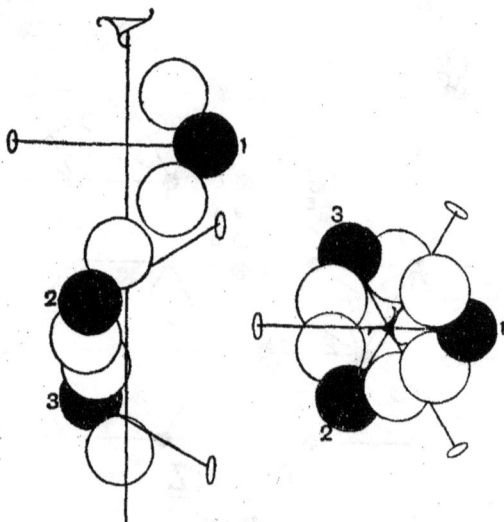

FIG. 2.

containing the three atoms in any SiO_2 group, such as that marked 1, 2, or 3 in Fig. 2.

The distance between the silicon atom and the trigonal axis, and the three quantities required in the description of the positions of the oxygens relative to the silicons remain, all four, unknown at the end of the first stage of the X-ray analysis. All that has been done is to determine the number of molecules in the unit cell and the dimensions of the cell : and to define the general arrangement, or space group as it is technically called. The height of the cell is 5·37 ($= 3 \times 1·79$) and the base is formed by any rhombus of minimum area which has four similar spots at its corners in any of the figures of Fig. 1. The length of the side of the rhombus is 4·89 Å.U.

To get past this point, and it is here that the major difficulties begin, it is necessary to bring fresh principles into play. We can examine, not only the spacings in the crystal as determined by the X-rays, but also the relative intensities of the various orders of reflection by each set of crystal planes. In the analogous case, when we find the angle at which light is diffracted by an ordinary diffraction grating we determine at once the spacing of the grating. This corresponds broadly to the first stage of the X-ray analysis of a crystal. The second stage of the same analysis in which the object is the determination of the details of atomic positions corresponds to the attempt to determine the form of the groove in the light grating, by examining the relative intensities of the different orders of reflection : and this is obviously a far more difficult task. In fact, we are not sufficiently advanced in our knowledge of the action of X-rays to complete the second stage of the analysis in this way, although it is theoretically possible to do so. We may, however, introduce physical and chemical evidence of various kinds in order to help us to the solution : and in the case of quartz several attempts have been made to complete the solution by such means. In no case, however, has the result been quite conclusive or satisfactory, in the sense that the attempt to explain one set of facts has not in general led to the explanation of others.

The present paper deals with a new method of attack which seems more promising, since the solution which it yields explains, qualitatively at least, some of the main properties of quartz in addition to those which have been considered in the argument. For example, the complicated twinnings of quartz can all be accounted for. It is based on two propositions, the first, that quartz above 575°, in what is called the high temperature or β-form, is so much more symmetrical than in the low temperature or α-form that the four unknown parameters reduce to one, and so a solution is attainable : the second, that the change from α- to β-quartz involves only small movements of the atoms which do not affect the design in most of its main features, so that the determination of the structure of β-quartz paves the way to the determination of the α-form also.

For the sake of brevity, the argument will not be given in full detail here ; the main points can be explained in a general statement.

As regards the first of the two propositions, there is supporting evidence from many quarters. The structure of quartz itself is trigonal ; for example, the six pyramid faces alternate in character, there being three of each of two kinds. The piezoelectricity and the pyroelectricity of the crystal are bound up with the lack of symmetry which is expressed by the existence of two kinds of

T 2

pyramid face. Above the critical temperature of 575° the piezo and pyro effects disappear : the six faces which we have taken as examples become all alike; the structure is now hexagonal. Optical activity remains, because the screw-like arrangement persists.

The X-ray Laue photograph taken by passing the X-ray beam at right angles through a plate cut perpendicular to the axis shows the change in a very simple and striking way. The figure is trigonal in the case of α-quartz, hexagonal in the other case.

If any two planes be chosen in the α-quartz crystal so as to make equal angles with the axis and to meet in a straight line perpendicular to the axis, they will in general differ in their properties. In β-quartz, no difference can exist. This has been illustrated in a very striking way by Mandrot.* Mandrot cut two sections parallel to the respective planes and measured Young's modulus in both cases at temperatures above and below the critical point. He found a large difference between the elasticities of the two sections below 575°, but none above that point. It was noteworthy that close to the critical point, when the most of the change may be supposed to be going on and the crystal may be considered to be in a temporarily labile condition, the elasticity drops to a very low value.

The increase of symmetry in passing from the α- to the β-form can be demonstrated in other ways, but the three that are quoted above are sufficient to prove the point.

The second proposition can also be supported in many ways. Mr. R. E. Gibbs has made X-ray photographs by the revolving crystal method, placing his quartz specimens in a small furnace which was mounted on the X-ray spectrometer. Photographs at 700°, well above the critical point, when compared with photographs taken at ordinary temperatures, showed the characteristic differences between the trigonal and the hexagonal forms, but otherwise were so similar in character as to show clearly the close connection between the two. The same spacings were found in the two cases, showing the existence of similar planes, with only such changes as would naturally follow from the temperature expansions. Long ago, H. Le Chatelier † showed that at the critical temperature there were changes in volume which, although definite, were quite small. There is a sharp change in the relation between specific heat and temperature ‡ at 575°; but the actual latent heat of transformation is very small. The optical activity changes

* " Elasticité et Symétrié du Quartz." Lausanne, Imprimeries réunies, 1924.

† *Compt. rend.*, 1889, **108**, 1046.

‡ A. Perrier and F. Wolfers, *Arch. des Sciences*, 1920, **125**, 372.

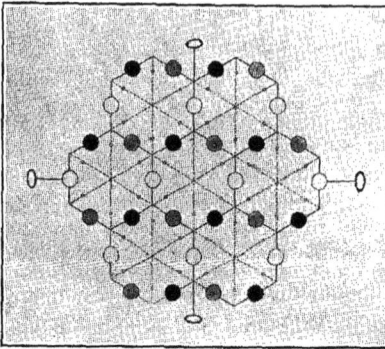

FIG. 3.

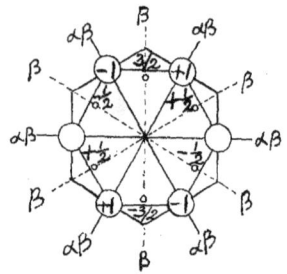

FIG. 4.

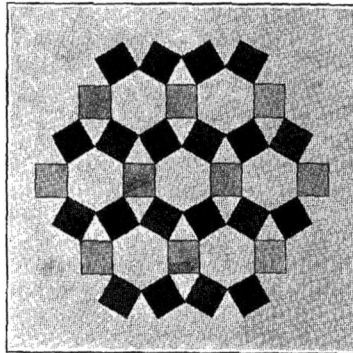

FIG. 5.

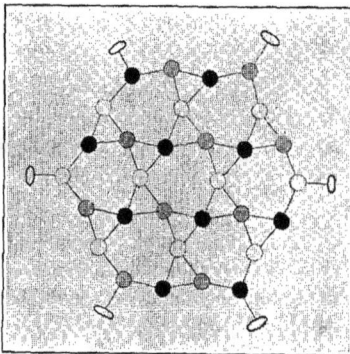

FIG. 6.

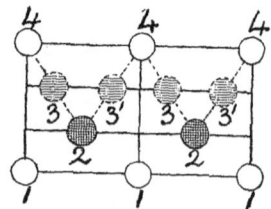

FIG. 7.

[*To face Trans.*, page 277

definitely at the same point : but the actual change is again very small (Le Chatelier). Goniometric measurements and determinations of refractivity show the same effect.

Assuming, then, the truth of these two propositions we proceed to apply them to the determination of structure. The gain in symmetry on passing from α- to β-quartz may be expressed as the addition of three new digonal axes at right angles to the first three, and also to the principal axis. If we refer again to Fig. 1, we see that, of all the various distributions of the silicon atoms which are illustrated in the four figures, only that which is illustrated in the last figure can satisfy the new conditions even as far as the silicons alone are concerned. It is to be remembered that β-quartz is to be looked for as a special form of α-quartz, since in both cases the unit cell has the same size and form (except for changes due to thermal expansion) and contains the same number of molecules, and the symmetry of β-quartz is the same as that of α-quartz, with additions. This particular limitation at once reduces the number of unknown parameters from four to three, which latter relate only to the relative positions of the oxygens and the silicons.

It is found on examination that the only possible position for the oxygens are such that in the projection of Fig. 3 a hexagon, formed by the projection of oxygen atoms, is associated with each of the hexagons in the figure. The centre of an oxygen hexagon coincides with the centre of a hexagon as drawn, and the projections of the oxygens lie on the lines drawn from the common centre to the corners. The positions of the atoms relative to the plane of the paper are indicated in Fig. 4, where a " 0 " implies that the silicon atom indicated is in the plane of the paper, " −1 " implies that it is at a depth $c/3 = 1.79$ Å.U. below the paper, and " +1 " the same distance above. The distances of the oxygens from the plane of the paper are indicated in the same way, and it will be observed that oxygen planes interleave the silicon planes.

The common distance of the oxygen projections from the hexagon centre in Fig. 4 is still undetermined. So long as they lie on the perpendiculars on the sides of the silicon hexagon, produced if necessary, the conditions of symmetry are satisfied. The four unknowns are now reduced to one.

It should be mentioned that in choosing these positions for the oxygens a somewhat similar set, in which the oxygen hexagon is turned round 30° with respect to the position actually adopted, will also satisfy the symmetry conditions. But it is rejected on the ground that the structure implied requires that oxygen and silicon atoms should be at distances from each other differing greatly from our experience so far. Again, the oxygen atoms might lie

in the silicon planes, but this would contradict the interpretations
of the X-ray photographs; for if this supposition were correct there
would be a series of planes perpendicular to the axis spaced at a
regular interval of $c/3$, each containing the same number of silicon
and oxygen atoms, with nothing between the planes. In such
circumstances, there would certainly be an intense reflection from
these planes (*i.e.*, from the 0001 plane of the crystal). On the
contrary, the first order reflection is very weak : the second and
third reflections are of an intensity which is much the same as that
of the first and this implies strong interleaving.

It now remains to use any available means for the determination of the remaining parameter; no more can be done from
considerations of symmetry. We have a long series of intensity
measurements at our disposal. We are not very expert at reading
them, for which reason little could be done when there were four
unknowns to be determined. But when there is only one, the case
is very different. Mr. Gibbs has made a full series of comparisons
between theory and observations for various values of the distance
of the oxygens from the centre of their hexagon (Fig. 4), and has
found that they must lie within the hexagon formed by joining
the projections of the silicon atoms, and that their distance from
the edge is between 1/8th and 1/16th of the distance between two
opposite sides of that hexagon. The value 1/12th gives satisfactory agreement. The distance across the silicon hexagon is
4·23 Å.U.; a twelfth of this is 0·35 Å.U.

If we place the oxygen projections accordingly, we find that they
very nearly form squares, of side 1·79 Å.U., and would exactly form
squares if a value 0·33 were chosen. It will be remembered at
once that this is also the value of $c/3$, and the height of one oxygen
plane above another. Hence the oxygen atoms lie at four corners
of a cube, the centre of which is occupied by silicon; in fact, the
oxygens are arranged about the silicon in a regular tetrahedron.
The structure is illustrated in Fig. 5, where the black, shaded, and
white squares are in successive layers. The result is remarkable
for its immediate adjustment to the known tetravalency of the
silicon atoms. It is to be observed that an oxygen atom does not
lie in the straight line joining the centre of its two silicon neighbours.

We may now take a backward view from the point we have
reached. If the structure of β-quartz can really be described by a
pile of cubes as illustrated in Fig. 5, the value of the crystallographic
" a " should be $S + S\sqrt{3}$, where S is the side of the cube, since it
is the distance between the centres of two silica neighbours in the
same plane. The value of " c " is $3S$: and therefore the ratio c/a
should be $3/(1 + \sqrt{3}) = 1·0981$. This is for β-quartz; to make the

value comparable with the known value for α-quartz, viz. 1·1000,[*] a small and rather uncertain correction has to be made for unequal contractions along and perpendicular to the axis. Le Chatelier's values [†] give, when used for this purpose, a value for c/a equal to 1·1025 nearly. The difference between the observed and calculated values is very small, but it is quite likely to be real and therefore to leave something unexplained. The calculation can be made in another way. The angle between two pyramid faces of β-quartz, not being immediate neighbours but having one face between them, is 85° 29′, according to Rinne.[‡] The simple structure described above makes this angle 85° 41′; again, a very slight but perhaps real difference.

The change from β-quartz to α-quartz implies small changes in the relative positions of the atoms within the unit cell. Changes in the positions of the silicons can only be of one kind, which would be illustrated in Fig. 1 by a movement from position 4 towards position 3. The drawing in Fig. 6 shows how the movement may be regarded (in the projection) as the result of rotations of triangles in opposite ways. How great that movement should be remains to be calculated. Further, the oxygens also move, no doubt, and their possible movements are varied to an extent implied by the increase from one unknown parameter to four. The consequences of various movements which seem probably of the right nature have been considered by Mr. Gibbs in connection with the consequent changes in intensity which can be measured experimentally : and his results will shortly be published. He finds that the triangles in Fig. 6 are turned about 8° from the symmetrical positions which they occupied in Fig. 3. For the moment, this description confines itself to certain general cases in which the structure arrived at can be considered in connection with the physical properties of quartz.

In the first place, we may take the twinnings of quartz, of which there are four kinds. Two of these are found in both α- and β-quartz, and are due to combinations of right-handed and left-handed portions. Two others are found in α-quartz only; their consideration may be deferred for the moment.

A twinning plane must have special characteristics. It must be a strong plane, containing a dense packing of atoms, for this is not only reasonable but in agreement with all cases examined. It must be such that if the crystal is built up to this plane, its continuance on the other side of the plane may follow one or two (or more) alternatives. One of the latter implies the continuation of the plan of structure across the plane, the other commences the

* Tutton, Vol. I., p. 371.   † *Loc. cit.*
‡ "Crystals and the Fine Structure of Matter," p. 167.

twin, and must not be quite so easy to follow, otherwise there
would be no regularity of structure at all.

Now if we take a section of β-quartz perpendicular to any of the
digonal axes or to the principal axis, so as to pass through silicon
atoms only, the atoms appearing in the section always show symmetry. For instance, the arrangement of silicon atoms in a plane
parallel to a prism face is as in Fig. 7; the full circles 1, 1, 1; 2, 2, 2;
4, 4, 4, being the silicons; it is symmetrical about a vertical line
in the paper. The shaded circles 3, 3, 3 represent silicon atoms in a
plane at a distance 2·15 Å.U. behind the paper. The spiral is
right-handed as the figure is drawn, the succession being 1, 2, 3, 4.
The layer of silicon atoms in the next plane above the paper should
be as at 3′, 3′. . . . But if there is a mistake, so to speak, and the
silicon atoms in this plane overlie 3, 3, . . ., the crystal above
the plane falls into the left-handed habit. There is nothing in the
arrangement of the silicon atoms in the plane to determine which
of the two alternatives is to be chosen : it is the influence of the
silicons in the plane below which should decide the choice. If
this fails for some reason, twinning takes place.

This state of things is found to hold also in all the other planes
referred to above, such as two planes at right angles to each other
and to the plane we have been considering; and because it can
occur in all these planes the twin forms can fit on to each other
in any way. Thus we have not only an explanation of the twin,
but also of its interpenetrating character.

The other form of twinning is infrequent and curious. It occurs
in crystals from the Dauphiné in France and from Japan. The
two portions are nearly at right angles to each other like the two
main strokes of L in italics. The twinning plane passes sym-
metrically through a pyramid edge. If a section of a quartz prism
is made parallel to this plane each of the two parts can be represented by one of the two portions shown joined together in Fig. 8.
Four silicons, AA, and two oxygens lie exactly in the section.
Two other silicons B and C lie very nearly in the section, as shown.
If they are transferred from one side to the other of the six-sided
section, the sense of the spiral in each of the portions is reversed.
A right-handed portion can be fitted to a left-handed portion as in
Fig. 8, thus giving the Dauphiné twin : a right joined to a right,
or a left to a left makes, of course, the regular prism.

We now consider the twinnings which α-quartz can show in
addition to the above. If we consider the section of Fig. 3 which
refers to a crystal with a right-handed spiral round the triangle
and imagine how it would change when the temperature fell below
the critical point, we see that it might go either of two ways, and

might attempt to do both in different parts of the same crystal. It might pass from the β-form of Fig. 3 to either Fig. 6 or Fig. 6 turned round 180° in its own plane. There is now an obvious polarity along each of the digonal axes, and it is possible and true that pressure along an axis develops positive electricity at one end and negative at the other. If, therefore, a β-quartz crystal, right- or left-handed cools down, it may develop these opposite arrangements in different parts of the crystal, and so form one of the well-known twin forms of quartz.*

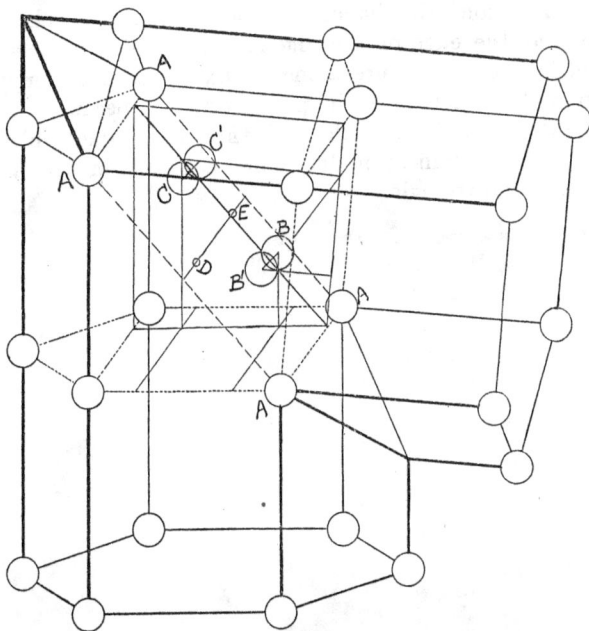

Fig. 8.

If the β-quartz crystal is already a twin and consists of right-handed and left-handed portions, then two different things may happen. Either the two portions develop their electric axes in the same sense or in opposite senses. In the former case, the plane of reflection or symmetry, across which the right-handed portions reflect into the left, contains an electric axis. In the latter case, it is perpendicular to one. The latter case is known as the Brazil twinning; the former is somewhat uncommon. Thus all four varieties of twinning are accounted for.

The piezo- and pyro-electric properties indicate a separation of

* H. A. Miers, " Mineralogy," p. 370. London, Macmillan, 1902.

electricities between the silicons and the oxygens. If we suppose the silicon to be quadruply positive and the oxygen doubly negative, we may suppose the arrangement of Fig. 6 to be divided up into separate triangles, each having half a silicon at each corner and an oxygen just off each side. The electric field of each of these triangles is very small at any distance, on account of symmetry; and when, as in β-quartz, there are equal numbers of triangles pointing opposite ways, the fields of the two sorts practically destroy each other. But when through cooling half the triangles revolve through a small angle in one direction and half in the opposite direction, the balance is destroyed; and charges are developed at the ends of the digonal axes, which are also called the electric axes. Pressure along an axis does not merely cause rotation of the triangle as in pyroelectricity, because the result would be the electrification along all three axes alike, which is not the case. The electrification due to pressure must involve an actual distortion of the triangle.

THE ROYAL INSTITUTION.
LONDON, W.1.

"*The Nature and Constitution of Glass.*"

DISCUSSION.*

I. On the paper by Prof. W. E. S. TURNER, entitled : " The Nature
and Constitution of Glass."

Prof. H. LE CHATELIER (*communication translated by* Mr. WALTER
BUTTERWORTH, jun.) :

Mr. Henry Le Chatelier does not believe that the investigation of
the physical properties of glasses can throw any light on their
chemical constitution or on the nature of the definite chemical combinations which occur in them, because in solutions the properties
are generally additive and therefore independent of the degree of
chemical combination. The attempts which have thus been made
to recognise the different hydrates of salts in aqueous solutions have
all failed.

There are, however, two physical properties which seem to
depend to some extent on the state of chemical combination : these
are the coloration or, more exactly, the absorptive power for different
luminous radiations, and the magnetic properties. When mercuric
iodide is dissolved in alcohol, a pale yellow solution is obtained which
deposits by evaporation the yellow variety of this salt. It may be
accepted that it is the yellow variety of the iodide which is in
solution in this case. As to the magnetic properties, Mr. Pascal
has shown that they could only be regarded as additive if to the
terms relating to each atom are added other terms relating to the
state of chemical combination. It is not impossible that this
method would lead to results in the study of glasses.

Mr. Le Chatelier has only observed one example of features in
a glass which could be connected with the existence of a definite
composition. It is the case of zinc borate. The mixture $2B_2O_3,3ZnO$
when melted gives a beautiful glass, but its composition cannot
be modified to any great extent by additions of boric acid or oxide
of zinc. In the former case, a layer of boric acid separates out and

* The discussion consists of communications received concerning the
papers contributed to the Symposium on " The Nature and Constitution
of Glass " at the May, 1925, meeting of the Society. See this J., TRANS.,
1925, **9**, 147.

comes and floats on the surface; in the latter, the mass crystallises on cooling and cannot maintain the vitreous state when cold.

Mr. Henry Le Chatelier is very interested in the fact that the crystallised compound $2SiO_2,Na_2O$ has been isolated. He is convinced that there must exist a very large number of silicates corresponding to this formula. His readers will recollect that the blue of the ancients consists of a perfectly crystallised silicate containing similarly two molecules of silica to one molecule of base $2SiO_2(0·5CuO,0·5CaO)$. He has obtained the corresponding compound with baryta replacing the lime. Its blue colour is still more beautiful. He had the notion that the black coating of Etruscan pots might be formed of the compound $2SiO_2(0·5FeO,0·5Na_2O)$ but he has never succeeded in demonstrating the existence of this body.

There should be a good chance of discovering new compounds containing the same number of molecules of silica by starting with its combinations with lithia. They crystallise very easily, but their double refraction is very small, and it is a delicate task to separate them from each other. Possibly even a silicate of formula $4SiO_2,Li_2O$ exists. The two corresponding compounds with boric acid have been obtained, $2B_2O_3,Li_2O$ and $4B_2O_3,Li_2O$, which crystallise very easily.

Finally, Mr. Henry Le Chatelier announces that the question of the point of transformation of glasses has just taken a considerable step forward as a result of research pursued in his own laboratory by Mr. Samsoen. The most characteristic change of properties accompanying the transformation is the sudden change in the coefficient of expansion, which in an interval of a few degrees may be increased tenfold. The point of transformation for boric anhydride is at 240°. Below this temperature the coefficient of expansion is 14×10^{-6} and, above it, it is 210×10^{-6}. This phenomenon is not connected at all with the allotropic changes of state of any particular compound, as has been thought hitherto. It manifests itself in exactly the same manner for all amorphous bodies, whether mineral or organic. And, moreover, it always takes place for the same value of the coefficient of viscosity, between 300° and 500° for ordinary glasses, that is, for the silicate, phosphate, and boracic glasses; between 0° and 100° for the resins, pitch, and concentrated acroleine, and below 0° for viscous liquids solidified by sufficient cooling, such as glycerine.

If we wish to conceive the phenomenon schematically, we may suppose that so long as the internal friction, or the coefficient of viscosity, is sufficiently high, the molecules swell under the action of heat but do not separate from each other. But once the internal friction has become weak enough the molecules begin to be spaced out and accordingly the coefficient of expansion greatly increases.

In any case, the anomaly of glasses is common to all amorphous bodies, and has nothing to do with any particular property of silica, still less with the transformation of quartz.

Dr. P. E. SHAW (*communicated*) : In recent researches in the surfaces of thin glass rods I have obtained evidence of " streakiness " which, as far as I know, is novel. From general reasoning, a glass surface, not much above the point of solidification, might be expected to be patchy. In the body of the fluid are several materials in a state of (imperfect) solution. Some would tend to be adsorbed to the glass surface and the process of adsorption would continue as long as fluidity existed. But the process would, on account of the great viscosity, never be completed. Further, it would be irregular. The fluid varies in composition and having bad conductivity the temperature gradient from place to place as the crust formed would be steep. All these factors and possibly the variable effect of furnace gases on the different surface materials would tend to cause patchiness of the glass surface when solidifying. When a patchy lump is drawn out we obtain a " streaky " rod. There are two ways of showing the streaks : (1) Boil two glass rods in chromic acid for a few minutes, then in distilled water. When dry and cold, at once test the coefficient of friction of the rods on one another. This coefficient is high; but it will be, say, 1·3 for one generating line, 1·2, 1·4, say, for neighbouring ones. The values repeat consistently as the rod is turned round. (2) Prepare two more clean rods as above. Wrap one with silk tightly. Rub this wrapped rod on the bare rod. It will often be found that for some generating lines we get a positive charge, for others a negative. Again this action will depend, consistently, on the place on the glass surface rubbed.

Prof. TURNER (*communicated reply*) said that he felt sure other workers besides himself were very much indebted to Prof. Le Chatelier for his interesting and valuable comments and suggestions. In regard to the study of the constitution of glasses by measuring their physical properties, he (Prof. Turner) did not think it desirable at this stage to exclude the possibility of important results being forthcoming in the future despite the apparent predominance of additive relationships in many of the properties of glasses. A somewhat similar position had existed at one time in connection with metallic alloys, but closer study had revealed quite a number of properties—even including the easily measured property of specific volume—through the measurement of which valuable light had been thrown on our knowledge of constitution. True it was that alloys were crystalline solids and glasses were not. On the other hand, the specific case of sodium disilicate, referred to in the course of his paper, provided encouragement for fuller investigation of glasses than had hitherto been given to them.

The most interesting remarks of Prof. Le Chatelier in regard to the striking phenomena observed in the softening range of glass and of other amorphous substances were of great interest to himself, especially as he (Prof. Turner) had conceived a similar idea three or four years ago, namely, that the heat absorption and marked change in thermal expansion, etc., were common to all amorphous substances; and as a test of the hypothesis Mr. A. R. Sheen had spent some time under his direction making experiments on resin, but no heat absorptions were found. These measurements were referred to in a paper in 1923.* He understood that Mr. F. W. Preston, during a study of pitch, had also not noticed any discontinuities. For these reasons, he looked forward with the greatest possible interest to an account of the investigations (that of Mr. Samsoen had already appeared) to which Prof. Le Chatelier had referred.

In regard to the observations of Dr. Shaw, he (Prof. Turner) thought his observations not only intensely interesting but suggestive. The patchiness of some samples of glass rod might be associated with the result of marvering the glass with the consequent formation of a layer in a toughened state. Some glasses were much more prone than others to develop a tough skin, as, for example, glasses containing considerable percentages of alumina and of boric oxide. When the marvered cylinder was drawn into rod or tubing, the skin, if hard and tough, broke up and might be distributed irregularly along the exterior of the rod or tubing drawn. In extreme cases they had found the rod or tubing had a lumpy or knotty appearance, most probably arising from such a cause. Patches or streaks of this character might give rise to the effects to which Mr. Shaw referred. His observations also had a bearing on the apparent lack of homogeneity revealed when, not only glass rods, but also glass slabs, which had been cast, were subjected to the action of corroding agents such as, for example, steam and water under pressure. The attack was not uniform; it began at certain points rather than appearing uniformly all over the surface. Some points, therefore, seemed to be less resistant and others more resistant. He trusted that Dr. Shaw would proceed with his researches on this subject and be able to give some account of his extended observations at a later date.

* "The Year in Review in the World of Glass Technology," by W. E. S. Turner, this J., TRANS., 1923, **7**, 139.

II. On the paper by Mr. V. H. Stott, entitled : "The Viscosity of Glass."

Prof. Le Chatelier (*communicated*) : Mr. Henry Le Chatelier recognises the soundness of the reservation made by Mr. Stott in regard to the non-homogeneity of the double logarithmic formula. It is merely a convenient empirical formula; the whole point is to know within what limits it sufficiently represents experimental results. It is difficult to discuss the examples adduced by Mr. Stott, in which a continuous curve of large radius of curvature would represent the experimental results better than the pair of straight lines, since he does not specify the glasses used by Dr. English. But in the majority of cases the radius of curvature is sufficiently large for the continuous curve to be replaced by two successive straight lines of different inclination.

This has the great advantage of allowing the establishment of a relation between the parameters of the two straight lines and the chemical composition of the glasses, which is impossible with curves not bearing any algebraic definition. Mr. H. Le Chatelier persists then in believing that it will be of practical interest to glass manufacturers to utilise the double logarithmic formula in discussing problems of glass manufacture.

He abandons, on the other hand, his notion that the point of intersection of the two straight lines corresponds to an allotropic transformation. The question of the transformation of glass has been completely cleared up by the researches of Mr. Samsoen. It is not a property peculiar to glass; it belongs to all amorphous bodies. Glycerine, below zero, shows the same anomaly as glasses at about 500°. Whenever an amorphous body, glycerine, resin, or glass, is heated to a temperature at which its viscosity is about 10^{13} poises, its coefficient of expansion suddenly makes an enormous change; it becomes sometimes ten times as great. Vitreous bodies at low temperatures have a coefficient of expansion of the same order of magnitude as that of solid crystallised bodies, and at high temperatures one of the same order of magnitude as that of liquids. This is a new fact which will have important bearings on all investigations into the constitution of matter.

Mr. V. H. Stott expresses his thanks to Prof. Le Chatelier for his interesting remarks, but has nothing further to add.

III. On the paper by Dr. F. Eckert, entitled : "Some Remarks on the Constitution of Glass."

Dr. F. Eckert (*communication translated by* Mr. A. Cousen) : Since my last communication, a series of facts have become

H 2

known to me, from literature as well as from discussions with a number of glass technologists. Moreover, several facts have come within my own experience, which appear to confirm in a wide degree the hypothesis advanced by me, of the dependence of the properties of glass upon its past heat treatment.

It is known that the viscosity of molten glass, particularly at the working temperature, is very dependent upon the type of the batch constituents, even when the glass has the same oxide composition. A glass made with soda ash is " softer " than the same glass (with an equal Na_2O content), made, however, from saltcake. When alumina is added as felspar the glass melts relatively quickly, but with slower refining. The glass itself in working is relatively of longer working range than when the alumina is added by means of aluminous sand, kaolin or the hydroxide. It has, however, in the latter case a somewhat higher temperature range of devitrification.

For estimating the change of the properties in the finished product it is necessary to consider three things :—

(1) The various decomposition temperatures and the various decomposition products of the batch constituents : for example, soda ash reacts directly with silica to form silicates at relatively low temperatures. Saltcake, on the other hand, decomposes temporarily producing caustic soda, and then forming a silicate (caustic soda at high temperatures explains, above everything else, the strong chemical attack of saltcake batches on the fireclay bricks and pots). The higher the decomposition temperatures, etc., the more will a change of the melting temperature, and of the duration of melting, be able to alter the properties of the finished glass. This effect appears to be particularly large in low-fusing glasses and enamels (according to a communication of Prof. Zschimmer, Karlsruhe).

(2) The effect of the raw materials (for example, felspar, cullet, etc.) on the resulting properties appears to be at most of a secondary character. Primarily, the cause of the change of the properties which is likely to be determinative is the high temperature and the duration of melting. The course of the melting and refining are in this way changed according to the materials employed in the batch. I take it that the action of the initial ingredients is not so much a direct one as an indirect one through the variation of the previous heat treatment. As an example, added cullet has not in itself an effect, but the fact remains that batches with a high proportion of cullet have a greatly shortened time of melting and a considerably increased refining period.

(3) The property which is influenced the most by the previous heat treatment, so that it is technically very noticeable, is the

viscosity of the molten liquid, and the properties connected with strength (brittleness) in the solid state. The two states (that is, the rigid and the fluid) are connected by a transition state, the characteristic of which is a change in the rate of expansion. It is probable that the change of the viscosity with temperature has, as a consequence, a change in the type of expansion, particularly through the influence of the adjacent layers, which, because of the poor heat conductivity of the glass in cooling, tend to a very varied condition of temperature. So far as the results are due to strain, whereby the mass as a whole becomes a doubly-refracting body, these physical anisotropes can be removed by cooling, in a known way. On the other hand, effects here mentioned, which arise from a varied molecular structure caused by a varied past heat treatment, cannot be removed by cooling to low annealing temperatures.

But it is also possible that reactions, which at high temperatures are relatively incomplete, lead to irregular heterogeneities, which, although in order of size greater than those of the molecularly disperse state, are yet so sub-microscopic that they can be made noticeable as optical homogeneities, particularly through double refraction phenomena. Such glass, according to ordinary ideas, is then apparently homogeneous in the sense of being a pure mixture; but there can be explained, however, in this way the large strains undoubtedly existing between these smallest heterogeneous constituents, which one must assume to be the cause of the increased brittleness of the glass. In some cases, I have undoubtedly been able to prove that a sample of glass which was formerly extremely heat-resistant showed such an increase of brittleness and decrease of breaking strength as to lower it closely to the level of ordinary glasses, although its chemical composition was identical with the original sample and its homogeneity and freedom from strain were proved by precise methods. This was in a glass with little soda and much boric oxide and zinc. The phenomenon was established by a large series of meltings.

My explanations have been surprisingly confirmed by the very noteworthy work of G. W. Morey and N. L. Bowen (see this J., TRANS., 1925, 9, 126). My hypothesis is that phases exist in equilibrium at any fixed temperature, having varying properties; practically this equilibrium is only more or less completely reached before new temperatures are applied and thus some of the high temperature condition is preserved up to the final cooling stage.

In a sense, glass in its fused state can thus be described, in reference to its state of equilibrium, as supercooled or superheated, with the exception that, in the case considered, the periods of time over which equilibrium is reached, due to the reduced viscosity, are not so enormously great as, for example, required for the crystallisation of the supercooled glass in the solid state.

I believe that to-day we need to consider it as an established fact that certain physical and also many chemical properties (in conformity with elementary analysis) can be very varied, according to the composition of the various mixed compounds or phases, of themselves completely homogeneous, which have been preserved in the solid state because of the varied previous heat treatment.

Prof. TURNER (communicated) stated that he welcomed Dr. Eckert's additional notes on the influence of heat treatment and other conditions on the properties of glass. This subject had formed the basis of much experimental work at Sheffield since 1922, and the results substantiated and extended Dr. Eckert's views.

XXXVI.—*The Viscous Properties of Glass.*

By VAUGHAN H. STOTT, M.Sc.

(*Communication from the National Physical Laboratory.*)

(*Read at the London Meeting, December 15th, 1926.*)

MEASUREMENTS of the viscosity of glass in the molten or the plastic state involve the special difficulties appertaining to all high temperature work in addition to the troubles associated with the manipulation of very viscous liquids. Molten glass has the further inconvenience of resembling, in its action on containing vessels, the universal solvent of the alchemists. The limitations imposed on the design of viscosimeters by these causes, and also by the very large range of viscosities to be measured, have prevented the achievement of what should precede the measurement of any physical property, namely, a complete investigation of the factors defining the property in question, so that the final results of the measurements shall in no way depend on the particular type of apparatus used.

A partial investigation of this nature conducted at the National Physical Laboratory has shown that the apparent viscosity of glass at high temperatures may be influenced in two ways by the method of measurement. In the first place, it is very probable that the apparent viscosity depends on the rate of shear, the former increasing as the latter diminishes. In the second place, the apparent viscosity is influenced by thermal treatment of the glass. When the viscosity is very high, as, for example, at temperatures below the annealing point, the viscous movements are complicated functions of the past history, both mechanical and thermal, of the specimen.

The above considerations suggest that two distinct problems require solution in connection with the viscosity of glass at high temperatures. The more urgent problem is to investigate thoroughly the relation of viscosity to thermal treatment, not merely for a particular glass, but for glasses of different compositions. Such an investigation could conveniently be carried out by means of the type of viscosimeter developed at the National Physical Laboratory. In this, a measurement is made of the mean thickness of glass which adheres to an iridio-platinum wire withdrawn at a known rate from the glass contained in an iridio-platinum crucible.* As an indication of the general nature of the information yielded by such work, we may recapitulate the conclusions to be drawn from measurements made in this way at the National Physical Laboratory on a particular soda–lime–silica glass.

It has been found that the viscosity–temperature curve of this glass is perfectly definite and reproducible at temperatures between 1200° and 1530°, provided that, by stirring at a suitable temperature (e.g., 1470°), the glass has been rendered homogeneous. At very high temperatures, the viscosity may be raised by the volatilisation of soda. Maintaining the temperature at 1624° for 24 hours results in an appreciable loss of soda, but the slope of the viscosity–temperature curve remains unchanged. If the glass be cooled below 1200° and then reheated, the viscosity may no longer be the same function of the temperature. Comparatively rapid cooling, even to room temperatures, followed by fairly rapid heating produces no change in the viscosity above 1200°. If, however, the glass be maintained too long within some critical temperature range lying below 1200°, then subsequently the viscosity at a temperature above 1200° may be found abnormally high. When the state of the glass has been changed in this way, the abnormality persists up to the highest temperature which can be reached without causing volatilisation of soda. At temperatures below 1200°, the glass can exist in more than one state. In a metastable form, it has the viscosity which would be expected from an extrapolation of the values found at higher temperatures. The determinations are repeatable to ± 3 per cent. of the viscosity, which is roughly equivalent to a temperature variation of ± 3°. In most cases, however, at temperatures below 1200° the measured viscosity values fluctuate very considerably from the mean, which is, nevertheless, the same as the value found when the glass is in a metastable, homogeneous state. The fluctuations may amount to as much as ± 30 per cent., and cannot be ascribed to experi-

* Stott, Turner, and Sloman, *Proc. Roy. Soc.*, 1926, A., **112**, 499.

mental error. The fact that these deviations occur about a mean point corresponding with a normal value is of great importance, since various possible explanations of the phenomenon are thereby ruled out. It is clear, however, that the peculiar behaviour of the glass below 1200° must be associated with some form of heterogeneity. It is natural to inquire whether the heterogeneity may be removed by heating the glass. When the glass has not been cooled below 1140°, the normal state may be regained at a temperature slightly exceeding 1200°. When, however, the glass has been cooled too slowly to room temperatures, a persistent increase of viscosity of about 10 per cent. at temperatures above 1200° has been found, as mentioned above, and this increase is associated with a slight reduction in the concordance of the measured values. The factor which determines the difference in the behaviour of the glass in these two cases is not known, although it may be noted that when the glass has been cooled to room temperatures the devitrification temperature (about 950°) is passed. No evidence of incipient devitrification has been obtained, however, nor does the microscope reveal any difference between the appearance of the " heterogeneous " and the " homogeneous " forms of the glass, nor between samples of "heterogeneous glass " of widely differing viscosity. This is not surprising in view of the well-known high sensitivity of the viscosity of very viscous liquids to small changes of composition or of constitution. The value of accurate viscosity measurements is, however, manifest, and it may readily be appreciated that most important results may follow the extension of such measurements to a variety of glasses. It is, perhaps, scarcely necessary to point out that brittleness and bad thermal endurance of glass may be associated with heterogeneity, and from this point of view further work on the lines suggested above may have important and immediate application.

Attention may here be directed to the desirability of investigating a problem which, although forming a part of the problem discussed above, may be of importance in connection with any theories of the constitution of glass. This is the question of what may be called " the scale " of heterogeneity involved in the phenomena described above. The thickness of glass deposited on the wire during a viscosity measurement at 1150° is of the order of $\frac{1}{8}$ mm. If we assume, therefore, that the heterogeneity is submicroscopic, the real variations of viscosity from point to point must be very large to yield residual effects on a scale relatively so great, since the mean effect of the heterogeneity is known to be zero. It is interesting to note that submicroscopic heterogeneity in glass at room temperatures has been postulated by others.* The deductions from viscosity determinations, however, form the first available evidence of such structure in glass at high temperatures.

In order to obtain further information, it would seem desirable to perform additional measurements of viscosity in which specially thin wires should be used, as in this way the effects of heterogeneity would be expected to be more in evidence. Unfortunately, considerable practical difficulties would have to be overcome, one of the most important being the great liability for bubbles to form in the glass on slightly lowering the temperature. When using the usual size of wire, a few small bubbles have no appreciable effect on the apparent viscosity, but their influence would undoubtedly be important in the case of exceptionally thin wires.

The second problem, less urgent than that discussed above, is the determination of the influence of a *very wide range* of rate of shear on the apparent viscosity of glass at high temperatures. This problem is very difficult from several points of view, but a solution is required in order to co-ordinate different methods of measuring the viscosity of glass. The most obvious method of attack would be to employ a Margules type of viscosimeter with a comparatively small clearance between the cylinders, in which the parts in contact with glass would be made of, or faced with, iridio-platinum. The elaboration of such an apparatus must be left to the future.

In this connection, attention may be directed to a point which has not hitherto received much consideration. No adequate mechanical theory of viscosity exists at present in the case of a liquid, although the viscosity of a gas can be explained quantitatively on the basis of the kinetic theory. For this reason, it has not been found possible to correlate with any accuracy the viscosity of a liquid with its other physical properties. Until this can be done, the intrinsic value of an absolute determination of the viscosity of a liquid is little greater from a theoretical point of view (*i.e.*, apart from mechanical considerations, such as the rate of flow of the liquid, the rate of rise of bubbles therein, and so on) than a relative determination. The latter is by far the easier to carry out, and it is for this reason that, of the two main problems just considered, the one relating to an extension to various glasses, of the viscosity measurements discussed above, and the use of thinner wires, has been considered more important than an investigation of the effect of rate of shear, which is essential from the mechanical point of view.

* F. Eckert, " Discussion on Constitution of Glass." This Vol., TRANS., p. 101; L. N. G. Filon and F. C. Harris, " On the Di-Phasic Nature of Glass." *Proc. Roy. Soc.*, 1923, A., **103**, 561.

THE CONSTITUTION OF GLASS   237

428   JOURNAL OF THE SOCIETY OF GLASS TECHNOLOGY.

Let us now consider viscosity at lower temperatures. As the temperature diminishes, the mechanical properties of glass exhibit to an increasing extent the properties peculiar to solids, whilst retaining to a diminishing degree the mobility peculiar to liquids. The phenomena of flow accordingly become very complicated when the stage is reached where the glass may be regarded either as a very viscous liquid or as a very soft solid. Fortunately, the solvent properties of glass are greatly reduced at this stage, and measurements of viscosity can be made comparatively easily with a Margules apparatus of china clay within the viscosity range of from 10^4 to 10^9 poises. The theoretical conceptions involved in the measurements are naturally intermediate between those involved at the higher and the lower temperatures and need not be further considered. The ease with which devitrification occurs within this range of viscosity is, however, a very serious inconvenience.

When the viscosity exceeds 10^9 poises, the glass flows with extreme slowness under the influence of its own weight, and with suitable experimental arrangements may be treated as a solid. Three methods have been used for measuring viscosity in this case. A cylindrical test piece, having in the middle a certain length of reduced diameter, may be subjected to tension, bending, or torsion. The first method has the advantage of imposing a uniform rate of shear throughout the narrow part of the specimen. The practical advantages of the last method, however, outweigh the theoretical disadvantage of non-homogeneity of strain. Since the form of the specimen is unchanged by twisting, a long series of measurements may easily be carried out on the same specimen when employing the torsional method. It is, moreover, a simple matter in this case to render the "end effects" negligible, as the resistance of a rod to torsion varies as the fourth power of the radius.

The viscous properties of glass are particularly interesting in the range between 10^9 and 10^{17} poises, since within this range other physical properties show discontinuities, the quantitative relationships depending somewhat on heat treatment. At very high viscosities, the velocity of deformation under load decreases with time to an asymptotic value. Measurements made at the National Physical Laboratory on a soft glass suitable for laboratory ware showed that if the viscosity be defined for a constant torque (using a torsional apparatus) in terms of the asymptotic velocity of deformation, a change in sign of the curvature of the logarithmic viscosity–temperature curve occurs at about the same temperature as the discontinuity in the thermal expansion curve. This result has recently been confirmed by a different method in the case of a lead glass.* It has been found by Samsœn,† working in the laboratory of Le Chatelier, that discontinuities in physical properties when the viscosity is about 10^{13} to 10^{14} poises are not confined to glass, but are a general property of amorphous materials. This similarity of behaviour of amorphous materials is very important, for advances in our knowledge of such substances as bakelite, ebonite, celluloid, etc., may be very suggestive in connection with phenomena occurring in glass which are difficult to investigate directly owing to the higher temperatures involved. These remarks apply with particular force to the study of " elastico-viscous " movements in which the similarity is very complete.

" Elastico-viscous " movements can be measured, not merely at temperatures near the annealing point, but even at room temperatures, in which case the movements, although small, may frequently be of considerable practical importance, as, for example, in the case of large mirrors or lenses, or in connection with the changes of volume of thermometer bulbs. Our knowledge of these movements is much less than a perusal of tables of physical constants would suggest. No general theory of the internal friction of solids which is not at variance with experiment has yet been proposed in a quantitative form. Moreover, great confusion of nomenclature exists which may easily lead to misconception. We have seen that whereas the word viscosity has a very definite sense when applied to the ordinary mobile liquids, its meaning becomes increasingly arbitrary as the viscosity of the liquid in question becomes greater. When the " solid " state is reached, the differences between the meanings attached to this word by different authors are so great that the grossest misconceptions may easily arise. Reference to a well-known compilation of physical tables tells us, for example, that the viscosity of steel at room temperatures is approximately 10^9 poises. If the definition of viscosity in this case were similar to that used in the present paper, we should be led to suppose that steel at room temperatures has a consistency something like that of glass at a temperature of more than 100° above its annealing point. This example has been brought forward because the outstanding feature of the phenomena of deformation of solids is their extraordinary complexity. Such complexity is not surprising in the case of metals which are usually composed of irregularly oriented crystals cemented together by thin layers of amorphous material; or in the case of a colloidal substance such as rubber. The behaviour of glass, an

* W. M. Hampton, *Trans. Opt. Soc.*, 1925—1926, **27**, 177.

† Samsœn, *Compt. rend.*, 1925, **181**, 354, and 1926, **182**, 517; Samsœn and Monvel, *ibid.*, 1926, **182**, 967.

amorphous substance *par excellence*, might, however, be expected to be somewhat simpler. In point of fact, the differences in this connection between glass and metals are quantitative rather than qualitative, there being few phenomena observable in the one case which do not have their counterpart in the other.

The analysis of the deformation of glass, under the action of a constant load, into an elastic movement, an elastico-viscous, or a retarded elastic movement, and a viscous flow has been discussed by the present author in a recent paper.* It is there shown that glass obeys Michelson's formula

$$S = S_0 + A(1 - e^{-\lambda\sqrt{t}}) + Bt,$$

where S is the displacement, t the time, and S_0, A, λ, and B are constants for a given load.

If the load be removed after a long time, a reverse displacement takes place which, measured from its starting point, in the reverse direction, is expressed by the formula

$$S' = S_0 + A(1 - e^{-\lambda\sqrt{t}}).$$

Michelson gives a further formula † to express the return flow at a time t after the removal of a load previously acting for a time t_0. This formula has the form

$$S' = S_0 + A(1 - e^{-\lambda\sqrt{t_0}})(1 - e^{-\lambda\sqrt{t}}).$$

It does not appear possible to reconcile this formula with the results of the present author's experiments on glass. In view of the difficulties of the experiments, and the abandonment of further work on the subject, little notice was taken of this point when the experiments were carried out, but it is interesting to note that the failure of this formula is confirmed by Filon's ‡ experiments on celluloid, although the previous formula which holds when $t_0 = \infty$ is obeyed accurately as in the case of glass.

In connection with the above formulæ, two problems are of particular interest. The first is the determination of the true relation giving the return flow after the removal of a load at a time t_0. The second is a determination of the variation with temperature of A, λ, and B for different glasses. (S_0 is practically independent of temperature.) At present it is known, for a particular glass, that A increases rapidly with temperature, whereas λ, at a temperature a little below the annealing point, has almost the same value as at room temperatures, although at a temperature a little above the annealing point, its value is doubled. B is, of

---

* This J., TRANS., 1925, **9**, 210.

† Michelson, *J. Geol.*, 1917, **25**, 405.

‡ Filon, *Phil. Trans.*, 1922, **223**, A, 120.

course, proportional to the reciprocal of the viscosity as previously defined.

Although the formulæ discussed above were established for torsional displacements, analogous phenomena are observed for other types of displacement. Thus Weidmann found that the after-effect due to flexure could be expressed by the equation

$$x = Ce^{(tm)}$$

where x is the ratio of the deformation after a time t, to the original deformation before the removal of the load. The values of m for three different glasses were 0·573, 0·552, and 0·417, respectively. This formula differs little from that of Michelson, in which $m = 0·5$.

Very interesting results have been obtained by Guye on the logarithmic decrements of various fibres undergoing torsional oscillations. In the case of " ordinary glass " * the curve representing the logarithmic decrement plotted against the temperature shows both a maximum and a minimum between 0° and 300°. A similar result was obtained with lead glass, but Jena glass gave a minimum only. Some kind of transformation is probably indicated by the curves for ordinary and lead glass, and further investigation by other methods is very desirable. It is interesting to note that neither the theory of Voigt † nor that of Boltzmann ‡ agrees with Guye's observations, although the former theory is obeyed approximately at low temperatures and the latter at high temperatures. Guye finally concludes that the decay of the oscillations is mainly due to " reactivity," or, to use Michelson's expression, " elastico-viscous flow."

It seems, then, that the conception of " elastico-viscous " flow is of considerable importance, and a realisation of its effects may, in many cases, throw light on little understood phenomena. It is well known, for instance, that when cutting glass with a diamond, the cut has a tendency to heal, and that the breaking should be done immediately after the cutting. This is probably a case of elastico-viscous flow. The problems of annealing are also intimately connected with elastico-viscous flow, and it should be noted that the usual theories of annealing, based on rates of viscous flow under load, are untenable except at temperatures at which the glass is

---

* C.-E. Guye and Mlle. S. Vassileff, " Archives des Sciences physiques et naturelles de Genève," March and April, 1914.

† Voigt's theory assumes a viscous resistance proportional to the velocity of deformation.

‡ Boltzmann's theory assumes that the material " remembers " previous deformations; the memory fades asymptotically with time. Further, the modification in the properties of the material due to the influence of a deformation is independent of the earlier modifications.

fairly soft. Moreover, measurements of viscosity which depend on bending, stretching, or compressing rods are untrustworthy at low temperatures, because the greater part of the motion is not viscous motion in the sense defined above, but elastico-viscous motion.

To turn to a rather different class of phenomena, we may consider the changes of zero of thermometers. These changes have been the subject of much attention, but no general explanation exists. Certain features seem closely analogous to the mechanical phenomena previously considered. It is definitely established that when the temperature of a thermometer bulb is abruptly changed, an abrupt change in volume takes place, which is followed by another change in the same direction, the latter gradually approaching a limit. The time required for this change is of the same order as that observed in elastico-viscous movements. There is also generally a secular change of zero which is probably analogous to viscous flow. It should be remembered that the very high sensitivity of a good thermometer enables movements to be observed which would be inappreciable in the case of the more usual types of mechanical experiment. All that can be said definitely about the closeness of the analogy between the mechanical and the thermal after-effects, is, that the order of the effects is similar, and that glasses which exhibit marked mechanical after-effects exhibit marked thermal after-effects. It has been objected that the analogy is inaccurate because the depression of the zero of a thermometer varies as the square of the difference of temperature which is required to produce it, whereas the mechanical after-effect, for a given time of application of a load, is directly proportional to that load. Such an objection is difficult to sustain, because the thermal effects involve many complications, which include the impossibility of finite instantaneous changes of temperature, strains, pre-existing, or set up during heating due to temperature gradients, etc., which are not present in the mechanical experiments.

The effects of strain in glass may be expected to be very important in thermometry, and some experiments, due to Winkelmann, on highly stressed glass are interesting. Winkelmann made a series of measurements of the expansion of cylindrical rods of glass which had been quickly cooled after having been drawn. These rods were, therefore, in tension in the interior and in compression in the exterior portions. The ends of one of the rods were made quite flat at the temperature of the room. After heating the rod for some time at 96°, the ends were found to be concave. The cylinder was then cooled and the ends were again ground flat.

THE VISCOUS PROPERTIES OF GLASS.　　　433

Three hours' subsequent immersion in boiling water produced no further change, but immersion in oil at 200° produced noticeable concavity in 5 minutes, which became constant in an hour. Similar results were obtained with other cylinders. It would appear that these effects are largely due to viscous flow. The glass is in tension in the central regions and in compression in the peripheral regions. On raising the temperature, the viscosity of the glass diminishes, and the central portions of the rod contract relatively to the outer portions. It is not certain whether the concavity produced at a given temperature is really constant after one or two hours, or whether flow continues with inappreciable velocity. Additional experiments are required to elucidate the phenomenon.

Some of the difficulties of investigating the thermal after-effects in glass might be avoided by observing the lengths of thin threads, the temperature of which could be changed with great rapidity, without introducing very large temperature gradients, by plunging them into liquid baths at different temperatures. The simpler geometrical conditions would also be advantageous in such a case. A difficulty is the possibility that the surface layers of the glass may behave in a manner different from that of the inner layers.

It is abundantly evident that a detailed study of the viscous properties of glass involves many problems of great theoretical and practical interest. At present, only the fringe of a vast field for research has been touched. The experimental side of such work is very difficult, and pitfalls abound for the unwary. The subject is essentially one in which anxiety for quick results, of practical application, should not be permitted to lead the investigator from the path of theoretical rectitude.

DISCUSSION.

MR. E. A. COAD-PRYOR said it was important to know if the phenomenon which Mr. Stott had observed in glass round about the working temperature of the latter depended on viscosity or on temperature. Was there a temperature below which they had to be careful in supplying glass from the furnace to forming machines ?

PROF. W. E. S. TURNER expressed his appreciation of Mr. Stott's thoughtful paper. The subject of anomalies in the working properties of glass in the temperature range between about 1100° and 1250° (or even higher) had been under investigation, particularly from the practical point of view, for several years in his Department at Sheffield. Moreover, the existence of anomalies had become apparent to some manufacturers, whether working glass by hand or by machine. Only during the previous few weeks he had come across two very interesting cases. One involved the continuous drawing of sheet glass, the other of automatic bottle-making. In the latter case, out of three machines of the same type, working under similar conditions at the same tank furnace and giving bottles which appeared homogeneous and satisfactorily annealed, two machines gave bottles resistant to heat shock, the third gave bottles which readily broke. In regard to Mr. Stott's measurements, as recorded in the recent paper in the *Proceedings of the Royal Society*, the fact which struck him (Prof. Turner) was the apparent smallness of the viscosity differences observed in the region of anomaly in comparison with the magnitude of the differences observed in the commercial working of glass. He hoped Mr. Stott would continue his work on this subject.

Mr. W. J. REES remarked that Mr. Stott's observations might have a bearing on the use of cullet in glass melting, and its selection.

Mr. STOTT, in reply to Mr. Coad-Pryor, said that recourse to experiment would be necessary in order to answer the question raised. With regard to the interesting question of the apparent smallness of the viscosity difference in the anomalous region raised by Professor Turner, Mr. Stott pointed out that the quantitative results discussed in the paper applied to a particular glass only, and that possibly the phenomena might be more evident in the case of the harder commercial glasses. He wished, however, to emphasise a point already mentioned, namely, that the smallness of the observed effects was due in part to the use of measuring apparatus having dimensions of considerable magnitude compared with those of a structure which may be submicroscopic.

VIII.—*The Structure and Constitution of Glass.**

By Walter Rosenhain, D.Sc., F.R.S. (of The National Physical Laboratory).

THE problem of the structure and constitution of glass has received a large amount of attention during the past fifty years, whilst quite recently our knowledge of the subject has been summarised in a symposium published in the Journal of the Society of Glass Technology. Careful study of that symposium, however, and of a number of other papers, leaves the reader with the impression that, beyond an elementary, but none the less fundamental generalisation, our knowledge is still vague. That fundamental generalisation is that glass must be regarded as an amorphous solid or under-cooled liquid, and although this view is more fully established to-day, it has been used by the author for purposes of discussion and research for more than 25 years. The problem, however, is thereby shifted rather than solved, since we are left with the wider problem as to the real nature and structure of under-cooled liquids or quasi-solid amorphous substances. Such a widening of the subject is further rendered necessary by the recent work of Le Chatelier and of Samsoen,† who have shown that certain phenomena believed to be typical of glass are met with in amorphous substances generally. This widening of the problem, however, implies also a certain narrowing of the range of investigation and of speculation, since the specific problem of the chemical constitution of glass loses much of its wider importance once it is recognised that, whatever its proximate chemical constitution, glass behaves much as any other super-cooled liquid of complex composition. It is, however, suggested below that the physical structure and proximate chemical constitution of such bodies are inter-related in a special manner.

In view of the fact that existing knowledge and ideas on the structure and constitution of glass have been so recently reviewed and discussed in the pages of the Journal of the Society of Glass Technology, it appears unnecessary to devote a large

---

* This paper constitutes the first Report prepared at the invitation of the Council under the Glass Research Association Trust Deed. After a survey of the existing literature on the Constitution of Glass the author has directed his efforts to framing a more general hypothesis of the inner structure of glass than has hitherto been attempted, with the intention of providing definite conceptions for all scientific investigators in the field of glass technology.—Editor.

† Le Chatelier and Samsoen, *Compt. rend.*, 1926, **182**, 519.

78   JOURNAL OF THE SOCIETY OF GLASS TECHNOLOGY.

amount of space to a recapitulation of known facts and ideas. On the other hand, the time seems ripe for an attack on the problem from a somewhat different angle. Our knowledge of the structure and constitution of truly solid, that is, crystalline matter has made considerable progress in recent years, and on the basis of new experimental facts, resulting largely from X-ray analysis of crystals, it has become possible to construct a fairly accurate picture of the inner structure of metals and alloys and of many other crystalline substances, including a number of the complex molecules of organic chemistry. Pictures of this kind cannot unfortunately be directly constructed by similar methods in the case of glass and other amorphous solids. None the less, it appears to the author possible to apply some of the knowledge gained from the study of the structure and behaviour of crystalline matter to the purpose of forming some idea of the nature of the structures which can and cannot be anticipated in amorphous matter. On this basis, it is, perhaps, worth while to seek to build up, as a tentative working hypothesis, some sort of picture of the inner structure of amorphous matter. In the existing state of knowledge it would be surprising if any such hypothesis could be constructed except as a rough first approximation, subject to future amendment or even entire rejection. On the other hand, if such an hypothesis furnishes fresh ground for attack on the whole problem, whether on the theoretical or the experimental side, it will have justified itself even if, in the course of time, it has to give way to better established views. That there is urgent need for further experimental research on the whole problem needs no emphasis for those acquainted with the subject, but one of the greatest obstacles to progress towards the solution of the whole problem is its vagueness and the difficulty of finding points of attack likely to furnish data of real value. On the experimental side as well as on the theoretical, glass has proved singularly elusive.

The modern study of crystal structure has led to the recognition of one general principle which, if not definitely formulated, yet tacitly underlies our whole interpretation of the data of X-ray crystal analysis. This is the fact that the atom rather than the molecule must be regarded as the structural unit. In some of the simpler types of structure, such as those of the metals and their mutual compounds, and in such substances as the alkali halides, we are obliged to carry this principle to the point of doubting whether such a thing as the chemical " molecule " has any real existence in the solid crystal. In the crystal of sodium chloride, for example, each sodium atom has precisely the same relation to each of six adjacent chloride atoms, and similarly each chlorine atom appears to be equally closely connected to each of six adjacent sodium atoms. It is clearly impossible to regard them as paired off into NaCl molecules. Where large and complex molecules are involved, matters may not always be so simple, but there is, even in those cases, some doubt whether there is any real distinction in kind between the inter-atomic bond or linkage which exists between the various atoms of the one molecule and adjacent atoms, belonging, nominally, to different molecules, but forming a cohesion-linkage between themselves. All that we can, perhaps, infer from the behaviour of complex molecules when forming aggregates—either as crystals or as absorption films—is that certain of what we may, for simplicity, term the exterior atoms of these molecules are more ready or able to form cohesion-linkages with corresponding atoms of other molecules. It is a slightly speculative step, but not a very long one, to conclude from this difference of behaviour that certain atoms in a complex molecule are left with more or stronger available " bonds " free to form outside attachments than are others the " bonds " of which are presumably more completely occupied by the intra-molecular linkages.

The bearing of these ideas on our views of the possible structure of an aggregate of complex molecules such as we suppose may exist in glass or other complex amorphous bodies lies in the inference that we may probably regard such aggregates as essentially assemblages of atoms in which certain molecular groupings may occur with greater or less frequency. The main point is, however, that the intra-molecular linkages need not be very different in kind or intensity from the other inter-atomic linkages which must undoubtedly exist or come into being as the substance is cooled down to the quasi-solid state. While we cannot, ultimately, ignore or neglect the existence of closer or more powerful inter-atomic linkages as between certain types of atoms present in such a substance, we may—as a first approximation—think of the whole substance as an assemblage of atoms linked together by " bonds " varying in strength, the strength of any given bond depending on the chemical nature of the atoms concerned and also on other factors among which the actual distance from atom to atom must—as we shall endeavour to show—be of very great importance.

In order to form a conception of the possible internal configuration of an amorphous body, therefore, we may begin by thinking of an assembly of atoms all of the same kind. This is, of course, an ideally simple case and one which, so far as experimental fact yet goes, cannot be realised except under very special conditions. For this latter fact, there is a simple explanation. Where the atoms in an assembly are all alike it is a comparatively easy matter for them to arrange themselves in a regular manner. Simple " close packing " is likely to occur almost automatically in a contracting assembly of such atoms and the immediate result is the formation of the simple type of crystalline structure usual in such substances. For that reason it has hitherto proved impossible to obtain a metal in an under-cooled amorphous form, although metal aggregates formed under conditions where the atoms are not free to pack or arrange themselves can be obtained in a condition which gives no typical crystal reflexion under X-ray analysis. Certain " sputtered " films of metal can be obtained in this condition, and there is, in the author's opinion, much evidence to suggest that more or less completely " amorphous " aggregates of metal atoms can be formed in other ways, although not by the super-cooling of the liquid. Broadly speaking, it is probably true to say that the more complex the molecule of a substance, and especially of a mixture or solution of different substances, the greater is the difficulty which the substances find in crystallising—the greater the ease with which they can be obtained in the vitreous or amorphous state. The experiments of Tammann, who has succeeded, by severe chilling, in producing the vitreous state in a variety of substances, bear out this generalisation, although the specific properties of certain substances naturally create apparent exceptions. In the range of actual glasses we find that the generalisation broadly holds. A simple silicate is much more ready to assume the crystalline state than is a mixture in which a variety of molecular groupings are likely to be present.

If we turn, then, to our ideally simple amorphous substance consisting of an assembly of one kind of atom only, how will such an assembly differ from the crystalline solid of the same composition? Our current conception and definition of an amorphous substance is in reality a negation. We think of a substance as " amorphous " when it does not possess the characteristics of crystalline structure, but this may be to a considerable extent a question of degree. The criterion of crystalline structure which we should probably apply to-day is the power of yielding sharp X-ray reflexions, and this in turn implies regular atomic spacing over a certain region of space sufficiently large relatively to the wavelength of the radiation to be reflected. It is believed that crystalline particles with not more than 40 atoms in line are sufficient to yield reasonably sharp reflexions. Beyond such a limit, therefore, we must conclude that the atomic spacing or arrangement in an amorphous substance must be irregular.

We next come to the question how irregular such an arrangement must or is likely to be ? Experiment shows quite conclusively that in crystalline substances a certain degree of irregularity is possible, but that it is strictly limited. Distortion of a crystal lattice can be brought about by several means, but the variations of inter-atomic distance always lie within narrow limits and these limits appear to be, if not identical, at least closely of the same order, whatever the method of lattice distortion adopted. For instance, a crystal lattice can be distorted and stretched in certain directions by the application of stress, whilst the same lattice can also be stretched (in this case expanded in all directions) by the application of heat, and it has recently been shown by one of the author's colleagues at the National Physical Laboratory that the maximum lattice parameter producible by strain is closely the same as that producible by thermal expansion. In either case, when this limit is exceeded, the lattice breaks down, under stress by fracture or plastic yielding, and under heating by fusion.

On these and similar grounds derived from other lines of experimental evidence, the author some years ago drew the inference that for any pair of atoms there is a maximum " parameter "— that is, a limiting distance apart—beyond which the forces which constitute the inter-atomic bond, cease to operate. This is not, in reality, a new principle, since it has long been tacitly recognised that there is a well-defined distance apart which constitutes what was formerly called the " sphere of molecular attraction." By thinking, however, of inter-atomic distances, whether within a crystal, an irregular atomic aggregate, or even a chemical " molecule," a series of interesting and important inferences can be drawn. These the author has previously applied to the explanation of a series of well-known phenomena in the behaviour of crystalline solids, and he now proposes to consider how far corresponding results can be obtained in regard to amorphous substances in general and to glass in particular.

If the principle of a definite limiting inter-atomic distance, beyond which inter-atomic linking is impossible, is accepted, it at once furnishes us with a law governing any random assembly of atoms. We find that the atoms must arrange themselves in such a way that, in the first place, their centres are never less than a certain minimum distance apart, whilst they will tend— in order to " satisfy " as many of their free bonds as possible—to place themselves within the above-mentioned limiting distance apart. Complete satisfaction of the bonds, and consequently minimum potential energy of the system, implies close packing and therefore crystallisation. The conditions under which the assembly is formed, however, are supposed to have prevented the formation of this ultimately stable grouping, except possibly in

small regions of a few hundred atoms at a time. These, for the
moment, we will ignore and direct our attention to the really
irregular regions. There is, however, another factor to be considered in the case of any irregular atomic assembly. It is difficult
to suppose that atoms which so persistently build themselves into
lattices of particular types behave towards each other as if they
were small spheres or points from which emanates a uniform spherical
field of force. In other words, it is difficult to avoid the idea that
inter-atomic forces are vectorial and that inter-atomic linkages
tend to be formed in certain definite directions, that is, that the
lines joining the centre of an atom to the centres of other atoms
to which it is linked tend to form certain definite angles with one
another and that linkages in other directions, if they can occur at
all, imply a certain distortion of the atom or its surrounding force-
field which implies the storage of energy. If this view is accepted,
it follows that even in an irregular assembly of atoms there will
be a tendency for the atoms to form linkages only in certain relative
directions or to turn round in such a way as to bring the angles
between linkages as close to the normal angle as possible. The
formation of linkages in an irregular assembly is thus limited by
two factors—inter-atomic distance and angular relations between
adjacent atoms.

The first inference which can be drawn from these considerations
is that in an irregular assembly only a portion, and possibly only a
small portion, of the possible inter-atomic linkages can be operative.
It is only where distance and angles are favourable that a linkage
can be formed. The consequence is that the latent heat of fusion,
which, in this way of regarding matters, represents the energy
stored in unsatisfied bonds, will be largely retained during the
cooling and " setting " of such an assembly; that is, the amorphous
body will show comparatively little absorption of heat during
melting and little or no noticeable evolution of heat during cooling.
This, of course, is one of the most striking facts about amorphous
substances generally. There is, however, another inference which,
if less obvious, is none the less important.

From the physical constants of a substance, particularly the
total heat of vaporisation, it is possible to calculate its theoretical
internal cohesion. The results obtained are, however, found to be
some twenty or more times higher than the cohesion indicated by
the resistance of the substance to rupture. A. A. Griffith, in his
remarkable paper on " The Phenomena of Flow and Rupture in
Solids," has endeavoured to account for this discrepancy by postul-
ating the existence within a solid of minute flaws or fissures, the
presence of which causes a great concentration of stress under load,

leading to rupture under an average stress far lower than that calculated from the physical theory mentioned above.* Whilst this view is not applicable to ductile crystalline solids, it fits in well with the general picture of the structure of amorphous aggregates which is here being presented, more especially in view of some suggestions developed below. At this point of the argument, however, it is interesting to notice that if only a fraction of the available inter-atomic bonds is brought into action in the ordinary "setting" of an amorphous body, we cannot expect it to develop the full theoretical strength, although on this basis we should not anticipate a discrepancy of more than at most 50 to 70 per cent. as against the actual discrepancy of the order of 90 per cent. or more. A. A. Griffith, however, has succeeded in producing one amorphous body—fused pure silica—in a special condition in which it approaches the theoretical cohesive strength. If our view of the structure of amorphous matter is, broadly speaking, correct, we should infer that in this super-strong silica the great majority of the inter-atomic bonds are in operation. If this inference is correct, we should expect the special form of silica to have a higher density than in the ordinary state and to show different thermal properties. We should not expect to find—for reasons elaborated below—any definite latent heat of fusion, but an apparently higher specific heat, due to the gradual breaking of inter-atomic bonds during heating. We should further anticipate that alternate heating and cooling would rapidly destroy the super-strength of the material. It must, however, be borne in mind that the super-strength of Griffith's silica may be only uni-directional; that is, that his process of rapid drawing into a rod or thread at a high temperature may make it possible for an exceptional number of inter-atomic linkages to be formed in a longitudinal direction, whilst in a transverse direction the linkages may be similar to, or even fewer than, those of the material in the ordinary state. This would imply that the total number of atomic linkages per gram need not be much larger than in the same material in the weak condition, and in that case the effect on the other physical

* A. A. Griffith, "The Phenomena of Flow and Rupture in Solids," *Phil. Trans. Roy. Soc.*, 1920, A 221, 163. Apart from the conception of flaws mentioned above, this paper contains many ideas on the inner structure of amorphous bodies which do not differ widely from those put forward here, and this the author wishes fully to acknowledge. Dr. Griffith, however, views the problem from an entirely different angle and not only arrives at a different picture of the structure of glass, but applies it to the explanation of a different set of phenomena. The two views are not irreconcilable, but it would have confused the present argument to refer specially to similarities of ideas where they occur.

properties may not be very marked. A further fact to be noticed is that the super-strong state is not stable and passes spontaneously into the weak state. This implies that the atomic arrangement of the super-strong state entails a higher potential energy content than does the weak state. On the present view, this might be interpreted as meaning that, while a larger number of linkages are formed during this special drawing process, some of them are left in a state of severe strain. Establishment of equilibrium would then occur by the breakdown of the severely strained linkages, possibly with the formation of other less strained linkages, perhaps in a transverse direction. It would be interesting to ascertain whether the change from strong to weak is accompanied by an evolution or absorption of heat, and how the other physical properties are affected.

Considerations of the kind which have just been discussed in regard to the super-strong state may be applied with perhaps greater certainty to the amorphous structure in its normal or weak state. Here the idea at the root of our considerations is that the total number of atomic linkages is materially smaller than in the stable crystalline arrangement, and that whilst in the undistorted crystal lattice all the effective linkages are, between given kinds of atoms, alike, those in the amorphous assembly will differ widely, both as regards inter-atomic distance and in regard to the angular relations between the atoms. In this sense, the amorphous structure is far less homogeneous than the crystalline. That this is really the case can be inferred from the fact that while the crystal breaks down completely at a single temperature, that is, has a definite melting point, the amorphous substance breaks down its atomic linkages gradually with rise of temperature and therefore has no definite melting point or even range. This, however, implies an interesting effect on another thermal property, namely, the specific heat. As temperature rises, at all events after a certain point, to which we shall refer below, has been passed, each increment of temperature implies that those pairs of atoms which are already near the limiting range of inter-atomic cohesion will be brought beyond that limit and their linkage will be broken. This entails a storage of energy which may be described, as is done by Griffith, as " surface energy " owing to the formation of a free surface where the bond has been broken, or—as the author prefers— as of the nature of local melting, with a corresponding storage of latent heat. The heat necessary to raise the temperature of an amorphous solid, therefore, must consist of two parts, one required to increase the amplitude of thermal oscillation of the atoms and the other to supply these small increments of latent heat. The apparent specific heat of a vitreous substance, therefore, at temperatures above the limit already mentioned, should be appreciably higher than that of a crystalline substance of the same composition. Unfortunately, satisfactory data for the verification of this inference are not available. The interpretation of such data as exist is also complicated by the differences between specific heat at constant pressure and at constant volume. Some data on the specific heats (at constant pressure) of vitreous silica and of quartz and cristobalite have been determined by W. P. White * and for the range of temperatures under consideration we find the following :

Temperature.	600°	700°	800°	900°
Mean atomic heat of vitreous silica	5·48	5·58	5·68	5·75
Quartz	—	5·46	5·58	5·65
Cristobalite	5·46	5·55	5·62	5·67

Since the coefficient of thermal expansion of vitreous silica is very small, while that of quartz and cristobalite is appreciably higher the correction for bringing these values to the corresponding figures for atomic heat at constant volume, will increase the differences between the vitreous and crystalline materials. As far as they go, therefore, these data bear out the inference suggested by our hypothesis. Further, in summing up the results of his measurements, White says : " Glasses show, in the main, a specific heat only slightly above the corresponding crystal forms, but with a tendency to increase at some rather higher temperature." It is this increase at higher temperatures, that is, in the temperature range where the process of " local fusion " or breakage of inter-atomic bonds is most active, that is to be anticipated from the present hypothesis.

Repeated reference has been made above to a limit of temperature below which thermal expansion alone causes few or no ruptures of inter-atomic bonds, whilst at higher temperatures such rupture occurs to a slowly increasing extent as temperature rises. The existence of such a limit, if it is accepted, need not imply that the vitreous substance is, below that limit, markedly more " solid " than above it. The limit referred to relates to the effect of thermal expansion alone, and must be conceived as applicable only under a given system of stress—which may, of course, be merely atmospheric pressure, or even zero. If the stress system is changed, that in itself may bring about ruptures of inter-atomic bonds which may continue to occur, at the same temperature, indefinitely. In that case, we should describe the substance as a viscous liquid. On the other hand, the stress system may not be sufficiently severe to cause the rupture of any bonds and in that case the material

* W. P. White, " Silicate Specific Heats," *Amer. J. Sci.*, 1919, **47**, 1.

behaves as an elastic solid. In an intermediate range, on the
other hand, rupture of bonds may occur for a time after stress
has been applied, but this may set up sufficient internal stress,
that is, lead to a storage of sufficient energy to put a stop, after a
time, to further rupturing of bonds. In that case, if the external
stress is removed there will be a gradual and incomplete return
towards the original configuration, the stored-up energy being
released and a certain number of ruptured bonds, or their equiv-
alents, will be restored. This is the typical behaviour of a substance undergoing visco-elastic * deformation. The range of stress
that brings about these different results will, naturally, depend to
a very great extent on the temperature. At relatively low temperatures it may happen that the stress required to bring about
true viscous flow may be so high that the material ruptures under
it. In that case, the material must be regarded as truly solid or
at least as quasi-solid, and not as a highly viscous liquid. There
is some reason for thinking that in many ordinary varieties of
glass at or near room temperature this is the case. Thus, in the
paper already referred to, A. A. Griffith finds that "below 730°
the glass was not a perfect viscous liquid," basing this conclusion
on measurements which showed the existence of a certain small
measure of true "strength" even at that relatively high temperature. The well-known phenomena of apparent flow and very
real deformation occurring slowly in glass kept under load are, on
this view, to be regarded as being visco-elastic and not truly viscous
in character. Attempts to measure viscosity in glass at low temperatures have led to the same view.†

In order to appreciate the further bearings of the matter just
discussed it is necessary to look into the mechanism of what we
conceive to occur in a vitreous substance under the action of either
stress or thermal expansion or of both. We have already noted
that in a heterogeneous assembly of atoms such as our hypothesis
supposes glass and other amorphous substances to be, atomic
bondings must exist which vary widely, both in regard to inter-
atomic distance and in regard to what may be termed the "angular"
factor. Neglecting the latter for the moment, we may consider
inter-atomic distance only. We must, indeed, simplify the matter
still further by leaving out of consideration all those strong inter-
atomic linkages such as exist between the atoms of well-marked
chemical compounds and confine ourselves to cohesion bonds such
as we must suppose to exist between atoms in an assembly consisting entirely of one kind of atom. In such bonding, we must

* Michelson, *J. Geol.*, 1917, 25, 405.
† Stott, this J., TRANS., 1925, 9, 210.

suppose that the inter-atomic force, that is, the net resultant attraction, is some function of inter-atomic distance which increases as that distance becomes greater or less than a given value. The increase with decreasing inter-atomic distance is simply another way of stating the very high resistance to compression or, in other words, the difficulty of forcing atoms too closely together. On the other hand, any attempt by external stress to pull the atoms further apart also meets with a resistance which increases the greater the stretch which is applied. If Hooke's law, which is known to apply to the resistance of an aggregate of atoms to extension in any one direction, could be applied to a pair of atoms, we could state that the inter-atomic force increases linearly with the distance apart of the atoms up to a certain limit—previously mentioned as the " maximum parameter " or inter-atomic distance of cohesion. We need not, however, seek to establish a law of inter-atomic attraction, but cannot escape the general conclusion that the stretching of an inter-atomic bond brings into play a resistance or attractive force which increases as the distance between the atoms increases.

If, then, we have an assembly of atoms irregularly arranged, in which inter-atomic distances vary from a minimum, corresponding to close packing, up to or very nearly up to the maximum range of cohesion, the application of either stress or rise of temperature will bring about results which can be clearly inferred. Rise of temperature will cause a general expansion and, therefore, an increase in the average inter-atomic distances. But those pairs of atoms which are already far apart will resist such further separation more strongly than will the pairs of atoms which lie more closely together. Whilst, therefore, some degree of further separation will occur everywhere, it will occur to a greater extent where the atoms are closest together. If the substance is very far below its normal melting point, we can infer that the inter-atomic distances will be, on the whole, well within the maximum range and a considerable degree of more or less uniform increase of inter-atomic distances will result. Little effect will be produced. But at a certain stage some of the atomic bonds will begin to approach the limiting range and at this stage the existing configuration will tend to preserve itself by concentrating the inter-atomic widening on the more closely-packed pairs of atoms, leaving the distances apart of the more widely spaced pairs almost unchanged. This will happen because, and so long as, the energy thereby stored in the system is less than that which becomes stored or latent by the rupture of inter-atomic bonds. As temperature rises, such an arrangement will more or less suddenly—depending on the rate of heating—become unstable. The strain-energy stored by the expansion of the originally closer-packed atoms will be relieved by the sudden rupture of a number of the more widely-spaced inter-atomic bonds. A certain amount of heat will be evolved by the release of the strain-energy in question, whilst some heat will be absorbed or become latent as the result of the rupture of inter-atomic bonds. The difference will make itself felt as a very slight thermal effect if the temperature of the glass is followed with sufficient accuracy and if the rate of heating is rapid enough to make the effect sensible. It must, further, be anticipated that the *rate* of heating will affect the temperature at which the break-down will occur. Factors of a statistical nature enter into the question, since the moment at which inter-atomic bonds will begin to rupture will be determined by the thermal oscillations of the atoms. Rup-ture will occur when the state of oscillation of a particular pair of atoms, the inter-linkage of which is near the breaking-point, results in an added impulse towards separation. Here a time element arises, and the shorter the time allowed at any given temperature, the higher the temperature at which rupture will begin. This effect is in many respects analogous, inversely as to temperatures, to the under-cooling phenomena the intensity of which, that is, the degree of under-cooling, is similarly affected by the rate of change of temperature.

It will be seen at once that this explanation fits the observed facts in regard to the slight thermal effect found during the rapid heating of glass, originally by Tool and Valasek * and subsequently confirmed by other workers. More recently still, Le Chatelier and his co-workers † have shown that this thermal effect is not a specific property of glass but occurs, at the corresponding viscosity, in all the amorphous substances as yet examined. Since the working hypothesis outlined above is equally applicable to any amorphous substance, it meets the necessity of offering an explanation not confined specifically to glass or its constituents. It also meets the second striking fact about this phenomenon, namely, that it can only be detected during the heating of the glass and then only if the heating is rapid. An instability of the kind suggested cannot occur during cooling whilst, if the rate of heating is slow, the breakdown will probably be local and the degree of " over-heating " upon which the effect depends very slight.

A further important fact in regard to these thermal effects in amorphous materials is the very minute amount of heat which is involved. This implies either that the total number of atoms

* Tool and Valasek, *Bur. Standards Sci. Papers*, No. 358.
† Samsoen and Monval, *Compt. rend.*, 1926, **182**, 967.

taking part in the change is very small or that the external effect is the difference between two effects occurring simultaneously within the substance. The explanation given above shows that, in a sense, both factors apply. The number of inter-atomic bonds likely to rupture suddenly when the limiting temperature is passed must involve only a fraction of the atoms of the substance, whilst the sensible heat-effect is, as has been suggested, a difference between strain energy released and latent heat of fusion absorbed.

Thus far the working hypothesis which is here put forward appears to fit the known facts satisfactorily, but the matter can be pushed rather further. During the stage of heating at all events immediately below the change-point which we are discussing, the effect of rise of temperature on thermal expansion is to some extent lessened by the storage of strain-energy to which we have referred. Far below this limiting temperature it would be expected that expansion would be normal, in the sense that it would produce purely statistical increase in the inter-atomic distances, but as the limiting temperature is approached and the storage of strain-energy commences, it would be anticipated that there would be a small but steady diminution in thermal expansion—possibly so small as not to be easily determined. At the point, however, where the thermal effect occurs and the breakdown of a relatively large number of inter-atomic bonds takes place, there should be a rapid and marked expansion, followed by a generally higher rate of thermal expansion, since each increment of temperature would provoke the rupture of some further bonds. This, again, fits the experimental facts. Samsoen has shown that this anomaly in thermal expansion is found in a number of amorphous substances accompanying the small thermal effect already discussed. The expansion effect, unlike the thermal effect, would not appear to be due to a difference between two opposed effects and would therefore be expected to be of greater intensity than the thermal effect, and the experimental evidence appears to justify that inference also.

Since the phenomena which occur in amorphous bodies in association with the thermal effect have not been more fully studied, it is not possible to test the validity of the present hypothesis by an appeal to additional experimental facts, but the hypothesis offers certain suggestions for further experimental investigation which may make a crucial test possible. One of these is the effect of pressure or stress on the thermal effect itself and on the accompanying expansion. On the present hypothesis, hydrostatic pressure, by lessening the average inter-atomic distance, must raise the temperature required to produce the instability which gives rise, by its breakdown, to the thermal effect. Heating-curves taken under high pressures could be used to verify this inference and the quantitative relation between pressure and the temperature at which the thermal effect occurs for a given rate of heating might furnish data bearing upon the actual changes of maximum inter-atomic distances involved. Further, if—as we should anticipate—the expansion effect is raised in temperature by the same amount as the thermal effect, the intimate connexion of the two would be thereby definitely established. Some effect on these phenomena may also be anticipated if heating-curves, for example, could be taken with the substance—not necessarily glass—under tensile or shear stress. The substances are, of course, appreciably viscous at the temperatures in question, but the maintenance of a constant stress would none the less be possible. As it is not possible to conclude directly from the applied stress what effect is produced on average or maximum inter-atomic distances, however, only qualitative results could be obtained.

In what has been said above, attention has been concentrated on the effects of temperature on an irregular assembly of atoms. If we consider briefly the effects of stress, some interesting inferences also emerge. It is not proposed here to enter far into the question of the mechanism of rupture in amorphous materials. This has been very fully discussed by A. A. Griffith in the paper already cited, although he approaches the subject from the "molecular" point of view and postulates, not an irregular assembly of atoms, but rather layers of molecules arranged in curved sheets. Yet many of the considerations advanced by him are almost equally applicable under the rather more generalised view of amorphous structure which is here put forward and it is not necessary to traverse the ground again. But the ideas here under discussion suggest an inference in regard to the elastic behaviour of amorphous bodies which it may be worth while to develop. In a crystalline body, even where the regular lattice structure is slightly distorted by the presence of stranger atoms or otherwise, the differences in inter-atomic spacing which exist cannot make themselves felt, so far as local rupture is concerned, in regard to elastic behaviour because the material—owing to its slip or cleavage mechanism—gives way in a plastic manner long before even the most widely-spaced inter-atomic bonds can be brought to rupture—or, rather, they are brought to rupture only with the production of slip or cleavage. In the amorphous structure such as is here postulated, however, the state of affairs is different. There is no slip or cleavage mechanism and it seems possible that the application of quite low stresses may bring about the rupture of some of the more highly-distended inter-atomic bonds. Under a uniform hydrostatic tension, it is probable that the higher resistance to further separation offered by the widely-distended bonds would lead to a piling-up of strain and of strain energy among the more closely packed atomic groups, in much the same way as we have supposed in the case of thermal expansion. But where a uni-directional stress acts on the substance, certain of the more distended bonds lying in the direction of maximum stress will not be able to escape the extension imposed on the mass, with the result that with each increment of stress a certain number of these bonds may be expected to be broken. Whilst the subsequent phenomena will be to some extent of the visco-elastic nature, yet the apparent modulus of elasticity (Young's modulus) must be expected to be decidedly higher than that of the same substance in the crystalline form where no ruptures of extended bonds will occur under similar stresses.

This inference, again, is verified by known experimental fact. Measurements of Young's modulus for vitreous silica and for quartz have been made by Auerbach * and these show values, in kilograms per sq. mm. for vitreous silica = 6970, for quartz parallel to the axis = 10,620, and for quartz at right angles to the axis = 8566. Verifications on other amorphous and crystalline substances, particularly of more complex constitution, are, of course, necessary before any very great weight can be attached to the apparent truth of this further inference, but the known facts, at least, are entirely consistent with the hypothesis here put forward.

A question of some difficulty arises in this connexion regarding the effect of rise of temperature on the elastic modulus, especially in tension. If increased inter-atomic distance (within the limits of possible cohesion) implies an increased inter-atomic attraction, this would at first sight seem to imply a corresponding increase in Young's modulus. But if rising temperature at the same time leads to a progressive rupturing of the most extended bonds, then the reduction in the total number of active bonds should counterbalance, and might more than counterbalance, the other effect. It would thus follow, on the basis of our hypothesis, that the change of modulus with temperature would be small and might be either an increase or a decrease. Experimentally, it appears to be a very small change, but decidedly a slight decrease.

In view of the apparent success of the working hypothesis put forward above in explaining a series of somewhat remarkable experimental facts which have been observed in amorphous substances, it may be interesting to inquire whether it can throw any light on other features in the physical behaviour of such materials. The phenomena of diffusion in solid and quasi-solid bodies are of

* Auerbach, Ann. Physik, 1900, (4), **3**, 116.

peculiar interest, since their explanation on the older lines of the kinetic theory of matter has proved unsatisfactory in regard to crystalline solids. The author has advanced a theory of diffusion in crystalline solids * which suggests that such diffusion occurs by movements rendered possible by the slip-mechanism which exists in plastic solids. The movement occurs, when favoured by thermal oscillation, in such a way as to relieve stresses set up in a crystal lattice where stranger or solute atoms are unduly concentrated. This theory at once accounts for the fact that the power of allowing diffusion is strictly associated with plasticity in crystalline matter; it is great in ductile metals, less in brittle metals and substantially non-existent in crystalline salts. It further accounts for the fact, established by Hevesey, that radio-active lead does not diffuse into ordinary lead, the explanation being that the atoms of the radio-active isotope are so closely similar to those of the rest of the lead that they cause no appreciable distortion of the lattice by their presence and therefore set up no stresses capable of causing movements in the slip mechanism. Neither of these facts can be explained on the "kinetic" view of diffusion in solids, since on that view the interchange of atomic positions would occur in any kind of crystal and without the need of local stress to set up motion.

If we admit that diffusion in crystalline solids depends on or is at least closely associated with the slip-mechanism, what is to be expected in a vitreous substance in which there is not slip-mechanism since there is no extensive crystal structure ? It is at once evident that diffusion of the type which occurs in crystals cannot occur in an amorphous substance and, indeed, it follows further that at relatively very low temperatures, where the super-cooled liquid is quasi-solid, so that such inter-atomic bonds as exist are all or nearly all formed between pairs of atoms at distances apart well below the limiting distance, there should be no diffusion at all. This is borne out by experience; glass shows no signs of diffusion phenomena at the ordinary temperature, whilst its high electrical resistivity also suggests the absence of diffusion, since electrolytic conduction, such as glass exhibits at higher temperatures, is dependent on translational movement of atoms in a manner akin to what occurs in diffusion.

At higher temperatures, however, where, on the present view a proportion of the inter-atomic bonds are near the possible limit of extension, conditions are different and our hypothesis would suggest that diffusion may occur somewhat on the lines of the "kinetic" theory, except that the details of the mechanism would be very

* Rosenhain, "The Inner Structure of Alloys," May lecture to the Institute of Metals. *Journ. Inst. Metals*, 1923, **30**, 3.

THE STRUCTURE AND CONSTITUTION OF GLASS.   93

different from what occurs in a mobile liquid or in a gas. Even in a glass of relatively moderate viscosity, such as would be encountered in ordinary medium-hard glasses in the neighbourhood of 400 to 500°, we cannot suppose that atoms—and still less molecules—are free to wander about through the mass, however slowly, in a practically free state, merely hindered or directed by successive collisions with their neighbours. On the contrary, our hypothesis suggests that at such a temperature a certain number of interatomic bonds must be so close to their limiting extension that, under the vagaries of thermal oscillation, rupture must occur from time to time between many pairs of atoms scattered through the mass. But where such rupture has occurred, the atoms concerned, and possibly some of their immediate neighbours, must receive, for a longer or shorter period of time, a certain measure of freedom of movement—a power to rearrange themselves. If the temperature continues to rise, or if there is an applied stress which tends to pull the atoms apart, there will be no or very few recombinations of the broken bonds, either as between the former partners or with other adjacent atoms. But where such disruptive forces are not at work, there must, on the average, be as many bonds formed per second as are broken. In the absence of any motive force, the result would probably be an oscillation about some mean configuration, although even then, by a process of bond-breaking and re-forming with other partners, any individual atom might wander for considerable distances through the mass. If, however, there is a motive force at work, such as an electromotive force due to application of external voltage, or a motive force due to the presence of some different type of atom producing a concentration gradient, then a general movement in the direction of that motive force must take place, that is, there will be conduction and electrolysis in the case of the $E.M.F.$ and diffusion in the case of solute concentration. Both will be very slow indeed at low temperatures, but will become more and more active at higher temperatures.

The mechanism of diffusion which we have just outlined on the basis of our hypothesis is, however, substantially identical with the mechanism of viscous flow and it would follow that in vitreous or amorphous bodies diffusion is associated with viscous flow in precisely the same way as diffusion in crystalline solids is associated with plastic deformation by slip.

For the purpose of simplifying the arguments as much as possible, the discussion of the hypothetical structure of amorphous bodies has so far been based, as was indicated at the outset, on purely atomic considerations and no attempt has yet been made to take into account the fact—which there is little reason to doubt—that a great part of such a substance must be occupied by atoms arranged in definite molecules, more or less complex in nature. It is also possible, so far as experimental evidence goes, that aggregates of atoms or even of molecules may be present in the form of very minute crystallites, too small to yield X-ray reflexions. The effect of the existence of such aggregates, in which, as a rule, inter-atomic bonding is probably too close and regular to be subject to the type of phenomena which have been suggested above, must be essentially that of the presence of neutral or "dead" zones, the phenomena we have discussed being confined to the remainder of the material. But other effects may have to be considered. The bonding between one atom which is an outer member of a molecule and another atom which is, perhaps, a "free" atom in the sense of not being part of any such organisation, will probably be some-what different from the bonding of two similar atoms both of which are "free." There is at present no means of knowing whether a bond between a "molecular" and a "free" atom will be weaker or stronger, or more or less easily stretched than the normal bond between "free" atoms. Further, the atomic bonding even within a "molecule," and certainly within any minute crystal, must be affected to a greater or less degree by the external bonding of the outer members of the organisation, and it is quite possible that, in some of the conditions indicated above, occasional rupture of bonds may occur even within the "organisations." A great deal must depend on the closeness of inter-atomic bonding within these molecules. As has already been indicated, there is probably, in a complex substance, a wide range of bondings, differing in closeness and strength. Yet there is evidence that it is possible, in certain circumstances, to bring about rupture of inter-atomic bonds which must be regarded as of the closest possible "chemical" type. Thus, when glass is subjected to electrolysis at temperatures above 400° it is possible to replace, electrolytically, one of the alkali metals by another within the mass of the glass, the ions or atoms of one metal entering a sheet of glass at one side, whilst those of the other metal leave the glass at the other side. Here, rupture of bonds between an alkali metal and oxygen must have occurred. Further, the changes of colour which are known to occur in glass containing oxides both of iron and arsenic under the influence of sunlight or of ultra-violet light can only be accounted for by "migration" of oxygen atoms, originally combined with iron, into combination with arsenic, or, at all events, by some internal chemical change of that order. This occurs even at room temperatures, and although it may not imply any translational movement of atoms through the glass, but only an exchange of bonds between pairs of neighbouring atoms, yet it goes to show that even the stronger " chemical " bonding is by no means invulnerable. In view of these facts, it may be regarded as reasonable to suppose that the general behaviour of glass in certain physical respects, such as have been discussed above, corresponds fairly closely with what might be expected of an assembly of independent atoms. The same considerations also account for the fact that many of the properties of glass, when looked at as a function of chemical composition, are not very far from being additive. If strongly-individualised molecules of definite compounds exerted a predominant influence in the structure of glass we should expect to find strongly marked maxima or minima, or at least breaks in the property-curves of glass corresponding with the compositions of those molecules. In fact, such features are only rarely encountered, principally in the simpler types of glass.

There are, however, possible ways in which the presence of well-defined molecules, particularly if they are of relatively large dimensions, should make themselves felt. Perhaps the most likely direction in which such effects might be sought is in the behaviour of glass surfaces and in thin films of glass. If we could think of a complex glass, in the mobile or semi-mobile liquid state, as consisting of such a type of molecule in solution in a liquid composed, for the rest, of heterogeneous congeries of atoms, it might be expected that these larger molecules might become concentrated at or close to the free surface of the liquid glass. Such a concentration of special molecules near the surface would readily account for the marked difference in chemical behaviour, especially, of " fire-polished " glass surfaces when compared with fractured or polished surface produced within the mass. These differences are well known in connection with the durability and chemical resistance of blown glass-ware. Where glass is to be used in the polished condition, as in the production of lenses, prisms, etc., a much greater degree of sensitiveness to surface deterioration is found than when the same glass is used in the fire-polished condition of blown ware. A glass relatively rich in alkalis and low in lime content, which in the form of a polished object would become rapidly dimmed, suffers very little deterioration when used as a bottle or a tumbler. In part, this is no doubt a question of the care with which the quality of the surface is regarded in the two cases, but even allowing for this factor, a considerable difference remains. A further fact of glass-making experience also appears to bear on these ideas. A glass which is perfectly satisfactory when used as blown ware of reasonable thickness becomes unduly sensitive to atmospheric deterioration when blown into very thin sheets. Such extremely thin sheets are produced by the rapid extension of a thicker layer of viscous glass, usually by the aid of centrifugal forces, and during this process, since the glass is cooling and "setting" very rapidly, there would be no time for the migration of the supposed molecules to the surface to form a special layer. The protective effect of a surface layer of well-defined molecules would thus be lost in the very thin sheets and their greater tendency to deterioration explained.

Were this hypothetical "adsorption" of layers of definite molecules from a vitreous solution really to occur, both in glass and in other kinds of amorphous matter, it would appear probable or at least possible that the consequent arrangement of molecules and of their constituent atoms should give rise to some form of X-ray reflexion effect. Even if such molecules formed a layer several molecules in thickness, and the molecules themselves were arranged in an approximately regular parallel manner, the regularity achieved in this way would scarcely be sufficient to yield reflexions of the sharp nature obtained from crystals, where great regularity exists over very much larger regions and much deeper layers. Glass, however, and many liquids yield X-ray reflexions which take the form of broad, indefinite bands which may, perhaps, be associated with the presence of surface layers of the kind here suggested. This is a matter which is capable of experimental investigation, since liquid solutions, and ultimately glasses, could be prepared which may be expected to contain large, well-defined molecules, and these could then be examined with a view to determine whether any correlation exists between the presence of such molecules and the X-ray reflexions obtained.

Other methods of studying the surface phenomena in glass suggest themselves. Thin films of many substances have been studied by Langmuir and others, films only a single molecule in thickness being obtained by the spontaneous spreading of suitable substances over prepared surfaces—such as films of certain oils on clean water. The preparation of glass films of such tenuity is obviously not possible or certainly very difficult if it is attempted to let molten glass spread over any solid body—even the surface of heated platinum would not readily serve the purpose. If, however, liquid glass were placed, in very small quantities, on the surface of a suitable molten metal, spreading over the surface might occur if the temperature were maintained. The formation of oxides of the metal, which would rapidly dissolve in the glass, would have to be avoided, and this might be done by using molten silver or gold for the substratum. Silver, however, probably as the result of chemical inter-action with the glass, rapidly passes into and colours molten glass in contact with it. Gold, being more inert, might behave more satisfactorily. Were it possible to overcome difficulties of this sort, the study of such thin films, when produced from glasses of different compositions, should serve to throw light on the problems here under discussion.

It is scarcely possible, in the existing state of our knowledge of the molecules present in glass, to carry the discussion of this aspect of the subject much further. Reference to it has, indeed, been made mainly in order to indicate the kind of bearing which the hypothesis as to the inner structure of amorphous matter may have on our views of the molecular structure of glass and more especially to indicate the probable relation of experimental data derived from the study of glass surfaces in particular to our hypothesis.

The attempt to formulate a theory of the atomic structure of amorphous bodies in general, and of glass in particular, which has been made in the preceding pages can claim nothing more than that the hypothesis which has been put forward appears to be consistent with views of atomic structure which have proved useful and in a large measure successful in their application to the theory of the inner structure of crystalline matter. On the other hand, it offers suggestive explanations of a series of well-established, if not very prominent, phenomena which have been discovered in amorphous matter. In the absence of any very acceptable or well-founded general theory, the hypothesis here discussed may perhaps prove useful not only as a basis for the discussion and explanation of known facts, but as a stimulus to further experimental investigation.

The author is indebted to his colleagues at the National Physical Laboratory, and especially to Mr. V. H. Stott, for valuable help in connexion with the preparation of this paper.

XV.—*The Structure of Glasses; The Evidence of X-Ray Diffraction.*

By J. T. RANDALL, H. P. ROOKSBY, and B. S. COOPER.

(Communication from the Staff of the Research Laboratories of the General Electric Company Limited, Wembley.)

Introduction.

MATTER in the glassy state has always excited curiosity, and speculations as to its constitution, based on its physical and chemical properties, have frequently been made.

A symposium published by the Society in 1925 indicates how widespread is the interest on this subject and emphasises how little is really known.

Glass has been described most generally either as an amorphous solid or as a super-cooled liquid.

It is obvious at the outset that glasses do not consist of crystals which can be recognised as such by the naked eye even when assisted by powerful microscopes; hence the word "amorphous," which has in the past been synonymous with lack of order and visible structure. On the other hand, it is equally obvious that glasses do not possess many of the properties of ordinary liquids.

During the last year or two experimental evidence has been accumulating on the subject of X-ray diffraction in liquids.* The results of this work, which have been corroborated by other workers, seem to show that even in liquids the atoms or molecules are always trying to arrange themselves in the regular fashion associated with the crystalline state. *A fortiori*, then, it is to be expected that the same thing would hold for glasses, which at any rate have a chance of growing crystals, however small. We therefore set out to examine a number of glasses by means of X-rays for evidence of incipient crystallinity. Should any such evidence be obtained, it is possible that it would be of help in the better understanding of some of the properties of glass.

It has been possible, in fact, to show that many glasses do consist of extremely small crystals, the order of size being 10^{-6} to 10^{-7} cm. For instance, vitreous *and* "amorphous" silica have been shown to consist largely of cristobalite crystallites of size about $1\cdot5$ to $2\cdot0 \times 10^{-7}$ cm. The vitreous and "amorphous" states are identical in structural unit, and this unit is really crystalline. This paper is an attempt to give a fairly brief account of the results obtained for several glasses. A preliminary account had already been given in

* G. W. Stewart, *Phys. Rev.*, 1927, **30**, 232; 1928, **31**, 174, etc.

Nature,* and a more detailed paper is shortly to appear in the *Zeitschrift für Kristallographie*, to which readers are referred for fuller details of the theoretical basis of the method used.

Experimental Arrangements.

The X-ray apparatus used in this work was of the Debye–Scherrer type. K–α radiation from the tube (which is of the hot cathode type) passes through a slit system, in front of which is placed the specimen. The diffracted radiation then passes into a cylindrical camera, round the circumference of which is placed a thin strip of film on which the X-ray pattern is registered. The apparatus is calibrated with standard substances, such as NaCl, so that for any line on the film, " d," the distance between planes of atoms, may be calculated according to the Bragg Law $2d \sin \theta = m\lambda$, where 2θ is the angle of diffraction and λ the wave-length of the radiation. Instead of a cylindrical camera, a plate camera may be used in conjunction with a pin-hole, through which the incident rays pass. This method gives rise to circular rings on the film; with the cylindrical camera lines are obtained. The " ring " films are better suited to visual examination, the " line " films to accurate photometric measurement. CuK–α and Mo–Kα radiations have been used in the present experiments. The former gives rise to openly-spaced lines for crystalline powders, the latter, being shorter in wave-length, crowds them together.

The diffraction pattern of a crystalline powder, the particles of which are, say, 1μ in size, consists of a series of comparatively sharp lines or rings. As the particles decrease in size these lines broaden out in simple cases according to the well-known law of Scherrer † until for particles of $1\mu\mu$ in size broad bands are obtained. The first examination of a glass by means of X-rays is recorded by P. Debye and P. Scherrer,‡ who obtained a broad diffraction band. Other workers, including Parmelee, Clark and Badger, and Clark and Amberg § have also obtained broad diffraction bands. Our own observations have also shown that good glasses invariably give rise to broad diffraction bands, superimposed lines being due to devitrification products.

The X-ray films have all been examined microphotometrically. A Moll microphotometer was used. In this way the positions of

* J. T. Randall, H. P. Rooksby, and B. S. Cooper. *Nature*, 1930, **125**, 458.
† P. Scherrer, *Nachr. Gesell. Wiss. Göttingen*, p. 98, 1918.
‡ P. Debye and P. Scherrer, *ibid.*, p. 16, 1916.
§ C. W. Parmelee, G. L. Clark, and A. E. Badger, this J., TRANS., 1929, **13**, 285; G. L. Clark and C. R. Amberg, *ibid.*, 1929, **13**, 290.

the bands could be determined with accuracy. It has been possible to show that the areas under the microphotometer curves, when re-plotted in terms of photographic density, were proportional to X-ray intensity. An X-ray film of a glass can thus be prepared ready for quantitative examination in a fairly short time.

The glass specimens were usually in the form of very thin slips, to allow the radiation to pass through without excessive absorption. Ground glass powder can also be used for chemically stable glasses, but the method is unsuitable for, say, B_2O_3, which quickly reverts to the acid when in powder form. Molybdenum radiation is useful in showing up extremely faint bands which would almost certainly remain undetected with Cu–Kα radiation, on account of the much greater angular width.

The glasses examined were usually made in Morgan pots, and the specimens used were always clear, exhibiting no signs of devitrification. Whenever sharp lines appeared on the photographs, indicating devitrification, fresh specimens were always made.

Results with Vitreous Silica, Wollastonite and Sodium Borate.

(a) *Vitreous Silica.*

Using Cu–Kα radiation, a single intense band at 4·33 Å. has been obtained. Recently Parmelee, Clark, and co-workers,* using radiation from a molybdenum target, obtained two bands, one at 7·1 Å. and another at 2·5 Å. In view of the discrepancy between this work and ours, we have repeated the measurements, using Mo–Kα radiation. We again obtained the strong band at 4·33 Å. together with a faint one at about 1·5 Å. At the time of our note to *Nature* we inclined to the view that this faint band was spurious; we believe now, however, that it is real, although it is difficult to establish definitely its existence. If the photographs be taken with both radiations and the distance from specimen to film be the same in each case, the bands using Mo–Kα are narrower and nearer the centre spot than those using Cu–Kα. It is possible, therefore, that the faint band is missed when using Cu–Kα, on account of its width. There is no reason why it should not appear when the distance from specimens to film is considerably reduced. We have actually arranged the experiment so as to reduce the distance in the ratio of 3 to 1. Calculation shows that the band ought to appear in these circumstances. Fogging of the film by the incident beam has so far made its detection impossible. The presence or absence of the second band does not, however, affect the main argument. Our

* *Loc. cit.*

222   JOURNAL OF THE SOCIETY OF GLASS TECHNOLOGY.

apparatus gives correct spacings for standard substances, and we cannot understand the values of 7·1 Å. and 2·5 Å. given by Parmelee and Clark.

In the preceding section it was pointed out that extremely small crystals give rise to diffraction *bands* instead of the *lines* common for

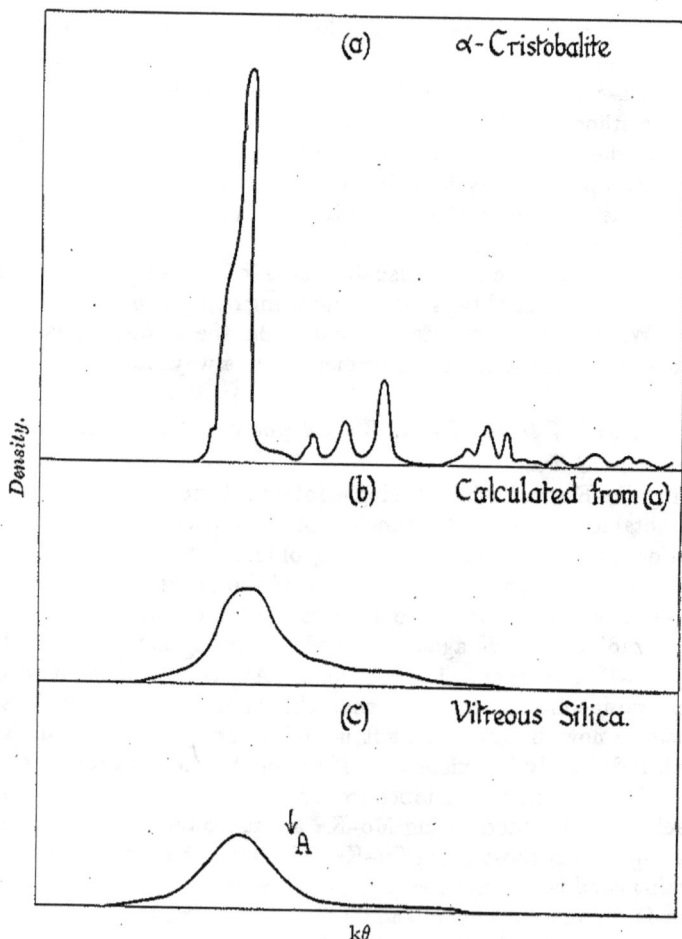

FIG. 1.

larger particles. In so far as vitreous silica gives rise to broad bands, there is no *a priori* reason why it should not consist of crystal-lites. It is to be expected, however, if the glass consists of crystal-lites, that some correspondence would exist between the bands obtained for the glass and the line pattern obtained for one of the crystalline forms of silica. It is found that such a correspondence exists between the sharp line pattern of cristobalite and the band pattern of the glass. The strong band at 4·33 Å. is in almost exactly the same position as the strongest line for α-cristobalite. This is shown in Fig. 1 (a) and (c). Further, it may be mentioned that it is possible to predict how Fig. 1 (a) would appear if the crystals which give rise to it were reduced to, say, 1·5 × 10⁻⁷ cm. Fig. 1 (b) shows this effect, and the resemblance of the observed band (c) and the calculated one (b) shows beyond reasonable doubt that the glass consists of cristobalite crystallites. It can be shown that the glass

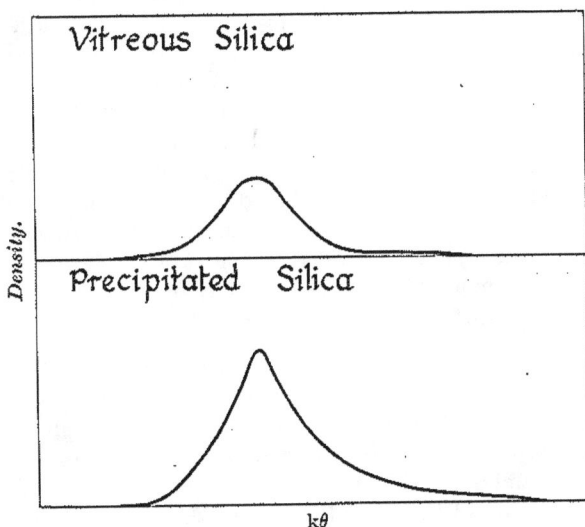

FIG. 2.

does not consist of quartz crystallites; if it did, it is to be expected that the band would have its maximum somewhere about the point A. One further possibility remains. The tridymite pattern is rather similar to that of cristobalite, and calculation shows that it would be difficult to distinguish between the bands for small crystallites of each. The facts that tridymite is a fairly rare mineral, and that cristobalite is invariably produced when the glass is held at 1,200° * for any length of time, would incline us to the view that vitreous silica is composed of cristobalite crystallites.

It is interesting to note that precipitated or " amorphous " silica gives almost exactly the same pattern, the only difference being

* All temperatures quoted in this JOURNAL, unless otherwise stated, are in Centigrade units.

that the crystallites in the precipitate are rather larger than those in the glass (Fig. 2). It follows, therefore, that there is little structural difference between these two forms of silica. In this sense the workers who hold to the " amorphous " theory of glasses are correct. The essential point is that the term " amorphous " has little meaning. The difference between the crystalline state on the one hand and the " vitreous " and " amorphous " on the other is one of degree, not of kind.

In order to get an idea of the number of molecules in the silica crystallites, we must consider the arrangement of the atoms in cristobalite, and not how many atoms of given " sizes " may be packed into a given volume. For purposes of calculation it is safe to assume that the arrangement of atoms in α-cristobalite is approximately the same as in β-cristobalite. This leads us to the conclusion that the average crystallite contains about 20 molecules. The arrangement and number of the SiO_4 groups would be almost identical with that of the tetrahedra shown in Fig. 3, the corners representing oxygen atoms or ions shared by neighbouring tetrahedra.

The X-ray results also show that the distribution of particle size is not simple. The crystallites probably consist of at least two groups, one set having a much bigger average size than the other.

(b) Wollastonite Glass.

It has been possible to show that the glass formed from calcium metasilicate consists, in all probability, of crystallites of pseudo-wollastonite, the hexagonal crystal which is formed at high temperatures. The order of size is the same as with vitreous silica. Thus, wollastonite and silica glasses contain crystallites of the high-temperature forms of the crystalline material.

(c) Sodium Borate, $Na_2B_4O_7$.

The glass was formed by rapid cooling of the fused decahydrate and the crystals by slow cooling. Examination has shown that the glass consists of small crystallites of $Na_2B_4O_7$.

Fig. 4 gives examples of the ring patterns of the crystals and glasses, and it is easy to see the similarities between the two sets.

Commercial Hard Glasses.

It has been found that the patterns of vitreous silica and a commercial hard glass containing 70 per cent. SiO_2, 17 per cent. B_2O_3 and other substances in minor proportions are almost identical. From this it is concluded that a considerable proportion of the silica remains in the form of cristobalite crystallites. This conclusion is borne out in several interesting ways. In the first place it is not

FIG. 3.

SiO₄ Groups in hexagonal formation as in Cristobalite. If 0·5 cm. be taken to represent 1 Å. the photograph gives a rough idea of the size of crystallites in vitreous silica.

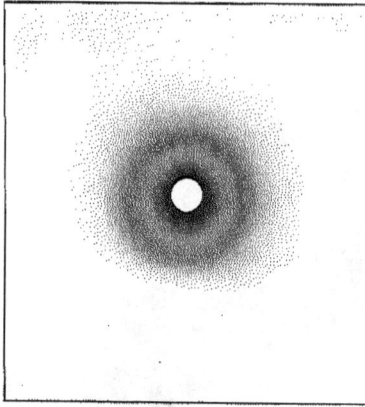

Vitreous silica.

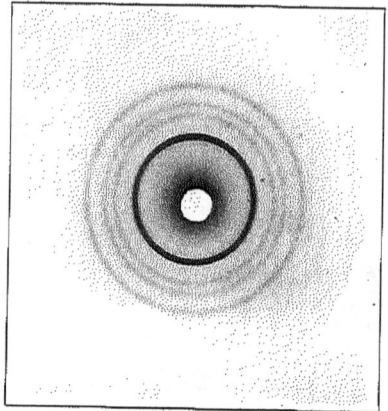

α Cristobalite.

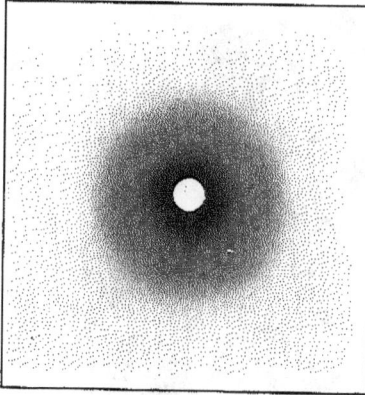

Wollastonite glass.

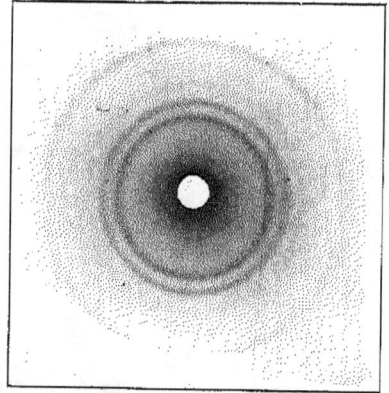

Pseudo-wollastonite.

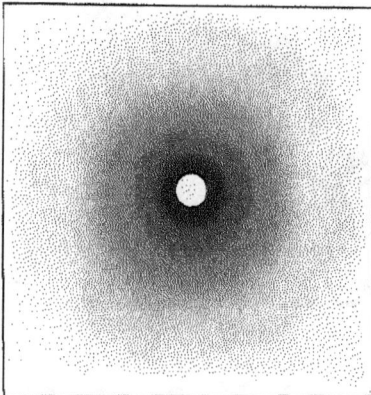

Borax glass.

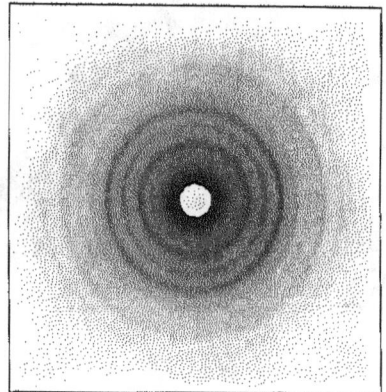

Sodium borate, $Na_2B_4O_7$.

FIG. 4.

X-Ray photographs illustrating the nature of the crystallites in various glasses.

to be expected that the B_2O_3 and SiO_2 would combine in any way, on account of their acidic nature.* (Curiously enough, the X-ray pattern of B_2O_3 glass has its strongest band very near to the strong silica band; this, combined with the weak scattering power of boron, makes it very difficult to separate the two.) Again, the patterns of felspar glasses are also very similar to that of the hard glass under consideration. The patterns show that the felspar lattice breaks down on fusion, and when the glass is formed the bulk of the silica remains more or less segregated from the other components in whatever form they exist. The existence of small crystals of cristobalite in the hard glass is confirmed by another phenomenon. When radiation of a frequency equal to a natural frequency of vibration of a crystal lattice is incident upon it, a marked absorption of the radiation takes place, and can be measured. These regions of the spectrum of the transmitted radiation are commonly called infra-red absorption bands. The vibrating atoms are also *radiating* energy of the same frequency, so that as well as strong absorption there is strong reflection. Hence, if a beam of radiation of several frequencies is reflected in turn from several quartz crystals, the residual rays or " Reststrahlen " from the last crystal will consist of rays of the natural frequencies of quartz. The " Reststrahlen " or infra-red frequencies of crystals are thus naturally dependent on the atomic arrangement. It was found some years ago, by Mr. J. W. Ryde of these Laboratories, that the infra-red absorption frequencies of the hard glass considered and those of crystalline silica were identical. The following observations make it clear that the presence of crystals of cristobalite in a glass depends not so much on the proportion of silica as on the nature of the other oxides. For example, another glass consisting of 70 per cent. SiO_2, 17 per cent. Na_2O, 5 per cent. CaO, 4 per cent. MgO, etc., gives no "Reststrahlen" corresponding with those of silica; neither is the X-ray pattern in any way similar. We have found that the oxides of boron and aluminium do not appear to enter into combination with the silica, whereas the oxides of the strongly electropositive elements calcium and sodium do so, to the destruction of the silica lattice and the disappearance of the corresponding X-ray pattern and infra-red absorption band. These observations show that X-ray diffraction will be a useful aid in determining the nature of the crystallites in commercial glasses.

Strain in Glasses.

We have been able to show during the course of the work that strain in glasses and glass-fibres, as exhibited by the strain-viewer,

* See A. Cousen and W. E. S. Turner, this J., 1928, **12**, 169.

does not appear to be associated with any change in the nature of the band systems. Strained and unstrained glasses give identical patterns. This result is in agreement with those of Parmelee, Clark, and Badger.* If the strain is the result of preferred orientation of the crystallites, there does not seem to be any reason why the diffraction rings should not be broken up in the usual way; it is, of course, possible that the extreme minuteness of the crystals would prevent any such effect being observed. On the other hand, the strain may be due to the orientation of *groups* of the crystallites, in which case the present technique would fail to distinguish between strained and unstrained material. It would be necessary to use longer wave-length *X*-rays to see

(*a*) if the groups exhibited any periodicity

(*b*) if they were oriented in any specific direction.

General Discussion.

It is natural to consider in what way the results of the previous sections can explain some of the physical properties of glasses and their variations under different conditions.

The Density of Glasses.

Before considering this question, it is necessary to go into the experimental results a little more fully. Although a close correspondence has been exhibited between line and band position for the *X*-ray photographs of powdered crystal and glass, it was never found to be perfect. The middle of the band was observed to be slightly to one side of the corresponding line. Usually the shift was in such a direction as to indicate a greater distance between the atomic planes in the glass than between those in the crystals. This means that the lattice of the glass crystallites is slightly bigger than that of large crystals, although the arrangement and number of the atoms in the unit of the lattice remain the same. It follows from this that the density of the glass should be rather less than that of the crystals from which it was made. This is actually the case ; the density of vitreous silica is 2·20, that of cristobalite 2·35. It is not possible to make more than an estimate of the density from the *X*-ray results, because, in the nature of things, the number of bands is insufficient to tell if the crystallites have expanded equally in all directions. The important result is that densities of simple glasses are generally less than those of the crystals from which they were made, and that the *X*-ray results confirm and explain this. Professor J. E. Lennard-Jones has kindly worked out for us one or

* *Loc. cit.*

THE STRUCTURE OF GLASSES. 227

two simple cases embodying certain assumptions, and he has shown that in a lattice of a non-polar type an expansion of the observed order is predicted. A simple cubic ionic lattice would be expected to contract on decrease of size. The details of these calculations and the assumptions on which they are based are given in a mathematical note to the paper in the *Zeitschrift für Kristallographie*. It is not too much to hope that the new technique will be of help in providing a new formula for the calculation of glass densities. While the Winkelmann–Schott formula is very useful as an approximation, it is well known to be inaccurate for glasses containing a preponderance of any given material.

The Melting of Glasses.

It is interesting to consider in what way the melting of an agglomerate of exceedingly small crystals will differ from the melting of a single crystal of the same total mass. Crystalline substances have, of course, definite melting points; glasses have only a melting range. It is to be expected on general grounds that the potential energy of the agglomerate will be greater than that of the single crystal which has had time to adopt the most favourable configuration. It therefore follows that the latent heat of fusion will be less for the agglomerate than for the single crystal. Also, since the agglomerate is made up of crystallites, some of which are smaller than the average, we may expect migration of atoms to take place at a much lower temperature in the agglomerate than in the single crystal. In Fig. 5, which is purely diagrammatic, $o \, b \, c \, d$ represents the fusion of the single crystal. The melting of the agglomerate will be represented by the mean of such curves as $ob'c'd'$, $b'c'$ being in general less than bc. On account of the probable existence of complicated particle size distribution, some particles will have a bigger latent heat than others. On these grounds it is not difficult to account for the indefinite melting of a glass as compared with a crystal.

Electrolysis in Glasses and Crystals.

In view of the results given in this paper, it is of interest to consider the electrolytic properties of glasses and crystals. Whilst electrolytic current is carried by both positive and negative ions in liquids and electrolytes, crystals and glasses are distinguished, in general, by having only one ion that is mobile. On the view that a glass is built up of extremely small crystals, it is to be expected that the mechanism of the passage of current would be the same in the two cases.

228   JOURNAL OF THE SOCIETY OF GLASS TECHNOLOGY.

Thermal Expansion.

It is possible that the discontinuities which are known to exist in the thermal expansion curves for most glasses may be explained on lines suggested by the results of the present work. At lower temperatures both the crystallites and the smaller units of the glass are held together by fairly strong forces, and the expansion will be characteristic of a matrix of crystallites cemented together by still

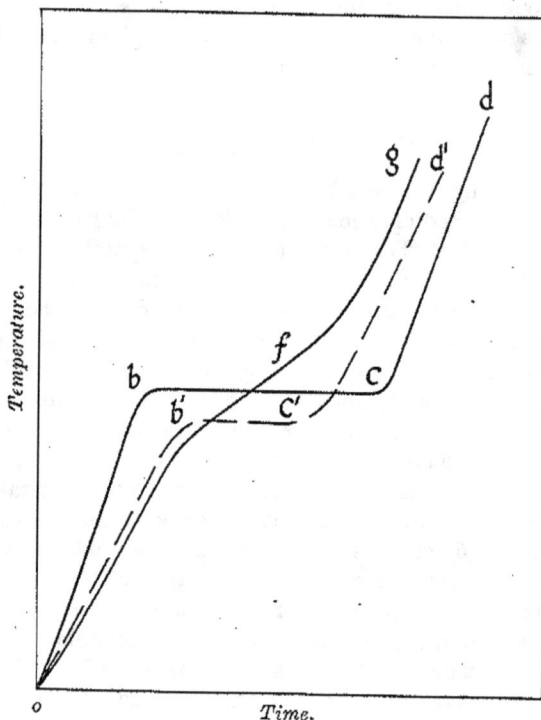

Fig. 5.

smaller units. As the temperature approaches the softening point, the glass is becoming more and more like a true liquid, and the discontinuity in the expansion curve is probably associated with the breaking down of the crystallite lattices; above this temperature the crystallites will only exist for small intervals of time.

Viscosity.

Recently S. E. Sheppard * and E. N. da C. Andrade † have shown

* S. E. Sheppard, *Nature*, 1930. **125**, 309.
† E. N. da C. Andrade, *ibid.*, 1930, **125**, 709.

independently of each other that the variation of viscosity, η, of a liquid with temperature, T, can be expressed by the formula $\eta = Ae^{-b/T}$, where A and b are constants for a given liquid. This formula was derived on the assumption that a liquid contains at any instant clusters of molecules distributed throughout its volume in the way suggested by Stewart's work. The simple formula above assumes that all groups in the liquid are equivalent. Their effect on the viscosity will, however, naturally depend on their size. Now the results given earlier in this paper indicated that the distribution of size of the crystallites in a glass is not simple. There appear to be at least two groups. The exceedingly small crystallites will be able to move above more freely at lower temperatures than larger ones, and their movements may account for the increase in fluidity near the annealing point. The change in slope of the viscosity–temperature curve above this point is no doubt associated with the fact that the glass is at this stage a true liquid. It is impossible to give a mathematical formulation of this idea at the present stage. The X-ray data, however, appear to give a plausible explanation of many of the physical properties of glasses.

Summary.

The diffraction patterns obtained by passing monochromatic X-radiation through glasses can be explained if it be assumed that they contain very small crystals, the order of size being 10^{-6} to 10^{-7} cm.

The experiments show that the term "amorphous" has little meaning. While glasses and "amorphous" bodies consist of units of about the same order of size, both are really crystalline. It has been possible to suggest explanations for many features of the physical properties of glasses on this basis. The densities of glasses, their melting, electrolytic, expansion and viscous properties are briefly considered in the light of the new evidence.

XX.—*The Structure of Silicates.*

By Professor W. L. Bragg, M.A., F.R.S.

(*Read at the Joint Meeting of the Society of Glass Technology and the Deutsche Glastechnische Gessellschaft, London, 3rd June, 1930.*)

The chemical composition of the very large class of compounds we term silicates has given rise to a great deal of controversy. This is largely due to the fact that it is so difficult to examine them by means of the chemical and physical processes ordinarily employed in determining composition. They are essentially bodies which exist in a characteristic solid crystalline state. If melted most of them break down into simpler compounds such as the constituent oxides, and the original compound does not recrystallise on cooling. When they are attacked chemically, it is not possible to isolate acids containing silicon, which may be supposed to combine with the bases to form the compounds. However, just because their characteristic form is crystalline, they provide ideal material for the new methods of Röntgen-ray analysis. The atomic arrangement in all the main classes of silicate structure has now been determined by these methods, and it is interesting to consider them in the light of this new knowledge. The conclusions arrived at often reflect well-known and widely accepted views of their nature, but the X-ray analysis has introduced a greater precision and many novel elements.

The distinguishing feature of silicate structures may be described as their intermediate position between salts of acid radicles on the one hand, and metallic oxides on the other. This has been fore-shadowed by the formulæ which have been used in the past to express their composition. It has been written as the sum of a number of oxides, an avoidance of the difficulty of expressing the atomic relationships which is really a confession of defeat; on the other hand hypothetical silicic acids have been invented, the weakness of this view being that no such acids have been isolated in the free state. The analysis of the structures has shown that both views have missed the mark.

We may consider the way in which oxygen is associated in crystal structures with the successive elements magnesium, aluminium, silicon, phosphorus, sulphur, and chlorine. The last three form acid radicles $(PO_4)^{3-}$, $(SO_4)^{2-}$, $(ClO_4)^{1-}$ in each of which the element is surrounded by four oxygen atoms at the corners of a tetrahedron. Such groups are stable, and can exist as ions in solution so that the corresponding acids can be recognised. More complex groups such as $(S_2O_7)^{2-}$ also exist as free ions, and all these self-contained groups unite with metallic ions to form salts. In certain silicate structures self-contained groups such as $(SiO_4)^{4-}$, $(Si_2O_7)^{6-}$, and more complicated forms, can also be traced; if we choose we may perhaps consider these groups as acid radicles like those in sulphates or perchlorates, though there is already the difference that the groups only exist in the solid state, and do not form free ions. In the majority of structures, however, an essentially novel feature is introduced by the manner in which silicon builds up *silicon–oxygen complexes with indefinite extension in space.* It is this feature which gives rise to the vast variety of silicates, and explains the difficulty of assigning formulæ to them as if they were ordinary salts. The rôle played by silicon in the inorganic world has been compared to that played by carbon in the organic world, but there is an essential difference between them. In organic chemistry the great variety of compounds is due to the possibility of continuing the link between carbon and carbon, so that more and more complex molecules may be built up. In the silicates there often is a similar indefinitely extended linking, but it is always one in which an oxygen atom is interposed between two silicon atoms. The extended silicon–oxygen linking is a transition from acid radicles such as $(PO_4)^{3-}$, $(SO_4)^{2-}$ toward the ionic lattices which metals such as magnesium form with oxygen. The transition is the more gradual because aluminium can replace silicon in a silicon–oxygen complex, and at the same time can replace a metal such as magnesium, in the common isomorphous substitution of natural silicates.

It is difficult to know what terms to use in describing these compounds, owing to their intermediate position. Our common use of the term " silicate " is some justification, however, for singling out the silicon–oxygen groups in the structure and comparing them to acid radicles. This has the further merit of providing a convenient basis for a classification of the silicate structures, though the way in which aluminium replaces silicon often makes the dividing line between the classes somewhat arbitrary. We have to decide in many structures whether aluminium is to be ranked with silicon or with the metallic cations and though its rôle is generally definite it is not so in all cases.

The fundamental feature of silicate structures, which has been found in all compounds hitherto analysed, is the *position of silicon at the centre of a regular tetrahedral group of oxygen atoms.* The oxygen atoms are about 2·6 A apart (1 A $= 10^{-8}$ cm.) and the oxygen-silicon distance is 1·6 A. Should the chemical composition not permit of each silicon atom having a complete group of four oxygen atoms to itself, these tetrahedral groups link up by sharing oxygen atoms. The linking is characteristic; *two tetrahedral groups only have one oxygen atom in common* (i.e. they are linked by a corner, and not by an edge or a face). The lower the ratio of oxygen to silicon, the greater is the extent to which this linking takes place. In this way

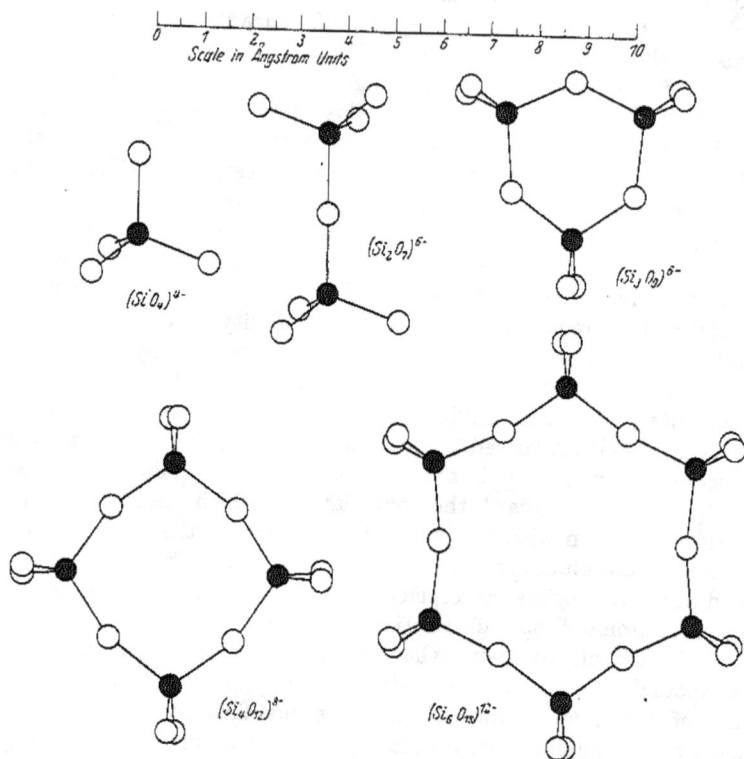

FIG. 1.

Types of self-contained silicon–oxygen groups in the silicates. The black circles represent silicon and the white circles oxygen. Silicon is in each case surrounded by four oxygen atoms at the corners of a tetrahedron, and these tetrahedra (if not independent as in the ortho-silicates) are linked by holding an oxygen atom in common. In these and subsequent groups, the oxygen atoms attached to the silicon atoms have only a slight residual attraction for the metallic ions. The oxygen atoms attached to one silicon atom, on the other hand, behave as if they had a single residual valency charge (–e) and an attraction to the metallic cations, thus binding the whole crystal structure together.

a range of structures is built up with a successive extension in space of the silicon–oxygen linking, represented at one end by the "ortho-silicates" with independent groups $(SiO_4)^{4-}$, and at the other end by the forms of silica such as quartz which W. H. Bragg and Gibbs first showed to be a structure of linked tetrahedra where *every* oxygen atom is shared by two silicon atoms. This is necessitated by the formula SiO_2, for if each silicon is surrounded by four oxygen atoms, it can only have a half share in each.

The following types of silicon–oxygen complex have been found.

(a) *Orthosilicates*, which have independent groups $(SiO_4)^{4-}$.

Examples :

    Olivine, $(Mg,Fe)_2SiO_4$ (Bragg and Brown).
    Garnet, $Ca_3Al_2(SiO_4)_3$ (Menzer).

(b) *Self-contained groups*, formed by linking a finite number of tetrahedral groups. Such groups are $(Si_2O_7)^{6-}$, $(Si_3O_9)^{6-}$, $(Si_4O_{12})^{8-}$, $(Si_6O_{18})^{12-}$. The latter three groups are formed by linking three, four, or six tetrahedral groups in a ring.

Examples :

    Melilite $Ca_2Mg(Si_2O_7)$ (Warren).
    Benitoite $BaTi(Si_3O_9)$ (Zachariasen).
    Beryl $Be_3Al_2(Si_6O_{18})$ (Bragg and West).
    Cordierite $Al_3Mg_2(AlSi_5O_{18})$ (Gossner and Mussgnug).

(c) *Silicon–oxygen chains.* These characterise a large group of minerals termed by the mineralogist " pyroxenes " and " amphiboles." A typical pyroxene is diopside $CaMg(SiO_3)_2$. In accordance with what has been said above, we do not find $(SiO_3)^{2-}$ groups in diopside like the $(CO_3)^{2-}$ groups in a carbonate. Instead there are endless chains of SiO_4 groups, each sharing an oxygen atom with its neighbour on either side so as to reduce the oxygen–silicon ratio to 3 : 1. In the amphiboles two such chains lying side by side have a further sharing of oxygen atoms, leading to a composition represented by the ratio $(Si_4O_{11})^{6-}$. In both these chains silicon can be partly replaced by aluminium. Such minerals often assume a characteristic *fibrous* form, asbestos being an example, and the fibres are parallel to the silicon–oxygen chains as we would expect.

Examples :

    Diopside, $CaMg(SiO_3)_2$
    Acmite, $NaFe(SiO_3)_2$ } (Bragg and Warren).
    Tremolite, $(OH)_2 Ca_2Mg_5(Si_4O_{11})_2$

(d) *Silicon–oxygen sheets.* If three oxygen atoms of each tetrahedral group are shared, the resulting ratio will be represented by $(Si_2O_5)^{2-}$ (since each silicon atom has one oxygen atom to itself and a half share in the remaining three). In this way a two-dimensional sheet of tetrahedral groups is built up, and the most direct way of linking tetrahedral groups into sheets leads to an arrangement which has hexagonal symmetry. This suggests at

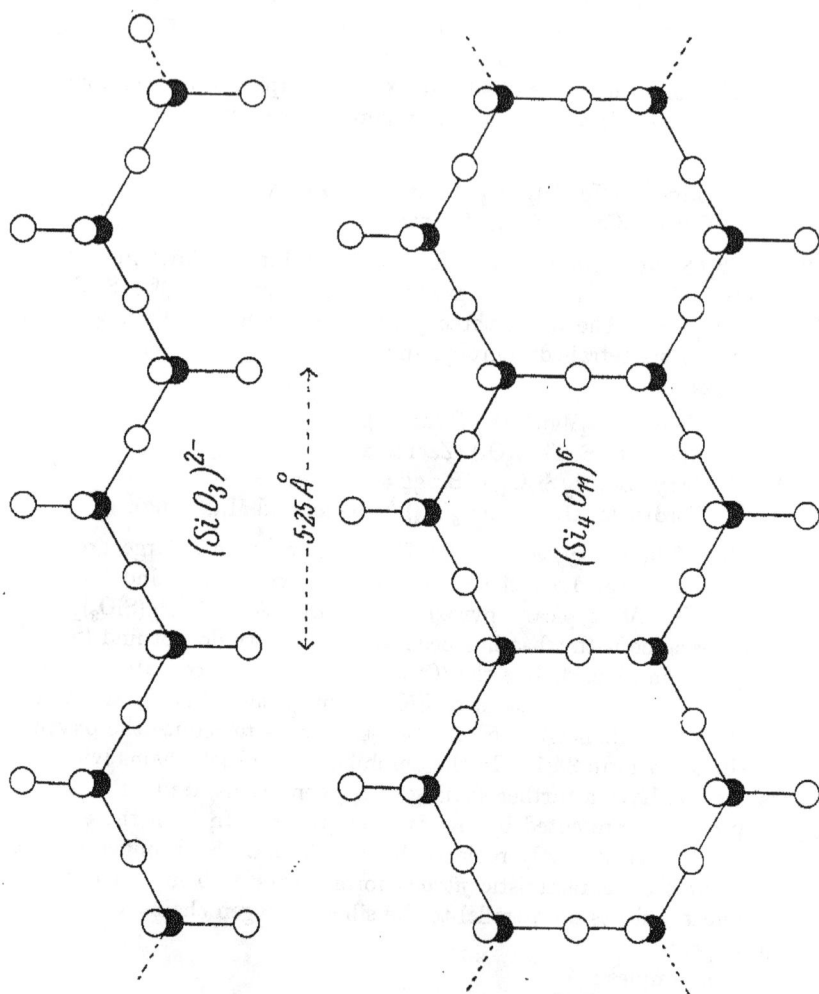

FIG. 2.

Silicon–oxygen chains of linked tetrahedral groups. Such chains are found in the pyroxenes and amphiboles amongst minerals. Owing to the strength of the binding between silicon and oxygen such minerals tend to have a fibrous form (*e.g.*, asbestos), the fibres being parallel to the silicon–oxygen chains. Aluminium replaces part of the silicon in the augites and hornblendes.

once that such a sheet is at the basis of such structures as mica, talc, kaolin, chlorite, and other scaly minerals. Pauling has recently analysed one of the mica structures and proposes structures by analogy for other compounds. Their relationships can be expressed by the following formulæ.

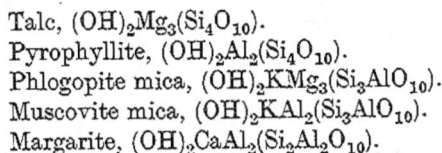

Talc, $(OH)_2Mg_3(Si_4O_{10})$.
Pyrophyllite, $(OH)_2Al_2(Si_4O_{10})$.
Phlogopite mica, $(OH)_2KMg_3(Si_3AlO_{10})$.
Muscovite mica, $(OH)_2KAl_2(Si_3AlO_{10})$.
Margarite, $(OH)_2CaAl_2(Si_2Al_2O_{10})$.

All these minerals are pseudohexagonal, with a marked basal cleavage, and are based on these two-dimensional sheets of linked tetrahedra around silicon and aluminium.

(e) *Three-dimensional silicon–oxygen networks.* If every oxygen of each tetrahedral group is shared between two silicon atoms, the resulting structure will have the composition of silica, SiO_2. Machatschki first pointed out that if a certain proportion of the silicon were replaced by aluminium, the result would be a silica-like arrangement of linked tetrahedra which had a total negative charge (owing to the lower valency of aluminium), and into which in consequence metallic ions would be incorporated. This is the basis of such compounds as ultramarine (Jaeger) of the *zeolites* (Taylor) and *felspars* (Schiebold). The linked tetrahedral groups in a typical zeolite form a very open, yet rigid network, with plenty of room for metallic ions to move from one part to another, and for water to escape from or enter into the structure. In zeolites the water can be driven off and replaced by other groups such as alcohol or carbon disulphide, and one cation can be substituted for another, without destroying the crystalline edifice. The whole crystal is a single acid radicle, into which the metallic ions are incorporated. This explains naturally the fact that the zeolites, almost alone amongst silicates, enter into chemical reactions.

Examples :
Analcite, $NaAlSi_2O_6 \cdot H_2O$ (Taylor).
Anorthite, $Ca(Al_2Si_2O_8)$ } (Schiebold).
Albite, $Na(AlSi_3O_8)$ }

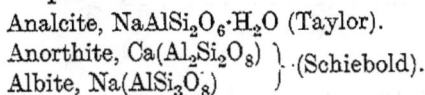

Choosing a few examples from these groups, we can see how the type of structure varies with the silicon–oxygen ratio.

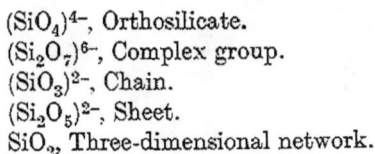

$(SiO_4)^{4-}$, Orthosilicate.
$(Si_2O_7)^{6-}$, Complex group.
$(SiO_3)^{2-}$, Chain.
$(Si_2O_5)^{2-}$, Sheet.
SiO_2, Three-dimensional network.

THE STRUCTURE OF SILICATES.   301

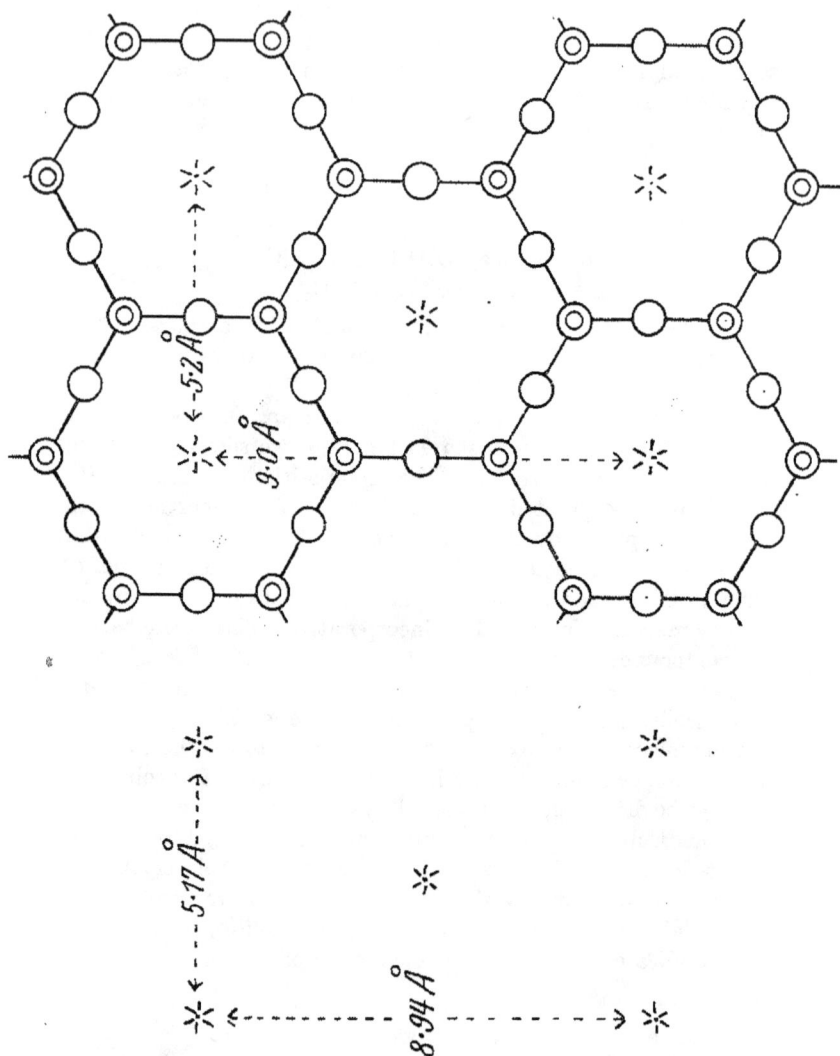

Fig. 3.

A sheet of linked tetrahedral groups. Such a sheet has a hexagonal outline, and forms the basis of the minerals with a platy texture, such as mica, talc, kaolin. The atoms within the tetrahedron may be either all silicon, or silicon and aluminium. The sheet is a further lateral extension of the double chain shown in Fig. 2.

Naturally the ratio in these groups overlap to a certain extent, and, as has been stressed above, aluminium can partly replace silicon in every group except the first. Nevertheless the classification is very convenient, and goes a long way to explain the properties of the structures. The rules for linking up SiO_4 groups show that certain groups which have been postulated in the past

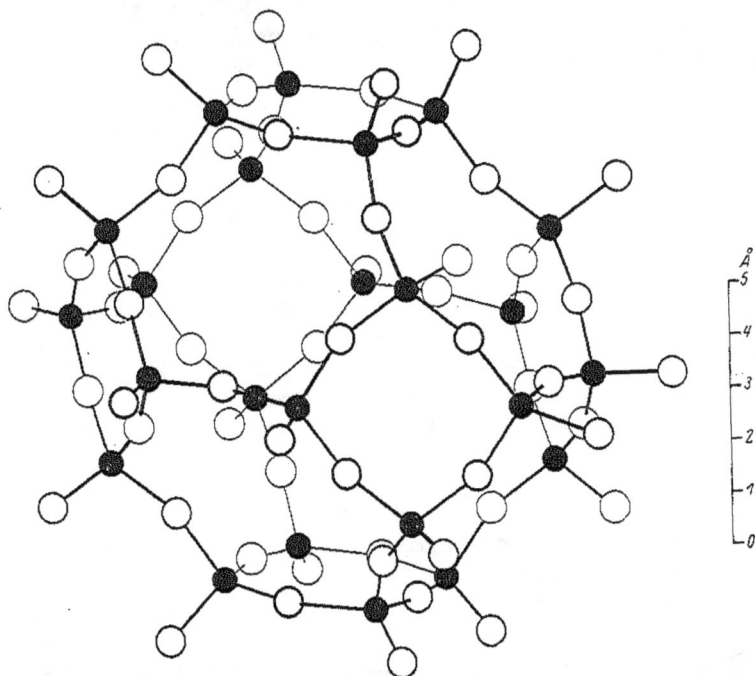

FIG. 4.

A three-dimensional network of linked tetrahedral groups. The network shown in the figure is continued in all directions in space, and represents the groundwork of the structure assigned by Jaeger to the group of compounds to which ultramarine belongs. It is typical of the networks in the zeolites. The black circles represent silicon and aluminium atoms, the whole network having a resultant negative charge owing to the partial replacement of silicon by aluminium. Metallic cations and other groups are incorporated into the spaces inside the networks.

are geometrically impossible, e.g., Si_3O_8. Three SiO_4 groups linked in a row give a group Si_3O_{10}, and in a ring they give a group Si_3O_9, but there is no way of linking them by corners only so as to form a group Si_3O_8.

We have next to consider how these silicon–oxygen complexes with their resultant negative charge are built up as a crystalline structure with metallic cations. The atoms fit together as if the oxygen atoms were comparatively large (diameter 2·7 A) and the most common metallic atoms relatively small. The author first pointed out the prevalence in silicate structures of regular groups of oxygen atoms around the metals, as if many cations such as $Be^{..}$ $Al^{...}$, $Mg^{..}$, $Fe^{..}$ fitted into close-packed groups of four oxygen atoms at the corners of a tetrahedron, or six atoms at the corners of an octahedron. Larger ions such as $K^{.}$, $Ca^{..}$, $Na^{.}$ have often more oxygen ions round them and the group is distorted. We may picture the silicon–oxygen complexes as coming together in such a way that they contribute oxygen atoms to symmetrical groups in the centres of which the metal atoms are imbedded. In the next place, these metallic ions appear to be attracted to the oxygen atoms which have only *one* link to silicon. Oxygen atoms linked to two silicon atoms have little external field, as if their valency were saturated (Fig. 5).

A very important principle holds which was first pointed out by Pauling in a general treatment of ionic compounds. The metallic atoms are so incorporated into the structure that there is a *local balancing of electric charge* between cations and the negatively charged oxygen atoms. Finally we always find that the silicon and metal atoms are dispersed in the structure, not crowded closely together in certain regions.

It is interesting to note that no case has yet been found where a group (OH) is attached to silicon. If hydrogen is present in the chemical composition of a silicate, it is either there as part of the water molecules (zeolites), or is incorporated into (OH) groups linked to metal ions only. This must be allowed for in reckoning the silicon–oxygen ratio. The formulæ of talc and serpentine, for example, are $H_2Mg_3Si_4O_{12}$, and $H_4Mg_3Si_2O_9$ respectively, yet both must have complexes with a ratio represented by (Si_2O_5).

West and the author, in a paper in 1927 on the structure of certain silicates, drew attention to the *importance of oxygen in silicate formulæ*. Oxygen atoms cannot be removed from the structure without breaking up the regular groups, and in most cases, owing to their relatively large size, additional oxygen atoms cannot be incorporated in the unit cell. On the other hand, Al can replace Si or Mg, Fe and Mn are interchangeable, Ca can replace Na, and so forth in the familiar way, provided a balance of total negative and positive valency is maintained. Hence for each type of silicate (mineral species) an analysis must be so expressed that the absolute number of oxygen atoms (ranking with them fluorine and the group OH) is correct for that type of crystal. This simple conception brings order into what has hitherto been a much vexed question, the problem of isomorphous substitution in the silicates. In past

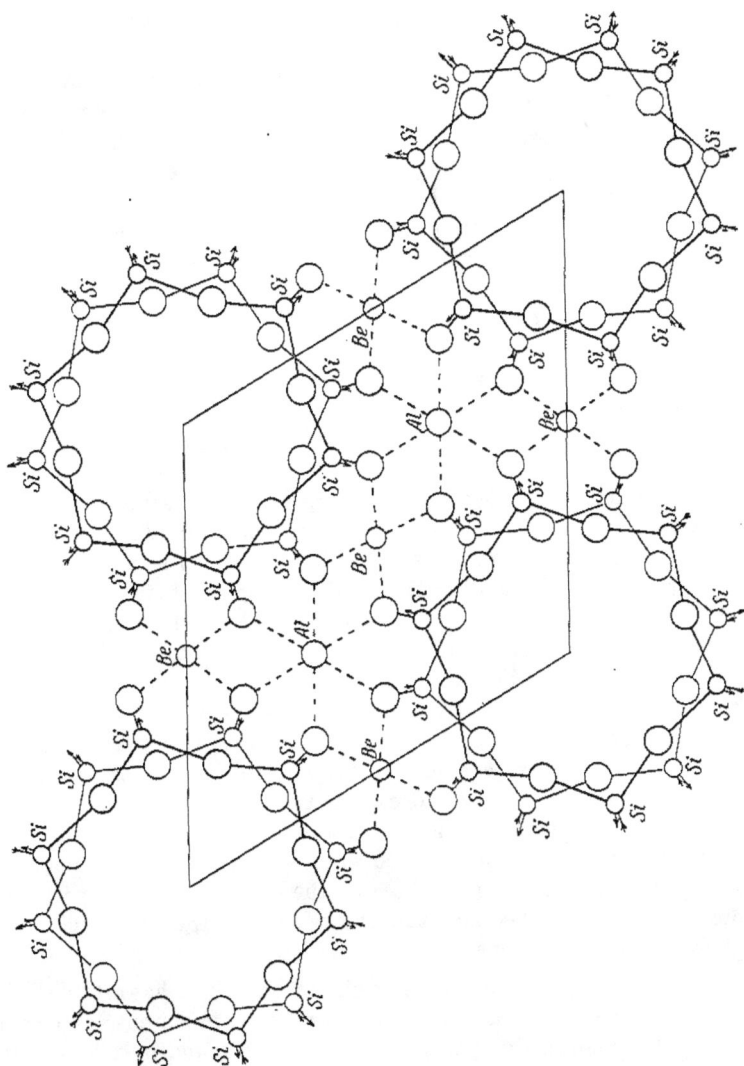

FIG. 5.

The figure represents the structure of Beryl, $Be_3Al_2Si_6O_{18}$, and is an example of the building of the crystal from the silicon–oxygen groups and the metallic cations. The groups in this case are sixfold rings of linked tetrahedra, of composition $(Si_6O_{18})_{12}$·. They are linked to each other by beryllium atoms between four oxygen atoms, and aluminium atoms between six oxygen atoms (it will be realised that the figure shows a projection on a plane, and that actually the atoms are at different heights so that the groups around the cations are tetrahedral and octahedral respectively). The metallic cations are linked to the " active " oxygen atoms which have only one link to silicon, the remainder being shared by two silicon atoms.

work the importance of oxygen appears to have been almost entirely overlooked.

The atom boron plays a rôle somewhat like that of silicon, but apparently does not replace it. It occurs either inside tetrahedral or threefold groups. Silicates in which boron is an essential constituent are not isomorphous with boron-free silicates, but have characteristic structures of their own. Aluminium, as we have seen, freely replace silicon isomorphously. The atom phosphorus does not replace silicon or behave like it. The vast number of natural phosphates are based on the simple $(PO_4)^{3-}$ group, and this group has not yet been observed to link up like the (SiO_4) groups into extended complexes.

The associations of silicon, metal, and oxygen have so far only been studied in their ideal crystalline form, but this is naturally a first step towards an understanding of the less regular grouping in a glass. X-Ray investigations into such so-called amorphous bodies have been made by Debye and Scherrer, Wyckoff, Clark, and others. We would expect a tendency towards regular arrangement to exist in the glass, and this is indeed found by experiments such as those recently made by Randall on X-ray diffraction effects in compounds which have both a crystalline and vitreous phase. The vitreous phase is not wholly amorphous, it gives indication of localised imperfect grouping like those in the ideal crystal. There can be no doubt both that a further X-ray study of the glasses will have most important results, and that the knowledge we have already gained about the atomic arrangement in the silicates points the way towards a systematic study and explanation of the relation between the physical properties of the glasses and their composition. Again in this case the X-ray method is ideal for a study of such compounds as a glass, porcelain, or glaze, because the body is examined without breaking down its texture, and the texture is the vital factor in giving it its useful properties.

NOTE.—The references to the original work which has established these features of silicate structures are very numerous, and can hardly be given in this paper. A list will be found in a paper by the author which will appear shortly in the *Zeitschrift für Kristallographie*.

THE PHYSICAL LABORATORIES.
    THE UNIVERSITY, MANCHESTER.

DISCUSSION.

PROF. TURNER said they were very much indebted to Professor Bragg for having addressed them. At a joint meeting of the English

Society with two other societies, held in London some time ago, they had the very great pleasure of having Professor Bragg's father telling them about the structure of quartz. It was so difficult to make out just what glass was that they were intensely interested in all this " approach work," if so he might call it, this work on the crystalline silicates. One supposed that the first step would be to ascertain that there was some sort of X-ray structure, and this was one of the things in regard to which glass was so tantalising. There were, of course, a number of analogies which could be cited as between glass and the crystalline forms of matter like the metals.

For example, metals like glass usually had a lower density when strained than when annealed. Glasses also were now known to absorb or evolve heat at a certain point during heating or cooling, in a manner analogous to the behaviour of certain metals.

The behaviour of the thermal expansion of glass was also specially interesting. They had discovered that even in the region in which glass was supposed to be rigid, breaks occurred in the thermal expansion curve. It did seem as if there was some kind of structure in glass which underwent changes with change of temperature.

. Types of structure had already been referred to as possibly existing in glasses such as the chain structure and the sponge structure. The sponge structure was suggested by the fact that they had found it possible at Sheffield to extract from glass containing high percentages of boric oxide all the constituents of the glass except silica which was left as a skeleton, showing the original form which the glass itself had at the outset.

He (Professor Turner) would like to ask if in the crystalline structures to which Professor Bragg had made reference, the sodium or potassium ions had a labile character. He asked this because it did seem that in glasses the alkaline oxides were the most loosely bound of the constituents and it was frequently possible to extract much more of these constituents than any of the others present in the glass.

PROFESSOR BRAGG, in reply, said that in regard to labile constituents, one could picture the mechanism by which silver can be substituted for sodium in the zeolite structures amongst minerals, or sodium and calcium could be interchanged in the artificial permutites used for softening water. The structures had a rigid skeleton of linked tetrahedral groups of oxygen around silicon and aluminium, and this enclosed sodium and calcium ions, water molecules, and more complex groups. These latter were very loosely attached to the framework and could be easily broken away, and also there existed open channels in the structure through which they could pass on their way in or out of the crystal. Presumably both factors must be present if an atom was to be mobile. In other silicates, such as $NaFe(SiO_3)_2$, for example, the sodium was surrounded by oxygen atoms and could not move from its place; hence the crystal was very stable.

PROFESSOR W. E. S. TURNER and PROFESSOR E. ZSCHIMMER asked if some sort of molecular crystalline structure existed.

PROFESSOR BRAGG said that the glasses showed diffuse diffraction patterns with X-rays which indicated an approach towards crystalline structure. Several writers had observed this, and some recent work had been done by Randal at the Research Laboratories of the General Electric Company which showed that some glasses gave haloes like those of corresponding crystalline solids, but much broader and more ill-defined. This might be explained by assuming a tendency even in glass to regular arrangement, which was interrupted at frequent intervals of about 10^{-7} cm.

PROFESSOR E. ZSCHIMMER asked if one ought not to express again in oxide form the compositions of the mineral silicates now one knew the important part played by oxygen atoms.

PROFESSOR BRAGG said he thought this was so. The unit crystalline cell of a silicate contained oxygen atoms, silicon atoms, and atoms of the metals. The metals could be substituted for one another, and aluminium could be substituted for silicon provided that ions of similar size were interchanged and valencies balanced. On the other hand, oxygen could not in general be taken from or added to the cell without breaking down the regular oxygen groups. Hence analyses should be so expressed that the number of oxygen atoms was correct for the particular mineral species in question. Analyses of silicates were much simplified if this was done.

54   JOURNAL OF THE SOCIETY OF GLASS TECHNOLOGY.

"The Structure of Glasses: The Evidence of X-ray Diffraction." *

By J. T. RANDALL, H. P. ROOKSBY and B. S. COOPER.

(Communication from the Staff of the Research Laboratories of the General Electric Co., Ltd., Wembley.)

DISCUSSION.

H. H. MACEY asked if any X-ray work had been done on rock-salt, and if so, if results could be given.

Mr. J. T. RANDALL : None, but rock-salt was one of the first crystals examined by Prof. W. L. Bragg many years ago. This crystal had since been used for much fundamental work on intensities, but he knew of no work on NaCl which had any relevance to the present work on glasses.

MR. S. R. HIND : The theory that vitreous silica really contained a high proportion of small crystals of cristobalite was difficult to accept in view of the great differences which existed between the thermal expansion curves of vitreous silica and of cristobalite. Not only was the rate of expansion very different in the two cases, but even a small proportion of cristobalite in fused silica might be expected to give indications of this in sudden expansion at around 200°, which it did not. He would like to suggest, tentatively, that the sensitivity of the X-ray method was such that a small proportion of minute crystals produced a large optical effect.

MR. J. T. RANDALL : Mr. Hind's question was a fair one, but difficult to answer at the present state of our knowledge. He had not tried to show that vitreous silica consisted entirely of cristobalite, but that it did so to quite a large extent. If it were truly amorphous, the patterns actually observed would not be got. There were difficulties which could not be explained yet, and he thought the point raised was one of them. The very tiny crystallites which had been isolated could be shown to have rather different properties from ordinary crystals. They had, for instance, a lattice *slightly* different from that of cristobalite. Therefore, it did not follow that one should expect complete correspondence between the properties of the glass and those of cristobalite.

Mr. W. J. REES : Bearing on the last remarks of Mr. Randall, he would like to ask if it was possible to obtain any basis of quantitative measurement. Was it possible from the intensities of the lines and bands to make an intelligent guess as to the proportion of cristobalite present in vitreous silica ?

* For the paper itself, see this J., TRANS., 1930, 14, 219.

MR. RANDALL : Yes. In ordinary vitreous silica he thought it quite reasonable to say that there must be some 80 per cent. of cristobalite. It was not yet possible, however, to give details of the nature and proportions of the various compounds which comprised commercial glasses. The sort of patterns to be expected from a soda–lime–silica glass, a lead glass, and so on were known; but they had not yet got down to finer details.

MR. A. T. GREEN : Since silica glass had a very small thermal expansion, whilst cristobalite had a very considerable expansion, had anything ever been noticed with respect to the amount of cristobalite present in glass ? Apparently they were to assume that cristobalite might have two essentially different types of properties when present in very minute crystals on the one hand and much larger crystals on the other. At the same time, manufacturers of refractory materials were afraid of the expansion of cristobalite at 200°. Therefore, one was disposed to inquire if anything of that nature could have any effect on the durability of a quartz glass. At 200° cristobalite changed, as they all knew. Apparently the speaker suggested that cristobalite crystallites were present in quartz glass. Were they then to assume that these crystallites were essentially different from the crystals encountered in, say, a silica brick ?

MR. RANDALL : Yes, he thought definitely so. As he had already said, in reply to one of the previous speakers, we had not 100 per cent. crystallites of cristobalite present in silica glass, and he was not prepared to predict what would happen if there were 100 per cent. of such crystallites in silica glass, or to explain at this stage why they should be different from those of cristobalite in silica bricks.

MR. V. STOTT : It seemed to him that there was an essential difference between cristobalite in reasonably-sized crystals and in such small crystallites as the lecturer had been referring to. One could conceive quite different properties in a material consisting of a great number of these tiny crystallites not bearing any specific relation to each other in their orientation, and a material like cristobalite which one might regard as composed of these minute crystals, but differently oriented in a specific direction. One would expect, he thought, quite different properties.

MR. RANDALL : The difference in size between the cristobalite crystals which one would get in an ordinary commercial product and those to which he was referring was something of this kind : if they could imagine an ordinary cristobalite crystallite as being equivalent to an ordinary brick in size, the crystallites would look like a small particle of dust.

MR. F. WINKS asked if the lecturer thought heat treatment

influenced the state of these glasses.   For instance, taking one glass as ordinarily prepared and another glass after heat treatment, could the authors tell the minimum temperature at which any change in the properties of the glass would be likely to occur? Was it possible for different properties to be realised in a glass through different treatment at the upper annealing temperature ?

MR. RANDALL : Such work as had been done on that point seemed to indicate no difference in the X-ray patterns after the glass had been subjected to heat treatment.   But, as he had remarked already, they were only, as yet, exploring the fringe of the subject.

PROF. W. E. S. TURNER thought they had to assume that if these minute crystallites existed their presence was masked by the amorphous medium in which they were situated.   For example, in the case of the expansion of a fused silica glass, there did not appear to be any break in the expansion curve until 1100—1200°, at which temperature the region of definite crystallisation was entered.   The idea propounded was really a striking one, and some phenomena had been observed to give countenance to it.   For example, he knew of a glass the expansion curve of which showed a very marked break below the critical zone, yet the glass appeared to the eye to be homogeneous and without trace of crystallisation.   When examined in ultra-violet light, minute crystallisation was observed to be present.   The theory of Mr. Randall and his co-workers impressed him favourably.   Yet, at the back of his mind lay the thought that the observations might be indicative of some other cause.   In Nature recently two investigators had reported a very definite break in the value of the dielectric constant of ethyl ether at − 105·4°.   And not only ethyl ether, but liquid helium, a simple substance, also showed something of this character.

MR. RANDALL : As a corollary to what Prof. Turner had just been saying, he might mention that some work had been done in America and in India by passing X-rays through liquids, and phenomenal although it may seem, the workers obtained patterns similar to those which they (Mr. Randall and colleagues) had observed with glasses.   The atoms or molecules in the liquid were trying to arrange themselves in the way they would in the solid body, but only succeeded in doing so in small groups over small intervals of time.   This was, however, sufficient to produce the similarities between the patterns for solid and liquid.

MR. B. P. DUDDING said he had watched the developments of the work described by Mr. Randall for ten years.   The work was first begun in connection with metallurgical problems, and he rather thought Mr. Green, in his suggestion, was getting back to where

they commenced. When they used the apparatus first they were
trying to discover what happened to a crystal when broken under
tension; and much of the work indicated that the so-called amorphous material had quite distinct properties from the crystalline
material. Two alternatives were suggested : that either extremely
small crystals result from breakage in the process, or the phenomena
might be explained by a stretch of the actual crystal lattice. The
latter theory was not generally accepted. He thought the Society
owed a debt of gratitude to the men who had been engaged in this
work, and who had been to very great pains over their work for a
period of very many months. He thought also they should be
congratulated on having tackled so many highly controversial subjects in order that people interested in such problems might go
away and scratch their heads about them.

APPENDIX

THE POLYMORPHISM AND ANNEALING OF GLASS*
(A Preliminary Communication)

A.A. Lebedev

Introduction

The problem of the influence of the internal stresses on the properties of glass, its refractive index in particular, is not very well understood or investigated. Frequently, phenomena that cannot be explained by a given chemical property of glass are attributed without any basis to these stresses. The basis of such a view is that glass is ordinarily represented as a supercooled liquid, *i.e.* as a molecularly disperse system in which the possibility of development of any profound structural changes in properties is difficult to visualise.

The work which I conducted on the suggestion of Prof. D. S. Rozhdestvenskii to study the problem of the most rational arrangement for annealing optical glass convinced me that the problem of stresses by far does not exhaust all factors which influence the properties of glass and that it is necessary to assume the possibility of a more profound (structural) change in the properties of glass not directly related to internal stresses. The results obtained are presented in a preliminary and schematic way in this paper.

It should be noted that in carrying out this study my main purpose was to elucidate only the basic character of the phenomena and to find methods that would allow one to determine in detail those characteristics that are found in glasses with the greatest ease and simplicity. Therefore, I did not strive for particular accuracy and care in the individual measurements but rather I attempted as much as possible to shed light on this problem from various directions in order that one then could select that method of study

---

* Translated from *Trudy Gosudarstvennyi Opticheskii Institut*, Vol. 2, No. 10, 1921, pp. 1-20, by the NATIONAL TRANSLATIONS CENTER.

which appears most advantageous. Therefore, the measurements had more of a qualitative rather than a quantitative character.

1.

It is well known that rapid cooling of glass from its softening point alters its properties markedly. In particular, my measurements of the change of the refractive index of glasses upon hardening performed according to the method of I.V. Obreimov* revealed that in very hard glasses the refractive index may change by 500 or more units of the fifth decimal place. In order to resolve the question whether these changes in refractive index are the result of internal stresses alone, the following experiments were conducted:

(1). A piece of hard glass 35×45×75 mm in size was crushed and from its fragments, 14 specimens were taken for which the refractive index was measured. Since studies of hard glass under polarized light had shown that the interior zones are under a tensile stress and the external under a compressive state of stress, one would expect that a great difference would be obtained between the refractive indices of specimens taken from internal and external fragments; however, it was found that the difference was relatively small: the average deviation from the mean was $6 \cdot 2 \times 10^{-5}$, with a range of -10 to $+15 \times 10^{-5}$, with no great regularity depending on the location from which the specimen originated, although on the whole, the refractive index was found to be somewhat greater on the edges. If this result is explained by the assumption that freed from the loads of the surrounding glass, the specimens were relieved of the internal stresses and at the same time of their preexisting difference in refractive index, then one would expect that if these fragments were annealed, no further changes in their refractive indices could take place; however, it was found that after heating to 570° and slow cooling (over a period of a week), the refractive index of all specimens rose on the average by 172×10^{-5} with mean variations of $\pm 5 \times 10^{-5}$; it follows then that both parts under compression and those under tension changed their properties

---

*   I.V. Obreimov, *Trudy Opt. Inst. Petrogr.* No. 1, **1** (1919).

to an almost identical degree, and consequently this change was not due to stresses but rather to other factors.

(2). If the variation of the refractive index in hard glasses depends only on stresses, then one would expect that in the finely ground form, annealed and hard glass has identical properties, in particular the same refractive index; however, experience shows that grinding has very little influence on the refractive index. I took a piece of annealed glass, divided it into two pieces of which one was ground into a fine powder and the other was first heated to the softening point and it was then hardened after which it was also converted into a powder; to these powders I added the same liquid (a suitable mixture of cedar oil and α-bromonaphthalene) with a refractive index close to that of glass to yield a viscous paste and these pastes from the two powders were clamped side by side between glass plates (the thickness of the paste layer was 1–2 mm). Such an assembly transmits only those rays for which the refractive index of the glass powder and of the liquid is identical and the remaining rays for which the refractive indices are different (owing to a large difference in dispersion of the glass and the liquid) will be scattered; in transmitted light such preparation will therefore have uniform colour. Christiansen* referred to such preparations as monochromes. It turns out that monochromes prepared from hard and annealed glass have entirely different colors (for example, one is red, the other is blue).

Knowing the dispersion of the glass and the liquid, one can rather accurately estimate[†] the difference between the refractive indices of the two powders and it is found to be of the same order of magnitude as for the fragments, *i.e.* it may reach up to 500 or more units of the fifth decimal place.

Finally, measuring the refractive indices of small glass fragments under the microscope[‡] also indicates that the difference in refractive indices of hard and annealed glasses is not eliminated by

---

* C. Christiansen, *Ann. Phys. (Leipzig)* **259** (1884), 298.
† See, for example, Obreimov, p. 1.
‡ K. Exner, *Repert. Phys.* **21** (1885), 555; S. Exner, *Z. Instrum.* **6** (1886), 139; F.E. Wright, *Amer. J. Sci.* **17** (1904), 385.

grinding. It follows then that through the same internal stresses it is impossible to explain directly, without resorting to certain more or less questionable means, the variations in the glass properties observed during different heat treatments.

2. Study of Birefringence of Glasses During Heating

A study of the pattern of change of birefringence of glasses during heating convinced me that the properties of glass do not vary continuously, but that certain characteristic features exist. I used a rectangular block of borosilicate-chrome yellow glass from the State Porcelain and Glass Plants in Petrograd*, 35×75×75 mm in size, placed it in a rather massive electrical furnace which was immediately short-circuited and was smoothly, first more rapidly, then more slowly, heated to 600–700° for 5–6 hours with an approximately constant current. Owing to the temperature gradient that developed in the glass, stresses were generated with birefringence that was measured by means of a compensator made of a ¼λ plate and a Nicol prism[†]. Since the experiments demonstrated that although the stresses in the glass are generally distributed in a rather complex manner, they change at different rates of heating in all parts approximately the same number of times, *i.e.* there exists a rather complete proportionality; I performed my measurements only in the centre of the glass. The results of this experiment are shown in the upper curve of Fig. 1 where the temperature is plotted on the abscissa and the phase difference φ produced by the glass in two mutually perpendicular (parallel to the sides of the plates) polarized rays on the ordinate. The lower curve of this figure shows the temperature difference between the centre and the outer layer of glass measured by means of a thermocouple (Ag-constantan), one junction of which was placed in the centre of the glass and the other at the outer (upper) edge; for this purpose the glass was

---

*   The numerical data presented in Section 1 as well as in Figs. 1, 2, 4 and 7 refer to the same type of glass.

†   For the theory of such a compensator, see for example, L. Chaumont, *Ann. de Phys.* **4** (1915), 61; *ibid.* **4** (1915) 101.

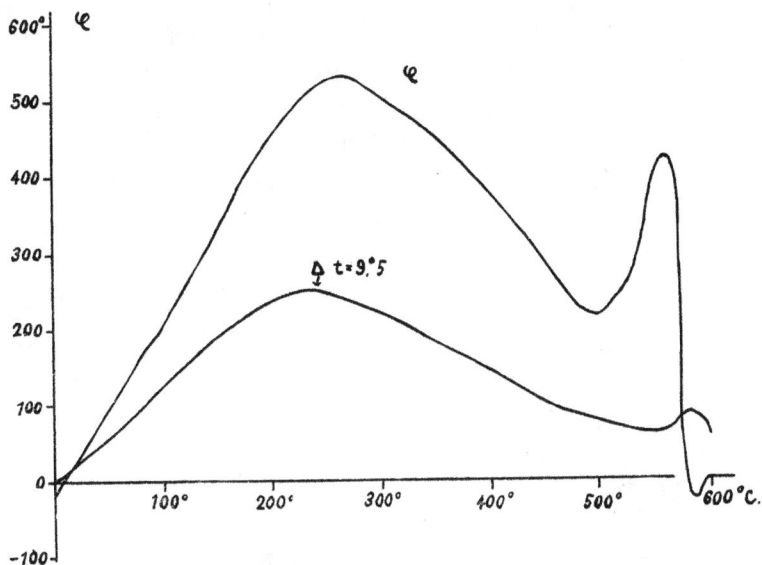

Fig. 1.

first split in half (perpendicular to the direction in which birefrin-gence was observed) and a small groove was made in the centre in which the thermocouple was inserted after which both halves were again closed and cemented. The experiment revealed that such a cleavage in the plate does not have a significant influence on the nature of the phenomenon. It is evident from the φ-curve of Fig. 1 that birefringence at first gradually increased with a corresponding increase in the temperature gradient (see lower curve), then began to drop because of the decrease in the heating rate, but, starting at approximately 510°, birefringence started to rise sharply, reached a maximum at 560° and then began to decrease even more rapidly; at a certain point the phase difference changed sign and finally at 600° birefringence vanished completely. The curve of the tempera-ture difference reveals that starting at 565°, the temperature dif-ference between the centre and the edge of the glass unexpectedly began to rise, reached a certain maximum and then began to drop.

These phenomena are readily explained if one imagines that beginning at 510° starting at the outer layers and then gradually moving toward the deeper layers, some type of structural change begins, accompanied by strong expansion. Actually, stresses develop in a rapidly heated piece of glass because the outer, more highly heated glass layers tend to expand but are hindered by the inner layers. It is obvious that if such a tendency toward expansion is caused not directly by a temperature rise but indirectly by a structural change which is a consequence of this temperature increase and is accompanied by an increase in volume, then the stresses must correspond both in character and sign to those observed during heating. Up to 510° we have a direct heating effect in Fig. 1: both curves φ and Δt rise and descend simultaneously. At 510°, the expansion of the outer layers resulting from structural changes is superimposed on direct thermal expansion and curve φ rises rapidly even though no noticeable change in the slope of curve Δt occurs. As the temperature increases, the structural changes are completed in the outer layers and begin in the inner layers. Therefore, the centre of the glass will now tend to expand but will be inhibited by the outer layers. The stresses as well as φ change signs. The effect of the structural change is subtracted from the direct heating effect and thus φ becomes negative somewhat below 600°. Curve φ, rising at 510° and having a negative sign below 600°, must have a marked maximum and then a rapid drop at the intermediate temperatures. At temperatures above 600° the glass is already so soft that it does not maintain stresses and therefore φ approaches zero. The maximum of the Δt curve somewhat below 600° can be readily explained if one assumes that the structural change of glass is accompanied by heat absorption. This phenomenon must be the cause for the decrease of the Δt curve when the structural change begins in the outer layer, *i.e.* beginning at 510°, and the cause of the rise when the structural change occurs only in the central parts and has already been completed in the outer parts at higher temperatures. Such a stage of the process begins when it causes the temperature in the centre of the glass to stop changing but it is still rising on the outside due to heating. The Δt curve must then begin to rise. And since it will undoubtedly have to drop again when all structural changes are terminated in the entire piece of glass, then a maximum must appear between these temperatures, as is actually the case.

One should note that the magnitude of the maximum of the φ-curve depends largely on the previous heat treatment, shape, size, and type of glass, *etc.*; therefore, such a sharp maximum as shown in Fig. 1 is never obtained but there is always a change in the phase difference φ before the complete extinction of birefringence.

In view of the fact that birefringence depends on very many factors, for example, the temperature gradient (consequently the rate of heating, thermal conductivity, heat capacity), the coefficient of expansion, shape, size, time, elastic properties, viscosity, and finally on the relationship between birefringence of the glass and stresses, it is obvious that a study of birefringence cannot yield a sufficiently clear and definite picture of the processes that take place in glass; I therefore turned to an investigation of other, simpler phenomena and properties of glass.

3. Heating Curves of Glass

As we showed in the preceding section, there was reason to believe that around 560° a certain process which is accompanied by heat absorption occurs in glass. In order to verify this I investigated the heating (and cooling) curves by the Roberts-Austen differential method.

Glass in a powder form was placed in a porcelain tube (15 mm O.D., 60 mm length), and another identical tube was filled with ground porcelain (or quartz glass); by means of an Ag-constantan

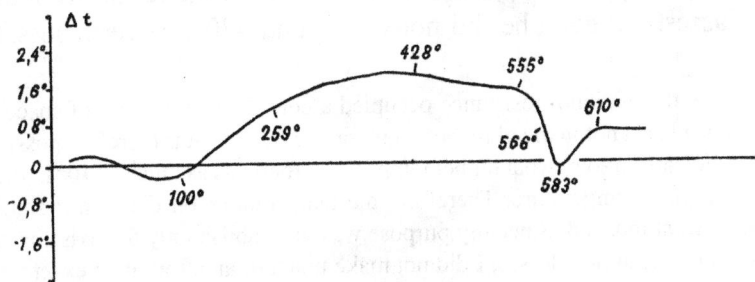

Fig. 2.

thermocouple and a recording galvanometer from the Cambridge Scientific Instrument Co., the temperature difference was recorded between the glass and porcelain powders placed side by side in an electric furnace, which as before was suddenly short-circuited and heated smoothly up to 650–700° over a period of approximately 6 hours. The temperature was recorded with a recording Callendar pyrometer with a platinum resistance*.

One of the curves obtained is shown in Fig. 2. It is evident from this figure that between 555° and 610° there undoubtedly occurs a process in glass that is accompanied by heat absorption, as indicated by the very characteristic inflection of the curve.

It must be noted that I observed the curve inflection near this temperature only during heating, while during cooling it is impossible to detect such typical inflections on any of the curves (of which I plotted more than 30). Although hard glasses also show an inflection at this temperature, it is considerably less prominent and less characteristic.

4. Change of the Refractive Index of Glass with Temperature

In Section 1 we presented considerations indicating that changes in the refractive index of hard glass cannot be explained by stresses alone and that they probably occur as a consequence of a certain structural modification of the glass at high temperatures that during rapid cooling is not completely evident in the reverse direction. If this is true, then we should expect a sharp change (decrease) in refractive index at high temperatures. To my knowledge Reed† attained the highest temperatures in his study of the refractive index of glasses, but even he did not go beyond 470°. Nevertheless, in

---

*   Since the platinum resistance occupied a considerable amount of space, it had to be placed not in the glass but rather on the outside; it is therefore possible that it indicated a somewhat higher temperature (perhaps as much as 10° higher) than the glass temperature. Therefore, the temperatures on diagram 2 should be somewhat reduced. Since my purpose was to establish only the existence of specific points in the glasses, I did not make much of an effort at an exact temperature determination.

†   J.O. Reed, *Ann. Phys. (Leipzig)* **301** (1898), 707.

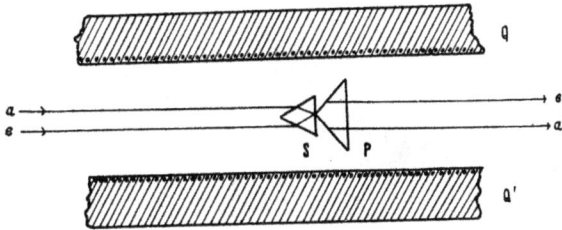

Fig. 3.

certain types of glasses he did actually observe a significant decrease in refractive index at high temperatures but it was smaller than the differences found here between hardened and annealed glasses; I therefore decided to make an independent study of the change of the refractive index of glasses with temperature while continuing heating up to noticeable softening of the glass.

The method used was a modification of the Pulfrich refractometer technique. It is especially suitable for measuring small variations in refractive index under the external influences. In an electric furnace QQ' (Fig. 3) I placed a rectangular prism P of the glass to be tested and another illuminating prism S was placed in front of it. This prism must emit glancing rays on the faces of prism P forming a right angle, and the refracting angles of prism S must be selected in accordance with this. In the two broad beams emitted by a monochromatic light source by means of a lens on prism S, we shall isolate rays a and b; after refraction through S have a glancing path on the legs of the triangle of prism P, they are refracted in prism P under a critical angle and then exit the prism almost perpendicular to the hypotenuse and parallel to one another* in the form of rays a' and b'. Both beams of the rays are observed in a telescope, the focal plane of which shows two bands with sharp boundaries facing one another, corresponding to the single band

---

*   Strictly speaking, exiting rays a' and b' will be parallel to each other only if the refractive index of prism P is equal to 1·414. If it is 1.514 (as was the case in my experiment), then the angle between a' and b' will be approximately equal to 11°; therefore for a' and b' to be parallel, it is useful to use a prism with a somewhat more obtuse angle rather than a right-angled prism.

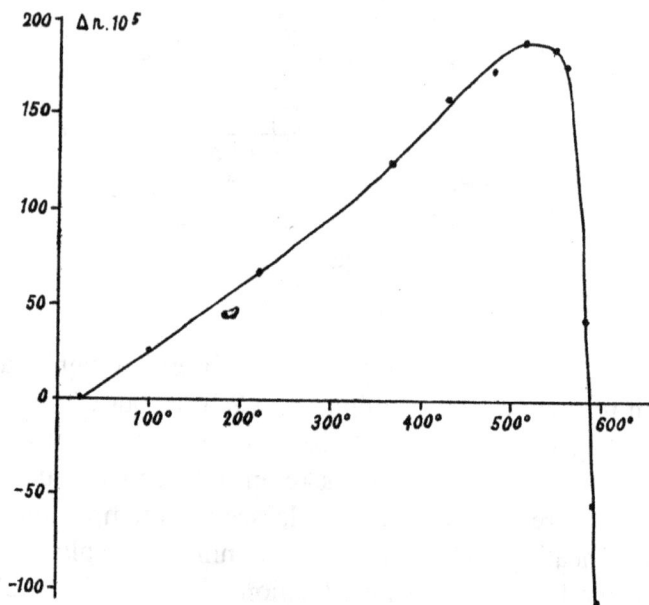

Fig. 4.

of the Pulfrich refractometer. The distance between the boundaries varies with a varying refractive index of prism P. Simple calculation reveals that the change in angle between a' and b' by 1 minute will correspond to a change in the refractive index of prism P (at $n=1\cdot514$) by $15\cdot4\times10^{-5}$, and consequently a change of the angle by 1 second corresponds to $\Delta n=0\cdot26\times10^{-5}$; thus, this method makes it possible to achieve rather significant accuracy.

Since a broad beam is used for illumination and the measurement is made on glancing rays, it is obvious that the change in refractive index and even a small random shift of the illuminating prism S do not affect the path of rays a' and b' and do not influence the measurements.

This method has the additional advantage that when using a Gauss ocular, it is possible to verify the direct angle in the usual way at any time without altering the setup.

The measurements made with this technique gave the following result (see Fig. 4): approximately up to 520°, the refractive index varies almost linearly, but beginning with that temperature it starts to drop rapidly so that during heating from 540 to 595°, n dropped* almost by 300×10^{-5}. Since at 595° the prism surfaces began to be noticeably deformed, it was impossible to raise the temperature further. In order to verify that it was not the distortion of the surfaces that caused the sharp change in angle between rays a' and b', this glass was then kept at approximately 450° for about 10 days, after which the refractive index was fully restored[†].

Thus, the hypothesis that at high temperatures there is a sharp reduction in the refractive index completely independent of stresses has been confirmed.

5. Variation of the Coefficient of Expansion of Glass During Heating

In the introduction I mentioned that one of my tasks was to establish a method that would permit the simple and clear detection of characteristic phenomena that originate in glasses, but the three methods presented are not very suitable for this since, for example, a study of birefringence gives only the total influence of very many factors which cannot each be evaluated separately; studying the heating curves only indicates that a process of some type takes place in the glass, but it can by no means furnish a representation of the influence of the ongoing processes on the properties of glass; finally, a study of the refractive index, although very instructive and important, unfortunately requires careful (optical) polishing of the specimens which considerably complicates its use.

In view of this I tested still another method, namely, a study of the changes in the expansion coefficient with temperature which, in my opinion is most convenient and has the potential of becoming at least at first, a fundamental technique for a study of processes in glasses.

---

* In my measurements I did not introduce a correction for the change in the refractive index of air, consequently I measured the relative refractive index.

† Unfortunately, it was not possible to check the angle by the above-mentioned method at the highest temperatures, since the surface deformation made it impossible to analyse the images of the filaments.

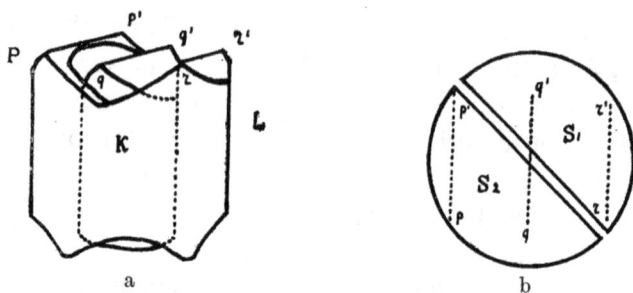

Fig. 5.

The method employed consisted of the following: A cylinder (L) was cut from a cube of quartz glass (16 mm O.D.) in the shape shown in Fig. 5a, *i.e.* at the bottom three projections were made and at the top two edges pp' and rr'; the height of the cylinder was 35 and 17 mm. The specimen K, having an edge qq' at the top and again three projections at the bottom, was inserted into the cylinder. Cylinder L and specimen K were placed on a quartz glass base, and two mirrors* S_1 and S_2 were placed at the top so that one mirror was lying on edge pp' and half of edge qq', and the other on edge rr' and the other half of edge qq', as shown in Fig. 5b. During heating specimen K expanded more than cylinder L and as a consequence the angle between mirrors S_1 and S_2 increased. By measuring the change of the angle between S_1 and S_2 it was possible to calculate the degree of elongation of specimen K, since the coefficient of expansion of quartz glass forming cylinder L is known from the literature[†]. This device was placed in the centre of electric furnace Q (Fig. 6). The measurement itself was carried out as follows: By means of tube S and right angled prism P it was possible to see the image of scale D in mirrors S_1 and S_2. If the mirrors are not completely parallel, then two superimposed images of scale

---

* As a mirror I used simply two glass plates polished at the top and frosted at the bottom. It is better to use mirrors made of quartz glass and to platinum-plate the reflecting surface, but even without platinum plating the amount of reflected light was rather high.

† See, for example, H.M. Randall, *Phys. Rev.* **30** (1910), 216.

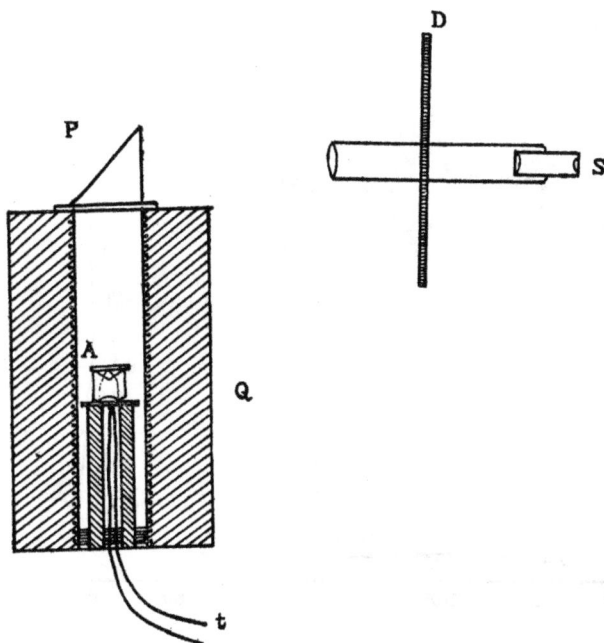

Fig. 6.

D are seen in the tube; upon heating the apparatus, these images begin to slide one along the other owing to the change in angle between the mirrors. In order that the images would not diverge to the side, it is necessary for scale D to be located in plane SPA, and for edges pp' and rr' to be perpendicular to it.

Noting the temperatures at which the divisions of one image of the scale coincide with those of the other, we obtain those temperature intervals Δt at which specimen K, on being raised to them, increases relative to cylinder L by the same constant amount Δl. It is obvious that the expansion coefficient will be $\Delta l / \Delta t \times 1/l_0$ where l_0 is the length of the specimen at $0°$ or, since Δl and l_0 are constants, $= K \times 1/\Delta t$, where K is a certain proportionality constant very simply dependent on the distance between edges pp' and rr' (edge qq' is assumed to be in the middle between these), on the distance between the scale and the mirror, the magnitude of the divisions of scale D, and length l_0.

Since my purpose was only to know the character of the changes of the coefficient of expansion, I did not calculate coefficient K;

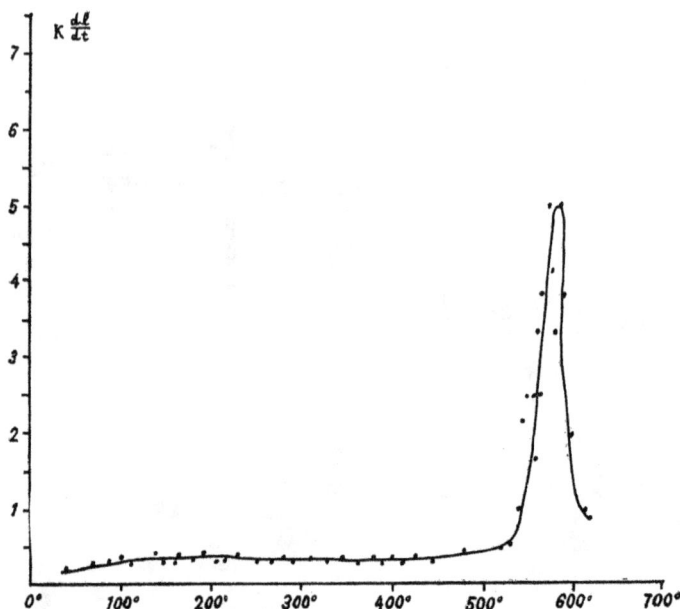

Fig. 7

moreover, I neglected the expansion of cylinder L since it is 20 times smaller than that of the specimen and in the interval up to 700° quartz glass shows no sharp changes in its coefficient of expansion.

The result of the glass study (Crown borosilicate) is shown in Fig. 7. It is evident from this figure that at temperatures up to approximately 520° the expansion coefficient of the glass varied smoothly, but then it started to rise rapidly, reaching a maximum at 570–580°, after which it began to drop again*.

Thus, the hypothesis that we expressed in Section 2 in explaining the anomaly in the change of birefringence has been confirmed.

In view of the fact that the preparation of the specimens is very simple and this method offers an idea about the change of a very

---

*   At the present time I have difficulty in determining whether this descent of the curve occurs actually because of a drop in the coefficient of expansion or whether the softening and deformation of the specimen already have an influence here; it is more likely that it is the decrease in the coefficient of expansion.

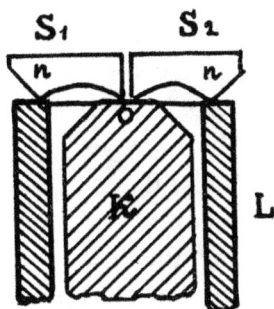

Fig. 8.

important characteristic; moreover, since the processes that are taking place in the glass are very clearly evident, I proposed to base the subsequent investigations on this method. I further note that the value of coefficient K in the system of the device shown in Fig. 5 depends on the position of edge qq' of the specimen with respect to edges pp' and rr'*, so that it is better to use another form of the device as shown in Fig. 8. Here L is again a quartz cylinder and K is the specimen to be tested. Their upper parts were polished to be plane. The two semicircular mirrors S_1 and S_2 (best made of quartz glass) have a pivot at point O resting on specimen K, and at points n – the edges on which they are supported on the edge of the cylinder[†].

6. Explanation of the Phenomena Observed Above

It is clear that presently it is not possible to provide a sufficiently well grounded answer to this question, but it appears to me that it will not be superfluous to present a hypothesis which appears to be very probable and which at any rate may, at least as a first attempt,

---

* A shift of rib qq' parallel to itself even by 1·6 mm at a distance between edges pp' and rr' equal to 16 mm causes an error in the absolute measurement of the expansion coefficient of only 1%, but more deleterious is the turn of prism qq' around the vertical axis, therefore it is useful to make edge qq' as short as possible.

† This device unfortunately has not yet been completed and I have not been able to test it in practice.

serve as a starting point for further investigations and thereby help to move this problem off dead centre.

The fact is that it becomes immediately apparent that even for very different glasses (only six types of glass have been investigated by any of the methods) the temperatures at which a sharp change in properties occurs differ little and are very close to 575°, *i.e.* the temperature at which the known polymorphic α⇌β transformation of quartz occurs, since quartz (in the form of sand) is an almost inevitable and in the majority of cases, predominant constituent of glasses; therefore the question arises whether these changes in the properties of glass are somehow related to this α⇌β transformation of quartz.

The following points can be cited in favour of this hypothesis: I studied three types of glass using the heating curve method and partially the birefringence method: Crown borosilicate, dense flint, and thermometric glass; these three types of glass differ greatly in composition and the only common component of all three types is silicon dioxide; however, in the investigation they all revealed an inflection of the heating curve at almost the same temperature (the difference was always less than 10°), although, for example, the temperature of noticeable densification of the Crown glass powder is 70° higher than for flint glass (for the former it is about 680° and for the latter about 610°).

Furthermore, the character of the change of properties of the glass fully coincides with that observed during the α⇌β transformation of quartz. Thus from the experiments of Le Chatelier* it is well known that the volume of quartz increases greatly (almost by 2%) during the transition from the α- to the β-modification. As shown by the curve of Fig. 7, the transition of glasses at this temperature is also accompanied by a strong expansion so that, when heated from 540 to 600°, the volume of glass increases by approximately 1·5%.

It is evident from the work of Rinne and Kolb† that the refractive index of quartz decreases greatly during the α⇌β transformation;

---

*   H. Le Chatelier, *Compt. Rend.* **108** (1889), 1046; **111** (1890), 123.
†   F. Rinne & R. Kolb, *Neues Jahrb. Mineral, Geol. Paläont.* **2** (1910), 138.

thus for the D_2 ray n_0 decreases by 400×10^{-5}, and n_e by 450×10^{-5} on heating from 550 to 580°.

As shown by curve 4, starting at 540°, the refractive index of glasses drops sharply and on heating from 540 to 595° it decreases by 300×10^{-5} so that even the order of magnitude of its value is the same as that of quartz.

If it is true that the observed characteristics in glasses are a result of the $\alpha \rightleftharpoons \beta$ transformation of quartz, one must conclude that the glass contains highly dispersed quartz crystals, because the conversion at 575° in all appearances[*] is a consequence of a change of the crystal lattice, and not a molecular change, *i.e.* in this case we are dealing with physical and not chemical isomerism. If the glass contains quartz crystals, then the question of their source arises: whether they are incompletely disrupted crystals of quartz (which is commonly used as sand in glass melting) or whether they were formed subsequently during the freezing of glass. In order to resolve this question I produced glass not from sand but from ground fused quartz (fused quartz does not have any special characteristics at 575°). The glass obtained is not significantly different from ordinary glass; consequently one must conclude that if the glass contains quartz crystals, then they formed during freezing, *i.e.* glass is not a supercooled liquid but rather an ordinary alloy consisting only of highly dispersed crystals. This is also indicated by the work of Wallace[†] (among others) who studied the Na_2SiO_3–$MgSiO_3$ system and found that good crystals are formed only in batches containing not more than 10% of any of the components; batches containing 20–80% $MgSiO_3$ solidify in the form of glasses; thus, if a melt with 80% $MgSiO_3$ is cooled slowly, then many spherulites with a size of about 0·5 mm are visible in the glass and he therefore explains solidification into glass by the slow rate of growth and not by the small number of nucleating crystals. It is difficult to assume that between the formation of visible spherulites observed by Wallace and the molecularly dispersed structure that some have attributed to glass, there are no intermediate stages of

---

[*]   See, for example, O. Mügge, *Neues Jahrb. Mineral. Monatsh.* (1907), 181.

[†]   R.C. Wallace, *Z. Anorg. Chem.* **63** (1909), 1.

submicroscopic crystals; nevertheless, it is most likely that glasses represent just such an intermediate structure of highly dispersed crystals.*

It seems to me that there is still another way in which one could illuminate this debatable problem. It is well known that in nature silicon dioxide occurs in three crystalline modifications: quartz, tridymite, and cristobalite; when heated, each of these modifications undergoes one (tridymite, apparently two) transformation. The modification that is stable below the transition point is designated by α, and that above by β. According to the study of Fenner[†] the transformations occur at the following temperatures:

	On heating	On cooling
Quartz:	$\alpha \rightarrow \beta$ at 575°	$\beta \rightarrow \alpha$ at 570°
Tridymite:	$\alpha \rightarrow \beta_1$ at 117°	(In cooling, the transition is not sharp in the case of tridymite.)
Tridymite:	$\alpha_1 \rightarrow \beta_2$ at 163°	
Cristobalite:	$\alpha \rightarrow \beta$ at 274·6°–219·7°	$\beta \rightarrow \alpha$ at 240·5°–198·1°

(The transition point of cristobalite varies depending on the previous heating history.)

Whereas these $\alpha \rightleftharpoons \beta$ transformations are completed very rapidly (especially in the case of quartz), the quartz$\rightleftharpoons$tridymite, tridymite$\rightleftharpoons$cristobalite, and quartz$\rightleftharpoons$cristobalite transformations go to completion extremely slowly. Sometimes they may extend over weeks or even months.

Fenner believes that the stable modification of silicon dioxide is quartz below 870°, tridymite from 870 to 1470°, and cristobalite above this.

In view of the above, one would expect that glasses which solidify below 870°, and glasses solidifying above this temperature will have different properties because in the second case silicon di-

---

*   Other considerations favouring a crystalline structure of glasses may be found in a number of papers by P.P. von Weimarn published since 1907 in *Kolloid-Zeitschrift.*

†   C.N. Fenner, *Amer. J. Sci.* **36** (1913), 331.

oxide must precipitate in the form of tridymite, and then no transition point can exist at 575°, but rather such a transition point must be located at about 117° and a less marked one at 163°. Unfortunately, the study may be highly complicated by the fact that, very frequently, at first not a stable but rather an unstable modification precipitates from the melts (Ostwald's rule), and in the case of silicates, this may be converted into the stable form only after a long time and may greatly complicate the study.

In any case, from this point of view it appeared of interest to study the properties of glasses with the composition $Na_2O + SiO_2$ for which Schaller* reports that devitrification still occurs at a higher temperature; for the investigated melts it varies from 730 to 1300°. One must believe that the transition point, *i.e.* the temperature at which highly dispersed crystals form on cooling, is close to this devitrification temperature and consequently, some of the glasses mentioned will solidify in the stability zone of quartz, and others in that of tridymite; in connection with this one could expect the well-known differences between the properties of these glasses.

Unfortunately, I have not yet been able to make much progress in the study of these melts; I made only one measurement of the change of the expansion coefficient for a glass with the composition $21 \cdot 5\%$ $Na_2O + 78 \cdot 5\%$ SiO_2 for which the devitrification point is at about 1200°. The curve obtained for this glass is shown in Fig. 9. At about 120° a small inflection can actually be seen which may be attributed to the $\alpha \rightleftharpoons \beta$ transformation of tridymite (occurring at 117°). The subsequent part of the curve beginning at 440° is still poorly understood: at 480° there is a maximum, followed by a very rapid drop in the expansion coefficient which even assumes a negative value, *i.e.* on heating from 520 to 610° the glass showed a volume decrease; then at 590° the minimum is at −10 on the scale of the graph, and subsequently the curve rapidly ascends to zero.

Because of the insufficient experimental material at this time, it is difficult to determine the reason for this shape of the curve: whether it is a consequence of the fact that, owing to the rapid

* R. Schaller, *Z. Angew. Chem.* **22** (1909), 2369.

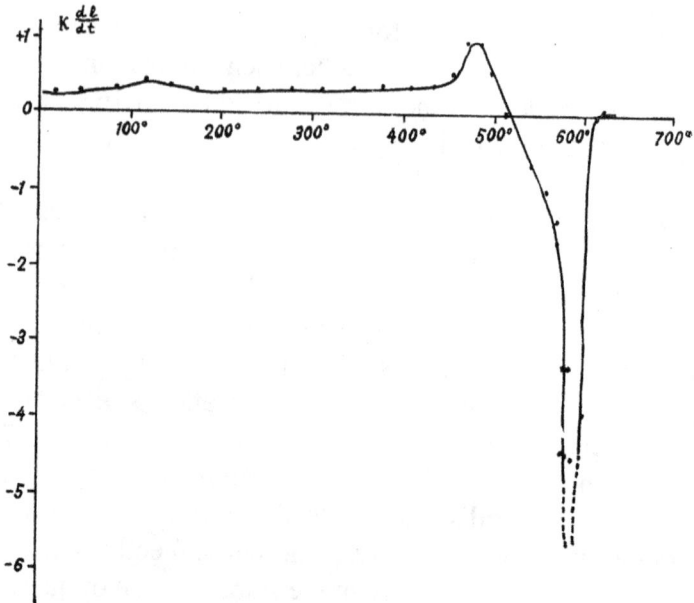

Fig. 9.

cooling, the glass had first been expanded and now, having softened somewhat, it began to contract; or whether not all of the silicon dioxide in the glass precipitated in the form of tridymite, a fraction being in the form of quartz, so that a polymorphic transformation must have taken place in it at about 570°, but due to the rapid cooling, this transformation could not be completed and the glass in supercooled form retained a modification of lower density which formed at higher temperatures and at about 500° began to be transformed into a higher-density modification that was stable at lower temperatures and also caused a volume reduction of the glass.

Although one cannot attribute much importance to this curve in the sense of solving the problem noted above, I still decided to present it because it can illustrate the fact that studying the expansion coefficient in all its simplicity may yield rather valuable information for the evaluation of processes that take place in glasses.

In addition, one may also recall the fact that Guye and Mlle Wassileff* who studied the internal friction of glasses (at up to 300–360°) found that the minimum was at about 120° for one glass, while for the rest, there was a sharp break in the curves at about 240° which may be related to the α⇌β transformation of cristobalite. Finally, according to Maxwell's theory the phenomena of aftereffects observed in glasses indicate the existence of inhomogeneities in glass; assuming a crystal structure of glasses, we can thus explain the aftereffects observed in them.

The final and indisputable solution of the problem concerning the existence of crystals in glass could be obtained by their X-ray study. A similar investigation is presently being conducted at the State X-Ray Institute in Petrograd under the direction of Prof. A. F. Ioffe, but as far as I know it has not been possible to obtain sufficiently sharp X-ray photographs because of the insufficient monochromaticity of the X-rays used.

7. Problem of the Annealing of Optical Glass

Although this study was undertaken with the purpose of investigating the methods for a rational annealing process of glass, *i.e.* to find such a method of cooling so that we would obtain the smallest possible fluctuations in refractive index in glasses (to the extent to which they are not the result of a chemical inhomogeneity) in the shortest time, it has not yet been possible to obtain any significant, concrete results sufficiently verified in practice in this direction.

If the problem were only to annul the stresses, it apparently could be solved relatively simply. Actually, the stresses are generated during hardening because the solid skin which forms on the outside prevents the inner layers from contracting during solidification, and consequently in essence, we are dealing with the same phenomenon as during the heating of a stress-free piece of glass when the outer layers tend to expand and are hindered by the inner layers. Experience shows that the stresses in hard glass actually are entirely of the same character (for the sake of simplicity we will talk of the "same sign" and call them "positive") as during the

---

* C.E. Guye & S. Wassileff, *Arch. Sci. Phys. Nat.* **37** (1914), 214.

heating of glass containing no residual stresses. It is easily understood that during cooling one obtains stresses directly opposite in sign ("negative") to those during heating or hardening as a result of the temperature distribution which is the reverse of that during heating. In this case it is easily seen that if a piece of glass which has not been very strongly annealed is heated to some degree and then prevented from cooling, the stresses begin to drop and if at a certain time the glass is almost entirely dark between crossed Nicol prisms, the stresses assume a "negative" character which can be readily verified by means of $\frac{1}{4}\lambda$ plates or any compensator; however, when the temperature equalizes in the glass, then of course the stresses which existed previously in the glass are restored. If this operation is conducted with glass not in the cooled form but at about 450–500° when it is still soft (but below the temperature where the expansion coefficient of the glass begins to rise), then we find that the negative stresses developing in the glass do not vanish without a trace but may remain even in the fully cooled glass in the form of "negative hardening." By using such a method of rapid cooling from relatively high temperatures, one can succeed in obtaining glasses in which the positive stresses will be compensated by negative ones and therefore they will produce almost no illumination of the field between crossed Nicol prisms.

A study of a flat sheet of a glass annealed in such a manner revealed that although the stresses were very low in this sheet, the variations of the refractive index in the different parts were nevertheless rather significant (almost up to 2×10^{-5}). It is evident then that the absence of stresses is still no guarantee of good glass quality.

In this manner the objective of annealing must be not to obtain glasses in which internal stresses are absent but rather glasses in which the heat treatment does not introduce any optical inhomogeneities.

The internal stresses that develop in glass are properly speaking only a secondary phenomenon that evidently does not play an important role in the properties of glass and therefore one should not consider them the sole criterion. It appears to me that the problem of annealing should be approached from another direction.

A study of the variation of the refractive index of hard glasses has shown that, at temperatures of 520–560° where the refractive index of annealed glasses decreases, it increases in hard glasses; this indicates that the polymorphic transformation in glass is essentially not an instantaneous but a slow viscosity change of the medium, and it has a character corresponding to the transformations of impure crystals and solid solutions, *i.e.* the transformation is completed in a certain temperature interval, so that the glass passes through a continuous series of equilibrium states and each temperature has its corresponding specific equilibrium state.

If the refractive indices of all of these equilibrium states of glass is known as well as the rate at which the refractive index in the non-equilibrium state will approach that of the equilibrium state at a given temperature, then the extent to which the refractive index of a given glass will finally differ from the equilibrium value if the glass is cooled by a specified program could be precalculated. Greater divergence from the equilibrium state will be allowed, the more rapidly the annealing process can be conducted, but then of course there will be a greater likelihood that the refractive index will have somewhat different values in different parts of the glass.

Summary

1. A study of heating curves, changes of birefringence with temperature, the refractive index and the expansion coefficient of glasses has shown that at 540–600° a sharp change in properties occurs in glasses which was detected with complete certainty with all of these methods.

2. The reason for these changes in glass properties is a polymorphic transformation that is assumed to be closely linked to the well-known $\alpha \rightleftharpoons \beta$ transformation of quartz.

3. The hypothesis has been advanced that glass is an aggregate of highly dispersed crystals which also include quartz crystals but in all probability not in the pure form but rather in the form of a solid solution with certain other components; therefore the polymorphic transformation does not occur immediately but in a certain temperature interval in which the glass gradually passes through a series

of equilibrium states. In view of the slowness at which the equilibrium states are reached, glass cannot reach these when cooled rapidly and it solidifies in a stage of more or less incomplete polymorphic transformation. This may also explain the differences in properties which are found in glasses subjected to different heat treatments and which cannot be explained by stresses alone.

4. Since the polymorphic transformation in glasses is accompanied by a great change in volume, it commonly entails the development of internal stresses, which are thus to a considerable extent a consequence of this transformation and, although they are not the main reason for the changes in properties observed in hard glasses, they accompany these in the majority of cases.

5. New methods are described which the author considers simpler than those available previously for determining the variation of the refractive index and coefficient of expansion with temperature.

In conclusion I wish to express my deep appreciation to Prof. D.S. Rozhdestvenskii and I.V. Grebenshchikov for their very valuable advice and unfailing assistance.

Petrograd. Optical Institute.
23 January 1921.

Name Index*

A

Abbé, E.   2
Académie des Sciences   80
Allen, E.T.   16, 179
Amberg, C.R.   266
American Optical Society   195
Anders-Gorczyza, L.   x
Armstrong, H.E.   11
Andrade, E.N. da C.   276
Auerbach, F.   258
Austen, -.   301

B

Badger, A.E.   266, 274
Baker, H.B.   101
Barlow, W.   9
Barus, C.   16
Bary, P.   20
Benrath, H.   94
Boam, F.J.   2
Boltzmann, L.   240
Bowen, N. L.   xii, 3, 5, 22, 33–5, 59–60, 66,
    93, 96, **166–204**, 206, 231
Bradford, S.C.   15
Bragg, Sir W.H.   xii, 2–5, 12, 22, 31–2, 52–3,
    55, **212–24**, 280, 289
Bragg, W.L.   10–2, 52–5, **278–88**, 281,
    289–90, 291
Breithaupt, -.   57
Brown, G.B.   281
Bryson, F.F.S.   5
Bureau of Standards (USA)   125, 183
Butterworth, W., Jr.   74, 225

C

Carson, C.M.   101
Christiansen, C.   297
Churchill College (Cambridge University)   66
Clark, G.L.   46, 48, 266–8, 274, 288
Clarke, J.R.   98, 129
Clay, R.S.   5
Coad-Pryor, E.A.   3, 5, **242**, 243
Consultants Bureau   66

Cooper, B.S.   265, 291, 40, 47–9, 54, 64,
    265–77, 291
Cousen, A.   207, 229

D

Daubrée, A.   93
Dauphiné (France)   222
Davidson, J.H.   106
Day, A.L.   16
Debye, P.   12, 266, 288
Department of Glass Technology (University of
    Sheffield)   2, 105
Deutsche Glastechnische Gesellschaft
    (DGG)   vii, x, 53
Dimbleby, V.   5, 105
Donnan, F.G.   5
Drane, H.D.H.   5
Dudding, B.P.   **293–4**

E

Eckert, F.   xii, 4, 27–8, 30, 38–9, 41, **207–12**,
    229–32
Eichlin, C.G.   19–20, 29, 211
English, S.   31, 33, 74–6, 80–1, 83–5, 99,
    120–1, 145, 147, 149, 159–62, 165,
    195, 199, 229

F

Faraday, M.   7
Faraday Society   1, 5, 15
Fenner, C.N.   179, 194, 312
Ferguson, J.B.   179
Filon, L.N.G.   5, 13, 41, 50, 88, 152, 201,
    239
Finn, A.N.   144
Foerster, F.   94
Foussereau, G.   134
Frankenheim, M.L.   4, 9, 50
Frink, R.L.   13, 93, 211

G

Gaskell, P.H.   66
General Electric Company   290
Geophysical Laboratory (Carnegie Institution
    of Washington)   166

---

* Bold page numbers refer to papers and/or discussion contributions included in this volume as facsimile copies. Only names occurring in the text are cited in this index, not those in author's addresses or confined to the references.

* Hilprecht is changed to Hilpert in Turner's
1927 volume.

* The misspelling of Oudemans' name in the original JSGT paper on page 95 is corrected in Turner's 1927 volume.

Chemical Formula Index

Subject Index*

---

* Bold page numbers denote facsimile papers or discussion sessions. Cross-references are indicated by italics, and those to the Chemical Formula Index are only provided when the latter has additional page numbers. Extra information is given in brackets in normal type.

www.ingramcontent.com/pod-product-compliance
Lightning Source LLC
Chambersburg PA
CBHW061621220326
41598CB00026BA/3838